T0213968

Lecture Notes in Computer Science 12732

More information about this subseries at http://www.springer.com/series/7408

Christian Attiogbé · Sadok Ben Yahia (Eds.)

Model and Data Engineering

10th International Conference, MEDI 2021
Tallinn, Estonia, June 21–23, 2021
Proceedings

 Springer

Editors
Christian Attiogbé 🄸🄳
University of Nantes
Nantes, France

Sadok Ben Yahia 🄸🄳
Tallinn University of Technology
Tallinn, Estonia

ISSN 0302-9743   ISSN 1611-3349   (electronic)
Lecture Notes in Computer Science
ISBN 978-3-030-78427-0   ISBN 978-3-030-78428-7   (eBook)
https://doi.org/10.1007/978-3-030-78428-7

LNCS Sublibrary: SL2 – Programming and Software Engineering

This Springer imprint is published by the registered company Springer Nature Switzerland AG
The registered company address is: Gewerbestrasse 11, 6330 Cham, Switzerland

Preface

The International Conference on Model and Data Engineering (MEDI) is an international forum for the dissemination of research accomplishments on modeling and data management. Especially, MEDI provides a thriving environment for the presentation of research on data models, database theory, data processing, database systems technology, and advanced database applications. This international scientific event, ushered by researchers from Euro-Mediterranean countries, also aims to promote the creation of North-South scientific networks, projects, and faculty/student exchanges.

In recent years, MEDI has taken place in Toulouse, France (2019), Marrakesh, Morocco (2018), Barcelona, Spain (2017), Almería, Spain (2016), Rhodes, Greece (2015), Larnaca, Cyprus (2014), Armantea, Italy (2013), Poitiers, France (2012), and Obidos, Portugal (2011). The tenth edition of MEDI was going to take place in Tallinn, Estonia, during June 21–23, 2021. Unfortunately, the third wave of the COVID-19 pandemic broke out this year and we had to move to a virtual conference without the opportunity to carry out face-to-face interchange with the MEDI community.

The Program Committee (PC) received 47 submissions from authors in 22 countries around the world. The selection process was rigorous, where each paper received at least three reviews. The PC, after painstaking discussions, decided to accept 16 full papers (34%), and eight short papers. Accepted papers covered broad research areas on theoretical, systems, and practical aspects. Some trends evident in accepted papers included mining complex databases, concurrent systems, machine learning, swarm optimization, query processing, semantic web, graph databases, formal methods, model-driven engineering, blockchain, cyber-physical systems, IoT applications, and smart systems.

Three keynotes were presented at MEDI 2021. Prof. Torben BACH PEDERSEN, from Aalborg University, Denmark, gave a keynote titled *"On the Extreme-Scale Model-Based Time Series Management with ModelarDB"* in which he presented concepts, techniques, and algorithms from model-based time series management and the implementation of these in the open-source Time Series Management System (TSMS) ModelarDB. Prof. Ali MILI, from the New Jersey Institute of Technology, USA, presented an invited talk titled *"Assume(), Capture(), Verify(), and Establish(): Ingredients for Scalable Program Analysis"*. Prof. MILI discussed the foundations and design of a programming environment that helps the programmer analyze her/his program by querying it through four orthogonal functions *Assume(), Capture(), Verify(),* and *Establish()*. The keynote by Prof. Marlon DUMAS, from the University of Tartu, Estonia, was titled *"From Process Mining to Automated Process Improvement"*, and was dedicated to the presentation of an overview of four key pillars of Process Mining 2.0: (*i*) predictive process monitoring; (*ii*) robotic process mining; (*iii*) causal process mining; and (*iv*) search-based process optimization.

In these troubled times, we are grateful to all the authors who submitted their work to MEDI 2021. In addition, we would like to thank the PC members and the external reviewers who diligently reviewed all submissions and selected the best contributions. Finally, we extend our special thanks to the Local Organizing Committee members who contributed to the success of MEDI 2021, even though they could not take pride in hosting the physical conference this year.

The EasyChair conference management system was set up for MEDI 2021, and we appreciated its ease of use for all tasks in the conference organizing workflow.

May 2021   Christian Attiogbé
                                                                                Sadok Ben Yahia

Organization

General Chairs

Ladjel Bellatreche   ENSMA Poitiers, France
Jüri Vain   TalTech, Estonia

Program Committee Co-chairs

Christian Attiogbé   University of Nantes, France
Sadok Ben Yahia   TalTech, Estonia

Workshop Chairs

George Chernishev   Saint Petersburg State University, Russia
Jüri Vain   TalTech, Estonia

Program Committee

El-Hassan Abdelwahed	Cadi Ayyad University, Morocco
Alberto Abello	Universitat Politòcnica de Catalunya, Spain
Yamine Aït Ameur	ENSEEIHT, France
Idir Aït Sadoune	Centrale Supélec, France
Christian Attiogbé	University of Nantes, France
Ahmed Awad	University of Tartu, Estonia
Ladjel Bellatreche	ENSMA, France
Orlando Belo	University of Minho, Portugal
Sadok Ben Yahia	Taltech, Estonia
Sidi Mohamed Benslimane	University of Sidi Bel Abbes, Algeria
Jorge Bernardino	Polytechnic Institute of Coimbra, Portugal
Drazen Brdjanin	University of Banja Luka, Bosnia and Herzegovina
Francesco Buccafurri	UNIRC, Italy
Antonio Corral	University of Almerià, Spain
Florian Daniel	Politechnico di Milano, Italy
Patricia Derler	Kontrol GmbH, Austria
Stefania Dumbrava	É.N.S.I. pour l'Industrie et l'Entreprise, France
Radwa Elshawi	University of Tartu, Estonia
Georgios Evangelidis	University of Macedonia, Greece
Alfio Ferrara	University of Milan, Italy
Flavio Ferrarotti	Software Competence Center Hagenberg, Austria
Mamoun Filali-Amine	CNRS, France
Enrico Gallinucci	University of Bologna, Italy
Matteo Golfarelli	University of Bologna, Italy

Anastasios Gounaris	Aristotle University of Thessaloniki, Greece
Marc Gyssens	Hasselt University, Belgium
Raju Halder	Indian Institute of Technology Patna, India
Irena Holubova	Charles University, Czech Republic
Luis Iribarne	University of Almerìa, Spain
Mirjana Ivanovic	University of Novi Sad, Serbia
Nadjet Kamel	University of Science and Technology Houari Boumediene, Algeria
Selma Khouri	ESI, Algeria
Adamantios Koumpis	University of Passau, Germany
Régine Laleau	Paris-Est Creteil University, France
Yves Ledru	Grenoble Alpes University, France
Sebastian Link	University of Auckland, New Zealand
Zhiming Liu	Southwest University, China
Oscar Pastor Lopez	Universitat Politècnica de Valencia, Spain
Ivan Luković	University of Novi Sad, Serbia
Hui Ma	Victoria University of Wellington, New Zealand
Yannis Manolopoulos	Aristotle University of Thessaloniki, Greece
Patrick Marcel	Université François Rabelais Tours, France
Dominique Méry	Université de Lorraine, France
Mohamed Mosbah	Bordeaux INP, France
Chokri Mraidha	CEA LIST, France
Yassine Ouhammou	ENSMA, France
Marc Pantel	ENSEEIHT, France
Jaroslav Pokorny	Charles University, Czech Republic
Giuseppe Polese	University of Salerno, Italy
S. Ramesh	General Motors, USA
Elvinia Riccobene	University of Milan, Italy
Jérôme Rocheteau	ICAM, France
Oscar Romero	Universitat Politècnica de Catalunya, Spain
Manoranjan Satpathy	Indian Institute of Technology Bhubaneswar, India
Milos Savic	University of Novi Sad, Serbia
Klaus-Dieter Schewe	Zhejiang University, China
Timos Sellis	Facebook, USA
Giovanni Simonini	Università di Modena e Reggio Emilia, Italy
Neeraj Kumar Singh	ENSEEIHT, France
Gàbor Szàrnyas	Budapest University of Technology and Economics, Hungary
Kuldar Taveter	University of Tartu, Estonia
Riccardo Tomassini	University of Tartu, Estonia
Riccardo Torlone	Roma Tre University, Italy
Goce Trajcevski	Northwestern University, USA
Javier Tuya	University of Oviedo, Spai
Michael Vassilakopoulos	University of Thessaly, Greece
Panos Vassiliadis	University of Ioannina, Greece

Organization Committee

Monika Perkman	TalTech, Estonia
Wissem Inoubli	TalTech, Estonia
Imen Ben Sassi	TalTech, Estonia
Chahinez Ounoughi	TalTech, Estonia
Siim Salin	TalTech, Estonia

Abstracts of Invited Talks

Extreme-Scale Model-Based Time Series Management with ModelarDB

Torben Bach Pedersen

Aalborg University, Denmark
tbp@cs.aau.dk

Abstract. To monitor critical industrial devices such as wind turbines, high quality sensors sampled at a high frequency are increasingly used. Current technology does not handle these extreme-scale time series well, so only simple aggregates are traditionally stored, removing outliers and fluctuations that could indicate problems.

As a remedy, we present a model-based approach for managing extreme-scale time series that approximates the time series values using mathematical functions (models) and stores only model coefficients rather than data values.

Compression is done both for individual time series and for correlated groups of time series. The keynote will present concepts, techniques, and algorithms from model-based time series management and our implementation of these in the open source Time Series Management System (TSMS) ModelarDB. Furthermore, it will present our experimental evaluation of ModelarDB on extreme-scale real-world time series, which shows that that compared to widely used Big Data formats, ModelarDB provides up to 14x faster ingestion due to high compression, 113x better compression due to its adaptability, 573x faster aggregatation by using models, and close to linear scale-out scalability.

Bio: Torben Bach Pedersen is a Professor of Computer Science at Aalborg University, Denmark. His research interests include Extreme-Scale Data Analytics, Data warehouses and Data Lakes, Predictive and Prescriptive Analytics, with a focus on technologies for "Big Multidimensional Data" - the integration and analysis of large amounts of complex and highly dynamic multidimensional data. His major application domain is Digital Energy, where he focuses on energy flexibility and analytics on extreme-scale energy time series. He is an ACM Distinguished Scientist, and a member of the Danish Academy of Technical Sciences, the SSTD Endowment, and the SSDBM Steering Committee. He has served as Area Editor for IEEE Transactions on Big Data, Information Systems and Springer EDBS, PC (Co-)Chair for DaWaK, DOLAP, SSDBM, and DASFAA, and regularly serves on the PCs of the major database conferences like SIGMOD, PVLDB, ICDE and EDBT. He received Best Paper/Demo awards from ACM e-Energy and WWW. He is co-founder of the spin-out companies FlexShape and ModelarData.

Assume(), Capture(), Verify(), Establish(): Ingredients for Scalable Program Analysis

Ali Mili

Ying College of Computing, New Jersey Institute of Technology, USA
ali.mili@njit.edu

Abstract. Despite more than half a century of research and development, the routine correctness verification of software remains an elusive goal, and software is often delivered with known failures but undiagnosed/unrepaired faults. Some of the reasons for this shortcoming include: inadequate training, lack of effective and efficient tools, lack of valid formal specifications, and the usual obstacles of scale and complexity. In this talk we discuss the foundations and design of a programming environment that helps the programmer analyze her/his program by querying it through four orthogonal functions: Assume(), which enables the user to make assumptions about program states or program functions; Capture(), which enables the user to retrieve properties of program states or program functions; Verify(), which enables the user to verify some properties of program states or program functions; and Establish(), which enables the user to alter the code to satisfy a requirement about program states or program functions. We envision that a user may start a session with vague/incomplete specifications and an incorrect program, and conclude, through successive iterations, with a valid specification and a correct program.

Bio: Ali Mili is Professor and Associate Dean at the Ying College of Computing, New Jersey Institute of Technology, and a member of the Tunisian Academy of Sciences, Letters and Arts. He holds a *Doctorat de Troisieme Cycle* from the *Institut Polytechnique de Grenoble*, a PhD from the University of Illinois in Urbana Champaign, and a *Doctorat es-Sciences d'Etat* from the University of Grenoble. His research interests lie in software engineering.

From Process Mining to Automated Process Improvement

Marlon Dumas

University of Tartu, Estonia
marlon.dumas@ut.ee

In the past decade, process mining has become a mainstream approach to analyze and improve business processes. The key idea of process mining is to extract event logs capturing the execution of business processes on top of enterprise systems, and to use these event logs to produce visual representations of the performance of the process. These representations allow analysts to understand how a given business process is actually executed, if and how its execution deviates with respect to expected or normative pathways, and what factors contribute to poor process performance or undesirable outcomes. For example, process mining techniques allow analysts to discover process models from execution data, to enhance process models with performance metrics, and to identify factors that influence long waiting times or rework in a business process. With hundreds of documented case studies and probably thousands of other success stories, process mining is now an integral part of the Business Process Management (BPM) discipline.

The first generation of process mining methods (Process Mining 1.0) focuses on descriptive analytics. It allows managers and analysts to get detailed pictures of their processes and to manually spot bottlenecks and sources of inefficiencies and defects. We are now witnessing the emergence of Process Mining 2.0: A new generation of process mining methods focused on automation and prescriptive analytics. In other words, we are moving from process mining as a way to produce insights to process mining as an instrument to recommend actions that drive continuous adaptation and improvement and improvement.

This talk will give an overview of four key pillars of Process Mining 2.0: (1) predictive process monitoring; (2) robotic process mining; (3) causal process mining; and (4) search-based process optimization.

Predictive process monitoring methods allow us to analyze ongoing executions of a process in order to predict future states and undesirable outcomes at runtime. These predictions can be used to trigger interventions in order to maximize a given reward function, for example by generating alerts or making recommendations to process workers. The talk will provide a taxonomy of the state of the art in this field, as well as open questions and possible research directions.

Robotic process mining seeks to analyze logs generated by user interactions in order to discover repetitive routines (e.g. clerical routines) that are fully deterministic and can therefore be automated via Robotic Process Automation (RPA) scripts. These scripts are executed by software bots, with minimal user supervision, thus relieving workers from tedious and error-prone work. The talk will present initial results in the field of robotic process mining and discuss challenges and opportunities.

Causal Process Mining is an emerging sub-field of process mining that seeks to discover and quantify cause-effect relations from business process execution logs. Such cause-effect relations may help process managers to identify business process improvement opportunities. For example, given an event log of an order-to-cash process, a typical causal process mining technique allows us to discover that when a customer is from Southeast-Asia, then assigning activity A to worker X or executing activity A before activity B (rather than the other way around) increases the probability that this customer will be satisfied by 10%. The goal of causal process mining is to identify interventions that make a difference in terms of specific performance metrics. The talk will show how causal machine learning methods are being adapted and extended to attain this goal.

Finally, the talk will introduce a gestating family of methods for search-based process optimization. These techniques rely on multi-objective optimization algorithms in conjunction with data-driven process simulation, in order to discover sets of changes to one or more business processes, which maximize one or more performance measures. The talk will present a framework for search-based process optimization and will sketch approaches that could be explored to realize the vision of a recommender system for process improvement.

Acknowledgments. The research reported in this talk is supported by the European Research Council (PIX project).

Contents

Machine Learning

Data Management - Blockchains

Databases and Ontologies

Stepwise Modelling and Analysis

A Refinement Strategy for Hybrid System Design with Safety Constraints

Zheng Cheng[✉] and Dominique Méry

Université de Lorraine, LORIA UMR CNRS 7503, Campus Scientifique - BP 239,
54506 Vandœuvre-lès-Nancy, France
zheng.cheng@inria.fr

Abstract. Whenever continuous dynamics and discrete control interact, hybrid systems arise. As hybrid systems become ubiquitous and more and more complex, analysis and synthesis techniques are in high demand to design safe hybrid systems. This is however challenging due to the nature of hybrid systems and their designs, and the question of how to formulate and reason their safety problems. Previous work has demonstrated how to extend the discrete modeling language Event-B with continuous support to integrate traditional refinement in hybrid system design. In the same spirit, we extend previous work by proposing a strategy that can coherently refine an abstract hybrid system design with safety constraints down to a concrete one, integrated with implementable discrete control, that can behave safely. We demonstrate our proposal on a smart heating system that regulates room temperature between two references, and we share our experience.

1 Introduction

Whenever continuous dynamics and discrete control interact, hybrid systems arise. This is especially the case in embedded and distributed systems where decision-making logic is combined with physical continuous processes. As hybrid systems are becoming increasingly complex, engineers usually start with different mathematical models (e.g. differential equations, timed automata) that are abstracted from systems. Then, the collection of analysis and synthesis techniques based on these models form the research area of hybrid systems theory (e.g. [4,13,18,21]). They play an important role in the multi-disciplinary design of many technological systems that surround us.

In this work, we focus on the problem of how to design discrete control such that with its integration, the hybrid system can behave safely (i.e. without specified undesired outcome). This is intrinsically difficult mainly due to the potential complex continuous behaviors of hybrid systems. However, even with simple dynamics, designing implementable discrete control for a hybrid system requires a significant amount of domain-specific knowledge that might be interleaved with each other [2,4,5,8]. The problem becomes more challenging when we try to specify the safe behaviors of a hybrid system, and reason whether it can be achieved with the designed discrete control [4,8].

© Springer Nature Switzerland AG 2021
C. Attiogbé and S. Ben Yahia (Eds.): MEDI 2021, LNCS 12732, pp. 3–17, 2021.
https://doi.org/10.1007/978-3-030-78428-7_1

We think that addressing these challenges in hybrid system design would first require a sound theoretic framework, which allows advanced reasoning and development to build on top of it. Our contribution is therefore a refinement strategy. Under such a framework, one can coherently refine an abstract hybrid system design with safety constraints down to a concrete one, integrated with implementable discrete control, that can behave safely. In our refinement strategy, the first milestone is to derive a discrete control design in the event-triggered paradigm, where the controller takes action when certain events occur. The keys to make this derivation sound and feasible are: 1) gradually refine safety property to an appropriate abstraction level (inductive). 2) explicitly modeling the time progression, and simultaneously modeling the safety property and plant progression inductively [3,20]. The second milestone is to refine the developed event-triggered design into a time-triggered one (where the controller takes action at ir/regularly occurring intervals) to make it more implementable. To make this refinement natural to establish, our strategy is to plan the event-triggered design as a specification of the time-triggered one. Then, following a worst-case analysis, the latter can be systematically implemented. We have instantiated our refinement strategy on the design of a smart heating system, which shows the feasibility of our proposal.

The paper is structured as follows. Section 2 motivates our approach. Section 3 introduces our refinement strategy to model and analyze the safety of hybrid systems. In Sect. 4, we detail how to instantiate our refinement strategy to design a smart heating system, and discuss the result. Section 5 discusses related research on modeling hybrid systems. Section 6 draws conclusions and lines for future work.

2   Motivation

To motivate our approach, we borrow an example of a hybrid smart heating system, which regulates the temperature of a house. The example is intentionally very small, so that it can be completely illustrated within this paper. However, we believe it to be easily generalizable by the reader to more complex scenarios.

The hybrid smart heating system that we consider can operate in two discrete modes: "ON" and "OFF". In each mode, the evolution of the continuous variable, i.e. temperature T, can be described by a differential equation (which we simplified for illustration purpose): when the mode of heating system is "ON", the value of temperature follows: $\dot{T} = 1$; when the mode of "OFF", the value of temperature follows: $\dot{T} = -1$.

Every δ (which is a constant, and greater than zero) seconds, the room temperature T is sampled by a sensor, and sends to a thermostat controller that controls mode switching. If the controller decides to switch mode, there is t_{act} (which is a constant and greater or equal to zero, but strictly smaller than δ) seconds of inertia, e.g. when switching from "ON" to "OFF", the evolution of temperature first follows the differential equation of "ON" mode for t_{act} seconds before following the dynamics of the "OFF" mode.

The safety property that we are interested in for this heating system can be expressed as an invariant and in a general form of: $\forall t \cdot t \in R^+ \Rightarrow T(t) \in [T_{min}, T_{max}]$, i.e. the room temperature T at any time start from 0 should be always greater or equal to T_{min}, and less or equal to T_{max} (where T_{min} strictly less than T_{max}). The safety property in the general form quantifies over the entire time span, which makes it ideal for compositional verification (by time-domain restriction).

Banach et al. develop a Hybrid Event-B modeling language (HEB) to design hybrid systems and verify them against safety properties in the general form [4]. The idea is to model the close-loop architecture by: 1) control logic (detect certain events and predict a feasible and appropriate trajectory for the system to progress), 2) plant progression (append predicted trajectory towards time infinity). For example, for the heating system, their controller design might detect that the temperature is about to exceed T_{max} in the "ON" mode at timestamp t, then it decides to switch and following "OFF" mode dynamics. Hence, after the prediction, the system agrees with its old state up until time stamp t(inclusive), but from t towards time infinity will follow the predicted "OFF" mode dynamics, i.e. $T := T \mathbin{\lhd\mkern-9mu-} ([t, +\infty] \lhd T_{off})$[1]. Intuitively, the system progression in this way would break the safety invariant (as when the temperature keeps decreasing under "OFF" mode, it will inevitably below T_{min}. Thus, to make the verification sound, Banach et al. depend on a so-called well-formedness assumption of the system runs, i.e. control logic can interrupt in the middle of plant progression (event-triggered paradigm). In other words, they define the time region where the safety invariant allows to break. However, **Problem 1:** to the best of our knowledge, there is not yet machinery to machine-check those safety properties in the general form based on the well-formedness assumption.

We think that the fundamental problem here is that the plant progression is inductively defined, whereas safety invariants are in the general form. To curb this mismatching, we first use refinements to formally establish the relationship between the safety property in the general form and its inductive correspondence. Then, we follow the idea of early action systems to define and prove inductive safety property [3,20]. Consequently, we not only allow the user to benefit from specifying safety properties in the general form for compositional verification, but also design a sound machine-checkable verification process to reason safety properties.

Event-triggered and time-triggered are two paradigms for control logic design. While the former one is easier to think of, the latter one is generally implemented by the hybrid system. Thus, many efforts were put into how to refine event-triggered control logic into time-triggered one to avoid repetitiveness. However, **Problem 2:** this is shown to be a complex process. For example, Loos and Platzer introduce differential refinement logic to determine whether two hybrid programs refine each other [17]. Using this logic they show how to establish a formal relationship between a time-triggered design and an event-triggered

---

[1] The notation we used in this paper are in traditional set theory. For example, $\mathbin{\lhd\mkern-9mu-}$ means relational override, and $\lhd$ means domain restriction.

design. To ease this process, we propose a pattern, which is still easy to think in terms of event-triggered paradigms, yet make their refinements to time-triggered designs more manageable.

3   A Refinement Strategy for Correct-by-Construction Controller

In this section, we propose a refinement strategy for hybrid system controller design with safety constraints. The strategy is based on the Event-B language, which we refer to [1] for details. Its input is an abstract hybrid system with its own safety constraints. The output is a concrete system, integrated with implementable discrete control, that can behave safely. Our refinement strategy from input to output is to consider the problem as the process of constructing a controlled safe time series. In each of the refinement steps, we show how to make specific part(s) of such controlled time series more concrete.

Specifically, we start with a machine M_specification, which consists of a variable x_a that represents the time series of a generic hybrid system. Such a variable is a model variable that only facilitates proofs but do not contribute to the final implementation [15]. Moreover, since M_specification is generic, we can only type time series x_a as a total function that maps from time-domain R^+ to an abstract domain D. Similarly, we encode a generic safety property $P_a(d)$, where $d \in D$, to generalize the main safety property that the system to-be-constructed needs to respect. This machine is general, which does not require the user to change. It acts as an interface in general programming languages, which provides an abstraction for all hybrid systems.

The next machine M_safety specializes M_specification by considering the safety of the input hybrid system. Such specialization is a refinement that consists of: a) a model variable x that refines x_a in M_specification by concretizing its type (which now maps time-domain to the concrete domain of the input system), and b) a predicate P that concretely specifies the safety property in the general form (i.e. $\forall t \cdot t \in R^+ \Rightarrow P(x(t))$), which quantifies system state over the entire time span). All events in this machine yield a "big-step" semantics for the input hybrid system, i.e. there exists some time series f that satisfies P at all times, which is assigned to x in one go. At this level of abstraction, both the safety invariant and plant progression are not inductively defined.

Next, the machine M_cycle refines M_safety to make the construction of x more precise by specifying that it needs to be safely and inductively constructed. We first introduce a model variable now, initialized at 0, to help tracking of cycles (as did by the early action systems [3,20]). now partitions x into two parts: the past cycles x_{past} up till now (inclusive), and the future cycles x_{future} that from now and beyond. The past cycles x_{past} is what we want to built inductively. The following glue invariant bridges between x and x_{past}: $\forall t \cdot t \in [0, now] \Rightarrow x(t) = x_{past}(t)$, i.e. the time series x and x_{past} agree on each other up till now. Thus, we can formally refine the safety invariant to be inductive (according to the glue invariant): $\forall t \cdot t \in [0, now] \Rightarrow P(x_{past}(t))$. Then, the "big-step" semantics of events is refined into "inductive-step" semantics for the input hybrid system: at beginning of the cycle (identified by now), there will exist some function f_n that satisfies P for the next cycle (identified by $now + tx$, where tx is an event argument to model irregular cycle duration), which is assigned to x_{past} in one go. In this way, assuming the inductive safety invariant at beginning of the cycle, by appending a safe trajectory f_n to x_{past} for the next cycle, then it is intuitive that the safety invariant is inductively preserved. The cycles recur towards time infinity to inductively construct the time series x as promised in the M_safety. The formal relationship between the safety property in the general form and its inductive correspondence, the explicit modeling of time progression, and this synchronization of inductive safety property and inductive plant progression is the key point to solve identified **Problem 1**.

Then, in M_closed_loop, we refine the inductive construction step, specifying that each piece in the safe time series to be constructed experiences the full cycle of a simplified but structured closed-loop architecture: at each cycle, the discrete control to be designed, as a black box, first predict a safe trajectory f_p (w.r.t. safety property P) over the next tx seconds. Then, the discrete control passes the initiative to the continuous system, where the predicted trajectory f_p is assigned to the dynamics of the current cycle f_n to model system progression. Finally, the continuous system passes the initiative back to discrete control for the next cycle and seamlessly close the loop. The intuition of this refinement is that, by the closed-loop architecture, we delegate the construction of safe trajectory f_n at each cycle to the discrete control, which is intended to be implemented on a computer and can be refined to be more implementable in the following steps.

In the refinement of M_control_logic, we aim to open the black box of discrete control, and more concretely specify when certain events are detected what actions should take (i.e. event-triggered) The key problem is what kind of events to detect. It is attempted to detect the precise threshold to react since they are easy to think of. However, it would require high overhead (e.g. continuous sensing). Besides, rewriting them into implementable is also a complex process [17]. We think this complexity is due to that the event-triggered design is not planned to be abstract enough to refine into a time-triggered design.

To curb this lack of abstraction, we propose a pattern of event-triggered design: events should be detected near the threshold to react. We can be imprecise about how near it is using abstract "buffer", but we need this level of abstraction to be presented. This is because, for implementation, we usually apply a different design paradigm, i.e. the controller takes action only every once in a while (time-triggered). The "buffer" that quantifies the distance to the threshold is the worst-case that the system can bear before the next time it can react, which is then concretely refined in M_worst_case_analysis.

In this way, we essentially see the event-triggered design as a specification of a time-triggered one. The logical relationship between the two becomes natural: the responsibilities specified in the abstract specification (e.g. "buffer"), need to be concretized during the refinement. This level of abstraction for the time-triggered design which cooperated with worst-case analysis is the key point to solve identified **Problem 2**.

In addition to time-triggered design, a worst-case analysis might involve other activities via separate refinements, e.g. disturbance approximation, sensor approximation.

Once the users are comfortable with implementation details for the designed discrete control, the final refinement M_implementation simply aims to merge designed discrete control cases together to be shipped for implementation.

4   Case Study

In this section, we apply our refinement strategy proposed in Sect. 3 to design the hybrid smart heating system shown in Sect. 2. We use the Rodin platform (v.3.4.0) to develop our Event-B program.

4.1   M_specification

Initially, M_specification (Listing 1.1) consists of a variable d that represents the time series of a generic hybrid system, which maps from time-domain R^+ to an abstract domain D ($type_d$). A generic safety property P_a is formulated to generalize the main safety property that any system to-be-constructed needs to respect ($safety_d$). Then, we have an $Update$ event that models a "big-step" semantics for any generic hybrid system, i.e. there exists some function f that satisfies the safety property at all times, which is assigned to d in one shot.

```
Machine M_specification
Variables d
Invariants
    type_d: d∈ R⁺→D
    safety_d: ∀t·t∈ R⁺⇒P_a(d(t))
Events ...
    Event Update ≙
        Any f
        Where
            grd₁ᵃ: f∈ R⁺→D
                ∧ ∀t·t∈ R⁺⇒P_a(f(t))
        Then
            act₁ d := f
    End
End
```

Listing 1.1. M_specification

```
Machine M_safety Refines M_specification
Variables Ta
Invariants
    type_Ta: Ta∈ R⁺→R
    safety_Ta: ∀t·t∈ R⁺⇒Ta(t)∈[T_min,T_max]
Events ...
    Event Update ≙
        Refines Update
        Any f
        Where
            grd₁ᶜ: f∈ R⁺→R
                ∧ ∀t·t∈ R⁺⇒f(t)∈[T_min,T_max]
        Then
            act₁ Ta := f
    End
End
```

Listing 1.2. M_safety

4.2   M_safety

The next machine M_safety (Listing 1.2) refines M_specification to be problem-specific, by considering the safety of the smart heating system. Thus, the main changes w.r.t. M_specification are:

– Ta, abstractly represents the time series of the room ($type_{Ta}$), refines the variable d of M_specification by concretizing its abstract domain D to the concrete domain of real numbers.
– the safety property is also refined w.r.t. our smart heating system, i.e. the room temperature should always be between two constant references T_{min} and T_{max} ($safety_{Ta}$).

Then, some induced changes fall into places, for example, the *Update* event is refined such that there exists some function f that satisfies the refined safety property at all times, which is assigned to Ta in one go.

To show the correctness of this refinement, we mainly need to prove that the missing elements in the abstract M_specification can find (witness) its correspondence in the concrete M_safety. A list of proof obligations is automatically generated by the Rodin to help us with this task. For example, for the *Update* event in Listing 1.1, a proof obligation is generated since the abstract guard grd_1^a is missing. To prove it, we forge a witness $grd_1^c \Rightarrow grd_1^a$, meaning the concrete guard grd_1^c is as strong as the abstract guard grd_1^a to replace it.

4.3   M_cycle

Next, the machine M_cycle (Listing 1.3) refines M_safety to make the construction of T more precise by specifying that it needs to be inductively constructed by a safe piece-wise time series. To achieve this, in this refinement:

– we introduce a variable now, initialized at 0, to help tracking of cycles. now partitions Ta into two parts: the past cycles up till now (inclusive), and the future cycles that from now beyond.
– we refine abstract variable Ta into concrete variable T to explicitly record the past cycles, where the following glue invariant bridges between Ta and T: $\forall t \cdot t \in [0, now] \Rightarrow T(t) = Ta(t)$, i.e. the time series Ta and T agree on temperature up till now.
– according to the glue invariant, we refine the safety invariant to be inductive: T is safe if the room temperature is bounded within the safe range up until now ($safety_T$)
– we also need to inductively specify that when now progresses, T remains safe. That is why we introduce an inductive event *Prophecy*: at the beginning of each cycle, the *Prophecy* event assumes the existence of a function f_n that can safely progress until the beginning of the next cycle ($safe_{fn}$). Under such assumption, we model the progression of the heating system: 1) time progresses for a sampling period of tx seconds (act_1), and T will follows f_n for the next tx seconds (act_2).

• The main proof in this refinement is to establish that the *Prophecy* event preserves the safety invariant $safety_T$, which can be proved by induction on T and now.

Consequently, in this refinement, now progresses towards time infinity, while T is safely built simultaneously. We, therefore, can construct a safe time series Ta as promised in M_safety.

4.4   M_closed_loop

We then refine the event *Prophecy* of M_cycle, modeling that the safe function f_n is constructed by experiencing the full cycle of a simplified closed-loop architecture. Specifically, as shown in Listing 1.4, we first introduce a system mode variable s ($type_s$) to distinguish *DECISION* and *RUN* modes in the closed-loop architecture. The *DECISION* mode corresponds to discrete control, and is modeled by the *Prediction* event. Its semantics is that it will predict a safe function f_n to progress within the next cycle ($safe_{fn}$), and assign the prediction as a candidate for the heating system to progress. How the prediction is done is a black box in this refinement, and will be refined in the next refinement. Once the prediction is finished, the mode is changed to the *RUN* mode. This mode corresponds to the system progression and is modeled by the *Progression* event, whose behavior is to follow the predicted candidate for the next cycle. Then, the heating system alternates back to the *DECISION* mode to predict for the next cycle, thereby forming a closed-loop.

```
Machine M_cycle Refines M_safety
Variables T now
Invariants ...
   safety_T:
   ∀t·t∈ [0,now]⇒T(t)∈[T_min,T_max]
Events ...
   Event Prophecy ≙
     Refines Update
     Any f_n  tx
     Where ...
        safe_fn:
        ∀t·t∈ (now,now+tx]⇒
           f_n(t)∈[T_min,T_max]
     Then
        act_1: now := now+tx
        act_2: T :=
           T ◁((now,+∞)◁ f_n)
   End
End
```

Listing 1.3. M_cycle

```
Machine M_closed_loop Refines M_cycle
Variables ... s fa ta
Invariants ...
   type_s: s ∈ SysMode
   safe_fa: s=RUN⇒ ∀t·t∈ (now,now+ta]⇒
              fa(t)∈[T_min,T_max]
Events ...
   Event Prediction ≙
     Any f_n  tx
     Where ...
        safe_fn: ∀t·t∈ (now,now+tx]⇒
           f_n(t)∈[T_min,T_max]
        grd_s: s = DECISION
     Then ...
        act_1: fa,ta := f_n,  tx
        act_s: s := RUN
   End
   Event Progression ≙
     Refines Prophecy
     Where ...
        grd_s: s = RUN
     Then
        act_1: now := now+ta
        act_2: T := T ◁((now,+∞)◁ fa)
        act_s: s := DECISION
   End
End
```

Listing 1.4. M_closed_loop

4.5   M_control_logic

Next, we aim to obtain an event-triggered control logic design that when applied to the heating system could maintain room temperature within two references to be safe.

We approach this by refining the *Prediction* event in M_closed_loop into a set of sub-cases. Each sub-case *Prediction$_i$* specifies more concretely when a certain event is detected how the system should react.

The event to be detected in each sub-case has 2 aspects to consider. One aspect is whether the system is increasing or decreasing or switching to increase/decrease the temperature. To allow each sub-case to access this information, we introduce a variable m for discrete mode control in this refinement. It can be increasing (ON) or decreasing (OFF) or switching $(OFFON$ and $ONOFF)$. The second aspect is whether the current temperature is **close** to the reference or not. Therefore, we assume that the true system state $(T(now)$ for current room temperature) is directly accessible by each sub-case, without external help (e.g. state estimators). Then, appropriate actuation command can be issued accordingly based on the considered sub-case.

For example, we consider *Prediction$_1$*, which detects that the system is increasing the temperature, and the temperature is not yet close to the reference. In this case, it is potentially safe and can continue to increase. This case is encoded as in Listing 1.5 for illustration purpose. The encoded *Prediction$_1$* event declares 3 parameters $buffer$, f_n and tx. It semantically means that: when the system is in the mode "ON" $(decision_1^m)$, and the current temperature $(T(now))$ is away from threshold T_{max} by "buffer" $(decision_1^e)$. This is considered to be safe. Then, the system can safely stay at the "ON" mode $(actuation_1^m)$, because there is going to be a safe trajectory f_n corresponding to this actuation decision that can safely progress for the next tx seconds $(safe_{fn})$.

Following the same parametrized pattern as shown in Listing 1.5, we develop another 7 sub-events, which together form an automata for our event-triggered control logic design.

4.6   M_worst_case_analysis

The event-triggered design developed in the previous refinement serves as a specification: each sub-case has the same set of abstract parameters. The concrete values of these parameters are yet unknown but their relationships are constrained by the guards of each sub-case. Next, we show how the worst-case analysis can be used to instantiate the abstract parameters of each sub-case, thereby refining our event-triggered design into a time-triggered one for our heating system.

We illustrate by still using the first sub-case, which is shown in Listing 1.6. In this case, the system is in the "ON" mode that monotonically increases the temperature. If we predict the temperature at $now + \delta$ $(T(now + \delta))$ is below T_{max}, then, we can deduce that the system can safely progress for another cycle. However, progression in this way does not guarantee that, when the next cycle starts, we have enough time to switch mode while remains in the safe envelope (since switching mode inevitably causes inertia that following the dynamics of "ON" mode for another t_{act} seconds). Therefore, the worst-case that the system can bear to increase the temperature is when $T(now+\delta+t_{act})$ equals to T_{max}. In other words, any prediction of $T(now+\delta+t_{act})$ that is below T_{max} $(decision_1^t)$, we can allow the system to stay at the "ON" mode. By expanding $T(now+\delta+t_{act})$

using the analytical solution of "ON" mode dynamics with initial temperature $T(now)$, and initial time now, we can concretize the abstract parameter "buffer" with $\delta + t_{act}$.

Then, since we decide to stay at the "ON" mode, we easily concretize f_n by the dynamics of "ON" mode T_{on}. Moreover, since the sampling time, tx has to be concretized by δ to obtain a periodic time-triggered design.

After the concretization of abstract parameters, our main job is to ensure the newly developed sub-case preserves the semantics of abstract sub-case. This means we have to prove the theorem $safe_{fn}$, which is implied by the following lemma: $\forall t \cdot t \in (now, now + \delta + t_{act}] \Rightarrow T_{on}(t) \in [T_{min}, T_{max}]$. The lemma holds for the first case due to the monotonicity of ON mode, and the fact that the boundary value of temperature on the time interval "$[now, now + \delta + t_{act}]$" is safely between the envelope T_{min} and T_{max}.

Using a similar worst-case analysis, we refine the other 7 sub-cases, and detect 2 bugs in our control logic design. Both bugs are related to a missing requirement on T_{min} and T_{max}, i.e. $T_{max} - T_{min} < 2*\delta$ in order to ensure a safe behavior when switching mode (otherwise, T_{min} and T_{max} would be too close, and mode switching will immediately cause the temperature out of the safe range within one cycle).

```
Event Prediction₁ ≙
  Refines Prediction
  Any buffer fₙ tx
  Where
    decision₁ᵐ: m = ON
    decision₁ᵉ:
    T(now)+buffer ≤ Tₘₐₓ
    safe_fn: ∀
 t·t∈ (now,now+tx]⇒
      fₙ(t)∈[Tₘᵢₙ,Tₘₐₓ]
    grdₛ: s = DECISION
  Then ...
    actuation₁ᵐ: m := ON
End
```

Listing 1.5. M_control_logic

```
Event Prediction₁ ≙
  Refines Prediction₁
  Where
    decision₁ᵐ: m = ON
    decision₁ᵗ:
    Tₒₙ(now + δ + t_act) ≤ Tₘₐₓ
    grdₛ: s = DECISION
  Theorem
    safe_fn:∀ t·t∈(now,now+δ]⇒
      Tₒₙ(t) ∈ [Tₘᵢₙ,Tₘₐₓ]
  Then ...
End
```

Listing 1.6. M_worst_case_analysis

4.7   M_implementation

The last refinement M_implementation will merge all developed control cases into a single event, which is the implementable discrete control as the output of our proposed refinement strategy. The Event-B language make the last refinement rigorous and simple. A snippet of the result for this merging is shown in Listing 1.7. As we can see in the merged event $Prediction$, each of its guard (whose name with the prefix $case$) is a conjunction of predicates that summarize a particular control case (including sensing, decision and actuation). To ensure this refinement is correct, we need to prove that the guards in $Prediction$ implies guards of $Prediction_1$ to $Prediction_8$ w.r.t. M_worst_case_analysis, which is trivially true by rewriting in our case. The full development of this case study developed in Event-B can be found in [7].

```
Event Prediction ≘
  Refines Prediction₁ ... Prediction₈
  Any a b
  Where
      case₁: m=ON ∧ decision₁ᵗ ∧ a=T_on ∧ b=ON
      ...
      case₈: m=OFF ∧ decision₈ᵗ ∧ a=T_off ∧ b=OFF
  Then
      actuation_T: fa := a
      actuation_m: m := b
      ...
End
```

Listing 1.7. M_implementation

4.8   Discussions

Table 1 summarizes our proof efforts for the case study. The second column shows the number of total proof obligations generated by Rodin. The last 2 columns show the number of automatic/manually discharged proofs. All proof obligations are machine-checkable.

Automation of Verification. From Table 1, we see that among 468 of proof obligations, 68% proof obligations can be automatically discharged. This is first thanks to the SMT solver plug-in, Atelier-B reasoner, and native reasoning support in Rodin. In addition, we find that a good way for proof automation is to encapsulate meta-theorems. For example, In M_cycle, we define and prove inductive safety property at an abstract level. This benefits the following refinements to prove similar properties by instantiation. Another good way for proof automation is to identify shared proof patterns in the proof tree to avoid rework of proofs. 32% of proof obligations are discharged manually. This is because that most of them are based on the domain theory of real numbers. We developed this theory using the Theory plug-in of Rodin [6], which currently does not natively support communication with the underlying solvers to encode decision procedures nor verification strategies. We agree that more engineering work is required in this direction.

Table 1. Overview of proof efforts

	Total	Auto.	Man.
M_specification	8	7	1
M_safety	14	11	3
M_cycle	16	9	7
M_close_loop	23	18	5
M_control_logic	42	27	15
M_worst_case_analysis	231	149	82
M_implementation	134	99	35
Total	468	320 (68%)	148 (32%)

Soundness of Veification. Our reasoning for hybrid system safety is essentially based on the axiomatization of hybrid system behaviors, and mathematical theories (e.g. real numbers). To ensure the axiom's consistency of hybrid system behaviors, we instantiate uninterpreted functions (dynamics of "ON/OFF" mode), then ensuring axioms are provable by instantiation (e.g. monotonicity). To ensure the consistency of the axioms of mathematical theories, we could use the realization mechanism by Why3 [9]: to ensure the theories in the Why3 framework are consistent, the developers clearly separates a small core of axioms, then build lemmas on top of the core.

Completeness of Verification. We do not claim that our verification approach for the hybrid system is complete. One of the key research in hybrid system verification is to be able to find appropriate algebraic invariants that quantify the progression of hybrid system states and is useful to prove assertions of interest at the same time. In our current work, we focus on the development of a theoretic framework for hybrid system design with safety constraints. We have not considered yet advanced techniques such as algebraic invariants generation. However, we believe that interact existing frameworks (e.g. Pegasus [19], HHL [16]) on top of our framework would be viable and beneficial for verification completeness.

5   Related Works

In this section, we review 3 categories of methods for hybrid system design.

Classical Engineering Methods. Matlab, based on model-based development, is one way to design complex hybrid systems. In model-based development, domain-knowledge are encapsulated in models, and rely on automated code generation to produce other artifacts. The benefits are saving time and avoiding the introduction of manually coded errors. Fitzgerald et al. propose a hybrid and collaborative approach to develop hybrid systems [10]. Co-modeling and co-simulation are collaboratively used. Classical engineering methods are usually based on a collection of collaborative tools, which do not necessarily have formal documents on their semantics[12,14]. This hinders the possibility to certify the translations among these tools and poses questions for the soundness of their hybrid systems design pipe-lines.

Formal Languages. Platzer designs KeYmaera X tool for deductive verification of hybrid systems [11,18]. Differential dynamic logic (dL) is designed as the back-end, which is a real-valued first-order dynamic logic for hybrid programs. Special rules are designed to ease the complexity of deduction reasoning, e.g. differential invariant (to reason hybrid systems without unique-analytic solutions), evolution domain (specify when to follow certain dynamics, which is equivalent to the role of our control events). Hybrid Hoare logic (HHL) has been proposed by Liu et al. for a duration calculus based on hybrid communicating sequential processes [16]. In HHL, the safety of a hybrid system is encoded as a Hoare-triple with historical expressions. HHL and dL are both deductive approaches. However, the verification process is very different, because of the way they model the message communication. These formal verification approaches are based on the formulation of the domain problem into machine-checkable logical statements. They are *a posteriori* verification process, where both the models and specifications need to be presented at the validation phase.

Stepwise Refinement. Loos and Platzer introduce differential refinement logic to determine whether two hybrid programs refine each other [17]. We both address that how to relate a time-triggered design to an event-triggered design, which is however quite complex in [17]. We think that the complexity stems from a lack of guidance on how to design event-triggered designs at an appropriate abstraction level. Therefore, we propose a pattern, which is still easy to think in terms of event-triggered paradigms, yet make their refinements to time-triggered designs more manageable. To make the Event-B language easier for developing hybrid systems, Banach et al. develop the HEB language [4]. The language proposes many user-friendly syntactic and semantic elements. For example, implicit control over time, encapsulation of domain-specific knowledge, and structured closed-loop architecture (as in M_closed_loop). In [8], Dupont et al. develop an architecture that is similar to HEB in Event-B. In addition, they demonstrate how to encapsulate domain-specific knowledge in Event-B. They also enrich the closed-loop architecture of Banach et al. with perturbation modeling. Both works of Banach et al. and Dupont et al. improve hybrid system designs in Event-B. However, they rely on a well-formedness assumption of the system runs, and have implicit modeling of time. These restrict the form of safety properties and how they can be proved mechanically. We show that this problem can be addressed by borrowing the idea of early action systems [3,20]. Our proposal can be used independently from [4] and [8]. When the integration is considered, we think that the front-end of [4] and [8], which with implicit time modeling, can be kept. Then, they can use separate refinements to link to our strategy for verification at the back-end.

6 Conclusion

The main contribution of this paper is a theoretic framework for hybrid system design with safety constraints. The framework is in the form of a refinement strategy that can coherently refine an abstract hybrid system design with safety constraints down to the concrete one, integrated with implementable discrete control, that can behave safely. In the process, we first derive a discrete control design in the event-triggered paradigm. Then, because the event-triggered discrete control is designed as the specification of a time-triggered one, it is then refined into implementable by worst-case analysis. We demonstrate our proposal on a case study of a smart heating system, which shows its feasibility.

Our future work will focus on: 1) developing more case studies. 2) integrating with advanced reasoning and modeling tools on top of our strategy for interoperability and automation. 3) technology transfer to other programming languages/frameworks for cross-validation.

Acknowledgements. This work is supported by grant ANR-17-CE25-0005 (The DIS-CONT Project http://discont.loria.fr) from the Agence Nationale de la Recherche.

References

1. Abrial, J.R.: Modeling in Event-B: System and Software Engineering. Cambridge University Press, Cambridge (2010)
2. Ameur, Y.A., Méry, D.: Making explicit domain knowledge in formal system development. Sci. Comput. Program. **121**, 100–127 (2016)
3. Back, R.J., Petre, L., Porres, I.: Continuous action systems as a model for hybrid systems. Nord. J. Comput. **8**(1), 2–21 (2001)
4. Banach, R., Butler, M., Qin, S., Verma, N., Zhu, H.: Core hybrid event-B I: single hybrid event-B machines. Sci. Comput. Program. **105**, 92–123 (2015)
5. Bjørner, D.: Domain analysis and description principles, techniques, and modelling languages. ACM Trans. Softw. Eng. Methodol. **28**(2), 8:1-8:67 (2019)
6. Butler, M., Maamria, I.: Mathematical extension in Event-B through the Rodin theory component (2010)
7. Cheng, Z., Méry, D.: The full development of smart heating system case study in Event-B (2020). https://github.com/zcheng05900/verihybrid
8. Dupont, G., Ameur, Y.A., Pantel, M., Singh, N.K.: Handling refinement of continuous behaviors: a proof based approach with event-B. In: 13th International Symposium on Theoretical Aspects of Software Engineering, pp. 9–16. IEEE, Guilin (2019)
9. Filliâtre, J.-C., Paskevich, A.: Why3—where programs meet provers. In: Felleisen, M., Gardner, P. (eds.) ESOP 2013. LNCS, vol. 7792, pp. 125–128. Springer, Heidelberg (2013). https://doi.org/10.1007/978-3-642-37036-6_8
10. Fitzgerald, J., Larsen, P.G., Verhoef, M. (eds.): Collaborative Design for Embedded Systems. Co-modelling and Co-simulation. Springer, Heidelberg (2014). https://doi.org/10.1007/978-3-642-54118-6
11. Fulton, N., Mitsch, S., Quesel, J.-D., Völp, M., Platzer, A.: KeYmaera X: an axiomatic tactical theorem prover for hybrid systems. In: Felty, A.P., Middeldorp, A. (eds.) CADE 2015. LNCS (LNAI), vol. 9195, pp. 527–538. Springer, Cham (2015). https://doi.org/10.1007/978-3-319-21401-6_36
12. Gleirscher, M., Foster, S., Woodcock, J.: New opportunities for integrated formal methods. ACM Comput. Surv. **52**(6), 1–36 (2019)
13. Landau, I.D., Zito, G.: Digital Control Systems Design Identification and Implementation. Springer, London (2010). https://doi.org/10.1007/978-1-84628-056-6
14. Larsen, P.G., Fitzgerald, J., Woodcock, J., Gamble, C., Payne, R., Pierce, K.: Features of integrated model-based co-modelling and co-simulation technology. In: Cerone, A., Roveri, M. (eds.) SEFM 2017. LNCS, vol. 10729, pp. 377–390. Springer, Cham (2018). https://doi.org/10.1007/978-3-319-74781-1_26
15. Leino, K.R.M.: Dafny: an automatic program verifier for functional correctness. In: Clarke, E.M., Voronkov, A. (eds.) LPAR 2010. LNCS (LNAI), vol. 6355, pp. 348–370. Springer, Heidelberg (2010). https://doi.org/10.1007/978-3-642-17511-4_20
16. Liu, J., et al.: A calculus for hybrid CSP. In: Ueda, K. (ed.) APLAS 2010. LNCS, vol. 6461, pp. 1–15. Springer, Heidelberg (2010). https://doi.org/10.1007/978-3-642-17164-2_1
17. Loos, S.M., Platzer, A.: Differential refinement logic. In: 31st Annual ACM/IEEE Symposium on Logic in Computer Science, pp. 505–514. ACM, New York (2016)

18. Platzer, A.: Logical Foundations of Cyber-Physical Systems. Springer, Cham (2018). https://doi.org/10.1007/978-3-319-63588-0
19. Sogokon, A., Mitsch, S., Tan, Y.K., Cordwell, K., Platzer, A.: Pegasus: a framework for sound continuous invariant generation. In: ter Beek, M.H., McIver, A., Oliveira, J.N. (eds.) FM 2019. LNCS, vol. 11800, pp. 138–157. Springer, Cham (2019). https://doi.org/10.1007/978-3-030-30942-8_10
20. Su, W., Abrial, J.R., Zhu, H.: Formalizing hybrid systems with Event-B and the Rodin platform. Sci. Comput. Program. **94**, 164–202 (2014)
21. Zhan, N., Wang, S., Zhao, H.: Formal Verification of Simulink/Stateflow Diagrams - A Deductive Approach. Springer, Heidelberg (2017). https://doi.org/10.1007/978-3-319-47016-0

Model-Based Approach
for Co-optimization of Safety
and Security Objectives in Design
of Critical Architectures

Kunal Suri[1]([✉]), Gabriel Pedroza[1], and Patrick Leserf[2]

[1] Université Paris-Saclay, CEA, List, 91120 Palaiseau, France
{kunal.suri,gabriel.pedroza}@cea.fr
[2] ESTACA, 12 Rue Paul Delouvrier, 78180 Montigny-le-Bretonneux, France
patrick.leserf@estaca.fr

Abstract. During the development of Cyber-Physical Systems (CPS) safety and security are major concerns to be considered as it has been established by various literature. Moreover, these concerns must be included early on during the System Development Life Cycle (SDLC). In this work, we focus on the design-phase of the SDLC to assist the engineers in conducting design-space exploration of the system hardware architecture w.r.t to both safety and security concerns. In this way, the engineers may perform simulations to find a set of quasi-optimal solutions before developing an actual physical prototype. To achieve this, our tooled method builds on our previous work [11] and supports a multi-concern analysis by leveraging Model-Driven Engineering (MDE) techniques such as SysML modeling along with the transformation of SysML models into representations which are finally optimized via constraint solvers. Overall, the method and framework shall support the design of the system architecture from a repository of components based on possible configuration alternatives, which satisfy the system objectives such as reliability and cost. Such functions can help to evaluate the effects of integrating safety and security features thus showing their interplay. The overall approach is illustrated via an automotive CPS case study.

Keywords: MDE · HW architecture · Optimization · Safety · Security

1 Introduction

Similar to information systems, the development of Cyber-Physical Systems (CPS) [10], such as autonomous vehicles [8] and industrial production systems [19] (Industry 4.0 [9]), involves a complex *System Development Life Cycle*

K. Suri–This work is the result of a collaborative project between CEA-LIST and ESTACA, a period during which Kunal was following a postdoctoral fellowship in ESTACA.

C. Attiogbé and S. Ben Yahia (Eds.): MEDI 2021, LNCS 12732, pp. 18–32, 2021.
https://doi.org/10.1007/978-3-030-78428-7_2

(SDLC[1]) [7]. Various stakeholders, namely systems engineers, software engineers, safety engineers, security engineers, are involved during the SDLC, wherein each of them is concerned with a specific aspect (or viewpoint) of the system. As CPS involves human users, it is essential to make them fail-safe by building them in a safety-aware manner. Likewise, these CPS must also be developed in a security-aware manner. This is because in the past many of the safety-critical CPS were constructed and used as standalone systems without giving much consideration to the security aspect. However, today most of these systems are highly interconnected and software intensive, which exposes them to the possibility of cyber-attacks which in turn may lead to safety related problems for the users [3]. Even with the recent increase in the use of Artificial Intelligence (AI) and other automation technology for CPS development, designing these systems still involves a lot of human experts/engineers. Thus, it is important to provide the engineers with methods and automated frameworks that ease the integration of non-functional aspects like, for instance, co-optimization of both safety and security objectives, especially during the development of the system hardware architecture [12,14]. For the same reason, the design phase is crucial within the SDLC, since errors or lacks introduced during the design time can propagate to other phases resulting in wastage of effort, time and resources [20]. This makes it critical to design a sound and well-formed system architecture, which satisfies all the functional and non-functional requirements (i.e., multiple objectives) and to analyze the system objectives via simulations for the desired outcomes. In this way, the engineers will be able to find a set of quasi-optimal solutions before developing a physical prototype of the hardware.

In general, different engineer teams participate during the design process and in particular when safety and security aspects need to be integrated. Thus, it is not expected a single stakeholder having expertise in both domains. Thus, in order to perform the aforementioned multi-objective system analysis, engineers need methods and automated frameworks that will support them to perform the analysis related to safety and security requirements along with the ability to discover or trace the link between these requirements and the lower-level criteria of the system. In the context of this work, our focus remains on supporting the engineers during the design phase following a model-driven approach wherein we build on our previous work [11] by extending it to handle the following: (1) multiple system components and (2) multiple objectives to be included in the optimization criteria for finding the quasi-optimal system architecture. Our method supports a multi-concern analysis by making use of Model-Driven Engineering (MDE) techniques such as SysML modeling along with the automatic transformation of SysML models into representations that can be optimized via relevant algorithms. We illustrate this method by reusing and extending the case study from [11] with the security perspective. Overall, this method shall support the design of the system architecture from a repository of components based on possible configuration alternatives, which satisfy the system objectives such as

---

[1] https://www.nist.gov/publications/system-development-life-cycle-sdlc.

reliability and cost which can impact or be related to both safety and security requirements.

The remainder of this paper is organized as follows: in Sect. 2, we introduce some preliminary information necessary to better understand various concepts used in this work. In Sect. 3, we detail our method along with the various steps (and sub-steps) and in Sect. 4, we detail the experimentation results and the automated framework developed for supporting this design-space exploration. In Sect. 5, we survey some related work and finally, in Sect. 6, we conclude the paper and provide some perspective on the future works.

2 Preliminaries

In this section, we briefly introduce various concepts and techniques necessary to understand the work detailed in this paper. These concepts are as follows:

- **MDE in systems engineering** has a large community of contributors and users along with the availability of mature tools (e.g. Eclipse Papyrus[2]). MDE uses models as the primary artifacts, which help to enhance the understandability of complex systems and aids in the reduction of the chances of error due to the use of principles, standards (e.g. work done by the Object Management Group (OMG[3])) and tools [11,15,16]. MDE supports creation of a coherent model of a system that may be augmented with relevant information for different stakeholders. This model when transformed into different formats allows representing various formalization relevant for different domains. In this work, we make extensive use of SysML 1.4 language along with model transformation technique, which is briefly detailed as follows:
  - **Systems Modeling Language[4] (SysML® 1.4)** is a general-purpose graphical modeling language developed by the OMG for systems engineering (SE). It reuses a subset of Unified Modeling Language (UML) and provides additional extensions to address the requirements in UML for SE. SysML is used to specify, analyze, design, and validate complex systems, which may include any type of system such as hardware, software and information. SysML comprises of the four essential diagrams that are referred to as the *Four Pillars of SysML*, which are: (1) Requirement, (2) Activity, (3) Block, and (4) Parametric diagrams. In this work, we mainly depend on the Requirements and the Block diagrams to model the system depicted in our case study.
  - **Model Transformation** plays a key role in MDE as it allows the generation, sometimes by refinement, of lower-level models from higher level (or abstract) models, eventually generating the executable codes (and viceversa, i.e., reverse engineering). Model transformation allows the mapping and synchronization of models, which may be at the same or different

---

[2] https://www.eclipse.org/papyrus/.
[3] https://www.omg.org/index.htm.
[4] https://www.omg.org/spec/SysML/1.4.

levels of abstraction. As per Czarnecki et al. [6], there are several major categories of model transformation. However, in our work, we use the model-to-text approach to generate executable scripts from our SysML models (based on a template). This template-based approach consists of target text along with the specific code that is generated based on the values provided in the model.

- **Constraint Satisfaction Problems (CSPs)** are mathematical problems, which are defined as a set of variables that have specific conditions that should never be violated while solving these problems (i.e., constraints) [4]. In other words, a CSP consists of the following: (1) a finite set of variables $(V_1, V_2, ..., V_n)$; (2) a non-empty domain having some values to be assigned to each variable $(D_{V_1}, D_{V_2}, ..., D_{V_n})$; and (3) a finite set of constraints $(C_1, C_2, ..., C_n)$, wherein each of the constraint C_i puts a limit on the values permitted or excluded for the variables (e.g., $V2 \neq V3$). CSP is used by researchers in AI to solve problems such as scheduling. The solution set can include a unique solution, all the solutions in the space, or an optimal solution. In this work, we support the engineers to find optimal solutions based on the values of the objective functions provided as input by the user in the SysML model.

- **Multi-Objective Optimization and Pareto Front** is an optimization problem involving multiple objective functions. It broadly falls under the area of decision making involving multiple criteria. These problems are concerned with scenarios wherein it is crucial to simultaneously optimize more than one objective function such as either cost, reliability or performance. These metrics are applied in various domains such as economics, logistics and engineering, where optimal decisions are needed by managing the trade-offs between two or more conflicting objectives. For instance, minimizing the cost of an automobile while maximizing its safety. Likewise, in a multi-objective optimization problem involving several objectives, an optimal solution is called *Pareto optimal* if there exists no possibility to improve an objective function without degrading the others. Thus, a Pareto optimal solution is an optimal trade-off between various objectives. The set of all the Pareto optimal solutions is called as *Pareto Front* (also known as Pareto frontier or Pareto set), which is graphically visible in form of a distinct front of points. In this work, we provide the engineers with a set of Pareto optimal solutions.

3   Method: Model-Based Co-optimization of Objectives

3.1   Overview

In this section, we present the overview of our method that consists of two main steps as detailed in Sect. 3.3 and Sect. 3.4. This method is illustrated based on the case study detailed in Sect. 3.2, which depicts an *Embedded Cognitive Safety System* (ECSS) (see Fig. 2 (source [11])).

This method is an extension of our previous work [11], and is envisioned to illustrate the possibility of integrating both safety and security related objective functions as a way to support decision making so as to avoid unnecessary trade-offs. Furthermore, in this work, one of our major focus has been on the readiness and automation of the framework presented in [11], i.e., to scale it for handling a dynamic range of hardware (HW) components to support the engineering needs. Figure 1 represent the steps (and sub-steps) of our method, which starts with capturing the safety and security based requirements from different stakeholders using SysML requirement diagram (see label 1 in Fig. 1). The model-based requirement gathering allows to capture different requirements and supports traceability between them. Such traceability enables the engineers to see which requirements are related to each other and how they can be tackled together [9]. Next, a SysML model is created, that is annotated with all the relevant information related to the system components along with the values of objective functions from both safety and security aspects (see label 2). The SysML model also contains the information related to the type of variability needed to explore the design-space (detailed in Sect. 3.2). Next, the SysML model is transformed into text (i.e., executable python scripts) using a template-based approach that involves the generation of mathematical optimization models based on the information about variability choices added to the SysML model (see label 3) [2]. Concretely, this model transformation step relies on Eclipse based technologies such as Xtext[5]. The variability choices are transformed into a set of 0-1 variables (integer programming). Then, based on the constraints defined between components of the SysML model, the problem can be solved as a CSP via some standard solvers. In our case, the generated Python scripts are executed using python specific CSP solver[6] (see label 4 in Fig. 1). The solver generates solutions that are visible as Pareto Front graphs. These solutions are based on the calculated value of the objective functions. In our case and to facilitate safety-security interplay, we rely upon basic objective functions which can be common to both safety and security, e.g., cost and reliability. An engineer can choose the best solutions among the trade-off solutions that fit their requirements. This method is implemented using the Eclipse Papyrus framework for both SysML modeling and model transformation.

3.2   Case Study

Figure 2 is sourced from our previous work [11] and depicts an Embedded Cognitive Safety System (ECSS). It is an integrated system on a chip (SoC) used in various domains such as automotive or drones. It supports line detection, obstacle detection and distance measurement with a stereoscopic view. The ECSS embedded hardware platform comprises of CMOS (complementary metal-oxide semiconductor), image sensors, processing elements (CPU), and vehicle interface networks (FlexRay, CAN). The CMOS image sensors are connected to the CPU via Digital Video Port (DVP), a type of parallel bus interface. CPU such as Cortex A9, iMX35 support image processing. The vehicle interface is integrated into

---

[5] https://www.eclipse.org/Xtext/documentation/index.html.
[6] https://github.com/python-constraint/python-constraint.

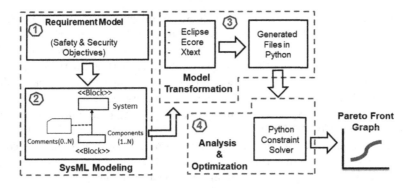

Fig. 1. Steps for finding optimal system architecture

the ECSS with a transceiver component, connected to the processing element
with a digital port (DP) which is a parallel bus interface.

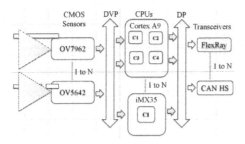

Fig. 2. Embedded Cognitive Safety System (ECSS)

The ECSS must be developed in both safety and cyber-security aware manner
as malicious attacks such as CAN bus attacks may expose the ECSS to component misbehavior or failure [3]. In this context, we consider that the ECSS can
be protected by integration of modules that include built-in security features like
specific encryption mechanisms, Public key infrastructure (PKI), Trusted Platform Modules (TPM) or Hardware Security Modules (HSM) [22,23]. As referred
modules may also require to be redundant so as to satisfy safety requirements,
an impact on the overall cost is expected due to security. Since built-in security
features are often HW greedy, a non-negligible impact on performance and real-
time safety constraints can occur, e.g., delayed braking CAN messages. Thus,
a basic but relevant interplay between safety and security can be studied via
the optimization of both cost and performance objective functions. In this work
towards multi-objective simulation, we mainly focus on the cost function.

Explaining the Design-Space Exploration: To illustrate some of the complexity faced by engineers, we represent a HW setup wherein there are three

slots available for each component of an ECSS (to allow redundancy). The Fig. 3 represents the different types of variability involved in selecting a set of HW components for the ECSS. Hence, the number of slots plays an important role, as the solution space grows exponentially on its number. To better illustrate it, let us consider a HW type such as CPU, wherein three slots shall represent a variability matrix of M×3. Indeed, each row of this matrix represents a set of non-distinct CPU elements from one specific manufacturer. For example, we may assume that row 1 represents a Cortex-A9 from Samsung, while row 2 from STMicroelectronics (ST), row 3 from MediaTek (MT), and so on. This is because there could be several possible configurations used to allow CPU redundancy in the chipset. This M×3 array can accept three main configuration categories which are (1) instance variability (IV), i.e., replicas of the same component of the same type, (2) component variability (CV), i.e., components from a different manufacturer, and (3) mixed variability (MV), i.e., IV + CV. Since for each column of size M, any subset of it is also a configuration, the solution space can grow up to $3.(2^M)$ configurations.

Fig. 3. Design-space definition for HW architecture

As each of the CPU incurs a specific cost (i.e., product cost plus the cost of security features) and provides specific reliability, an engineer must find the optimal redundancy values and variability type for designing an architecture that satisfies the system requirements. Furthermore, this type of variability issue shall exist for each HW component of the system to be designed, making it quite a complex process to find the optimal set of solutions considering all the components and all the different variability configurations.

3.3   Step 1: SysML Modeling

Our method starts with developing a model to capture requirements in textual form and link them to the various modeling elements via relationships such as verify or derive. Figure 4 represents a SysML based requirement model consisting of standard requirement elements. In [11] the authors introduced a new requirement type called <<ArchRequirement>> to model a specific type of architectural requirement by extending the standard requirement element using a stereotype. Likewise, in this work, we assist the engineers to model safety and security requirements by introducing <<SafeReq>> and <<SecReq>>

respectively. The overall system requirements are composed of the aforementioned requirements, which shall be further refined based on the system needs. The $<<ArchRequirement>>$ requirement is evaluated via an objective function $<<HWCostEvaluation>>$, which we introduce as a stereotype by extending the SysML Constraint Block. This objective function is related to the requirement via a dependency called *Evaluate*, depicting a design-time relationship between the elements. The architecture requirement called *MaxRedundancy* introduces a maximum limit to the redundancy of a component (e.g., max number of sensors $= M/2$). These types of restrictions are needed as the hardware redundancy incurs a cost that directly influences the overall system cost.

To better visualize and integrate the Multi-Objective Optimization (MOO) context into system modeling, i.e., a type of analysis context, a SysML Block Definition Diagram (BDD) is modeled (see Fig. 5). This BDD consists of Constraint Blocks along with their relationships having a top-level block called *ECSS MOO*, which references the ECSS system block. This BDD also contains the objective functions that are a representation of the optimization model. The Pareto front is a result of the MOO context and provides various alternatives to the engineer. In this approach, the MOO context shall be passed to an external CSP solver. The results from the solver are provided to the engineer in form of the values of the Pareto front. The objective function extends the standard SysML Constraint Block and consists of the optimization goal (maximize or minimize). The ECSS MDO Context constraint block in Fig. 5 represents the Pareto front via two value vectors, i.e., *BestCost* and *BestRel*, which are produced by the two objective functions. The BDD has constraint properties, i.e., *HWCostEvaluation* and *SystemReliability*, which are typed by the objective function. The *SystemReliability* function, represents the calculation of system reliability (R) based on the parameters received from the ECSS system (the component reliability) and the Zero-One model. Likewise, the *HWCostEvaluation* gets its values from the ECSS including both safety and security components. The Zero-One model represents the optimization model described in Sect. 3.4. It has a parameter and a set of constraints deduced from the ECSS and from the model itself.

As the main part of this step, a SysML BDD is developed for modeling the ECSS architecture (see Fig. 6). Based on the requirement, it shall contain all the information about the underlying hardware resources and their composition.

In our method, the optimization problem to be solved involves finding the right balance between the optimal level for redundancy of each component and the cost of the component. During the first step, only the composition is known and not the redundancy level. We already detailed different types of variability configurations in Sect. 3.2. They help to explore the design space by permuting the redundant elements (including security features) to reach minimal global cost and the maximum redundancy allowed.

3.4   Step 2: Model-Transformation and Optimization

This section details the second step of the approach wherein a SysML BDD model (see Fig. 6) is transformed into a mathematical formalism susceptible to

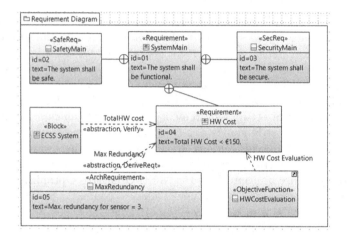

Fig. 4. Requirement model for ECSS architecture optimization

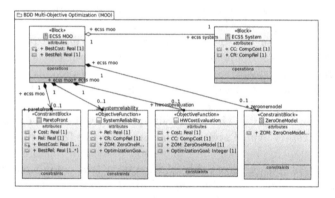

Fig. 5. BDD for MOO context modeling

optimization via a space search and constraint solver algorithm [4]. The representation is based upon zero-one variables that can be solved via a CSP solver. The ECSS is composed of subsystems S_i, wherein each S_i is associated with a given block. A subsystem S_i is nothing but a component slot set (vertical) to be configured by components of the same type selected from repository C_i (provided by the same or different manufacturers). C_{ij} represents the jth component in the repository C_i: each selected component C_{ij} shall have a position j in the repository C_i and can be used at a position k in the subsystem S_i. We define the following sets and parameters:

- $cost_{ij}$ is the cost of the component C_{ij}, while θ_i is the interconnection cost for any component
- rel_{ij} is the reliability of component C_{ij}
- α_{ij} and β_{ij} is the number of input and output ports of component C_{ij}. For instance, in an ECSS, first the video sensor is activated and then the data is

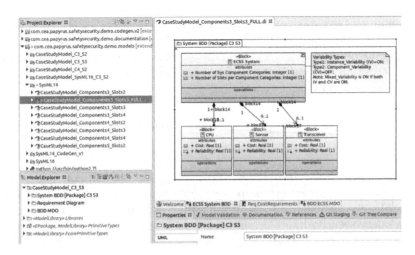

Fig. 6. BDD for ECSS composition

processed, thus the video sensor has no input port and only one output port, i.e., $\alpha_{ij} = 0$ and $\beta_{ij} = 1$.

To find optimal solutions, we first find all the possible solutions and then evaluate each different solution relying upon the mathematical representation of the objective function. For achieving this, we make use of the Python-based CSP package called *Python-Constraint* (version 1.4.0) as it is simple to use and is widely used in various research works.

For the use of zero-one programming, we assume that the range of components C_{ij} will be a M×N matrix, where M is the number of slots for each component (vertical size), and N the number of components (horizontal size). Thus, the problem is defined as follow:

$$\forall i \in S, j \in C_i, k \in S_i$$

$$a_{ijk} = \begin{cases} 1, & \text{if } C_{ij} \text{ is used in system } S_i \text{ at position k} \\ 0, & \text{otherwise} \end{cases} \qquad (1)$$

Constraints: Each system has some default constraints that are derived based on the decision variables. Even if the introduction of new constraints may not be straightforward, some of them have been previously prioritized in different application domains and can be reused, for instance, performance and energy constraints [17]. In our case study, the following constraints are modeled, such as at any position k in the final sub-system, there shall be only one component:

$$\forall i, j, \sum_k a_{ijk} \leqslant 1 \qquad (2)$$

Based on the requirements, several other constraints can be included in the CSP program. This shall allow the solver to provides a better solution targeted to the specific objectives of the system design. For instance, the constraint on a digital connection (system BUS) that connects each sensor to a CPU and each CPU to a transceiver can be represented as:

$$\sum_{j,k} a_{1jk}\beta_{1j} \leqslant \sum_{j,k} a_{2jk}\alpha_{2j} \tag{3}$$

Objective Functions: In our case study, we have a system cost and system reliability which are based on component redundancy and extended from functions in [11]. As previously discussed, the total cost is an objective function that allows a safety-security interplay since components can include features impacting both. Thus, $Cost = Cost_{safety} + Cost_{security}$ and the total cost is given by:

$$TotalCost = \sum_{i,j,k} Cost_{ij} \left[a_{ijk} + exp\left(\theta_i \sum_k a_{ijk}\right)\right] \tag{4}$$

The system reliability (R) is calculated by using the serial-parallel interconnection model, which is given as:

$$R = \prod_i \left[1 - \prod_{j,k}\left[1 - a_{ijk}rel_{ij}\right]\right] \tag{5}$$

The goal of the optimization is to minimize the cost and maximize the reliability, i.e., $min(TotalCost)$ and $max(R)$ by using a different configuration of components based on variability types.

4   Experimentation Results

A proof of concept for our method is implemented using the Eclipse Papyrus framework for modeling SysML models and performing model-transformation (i.e., code generation). The code generation is based on the API's provided by *Papyrus Designer*[7] and the experimentation is performed using multiple SysML models (visible in top left-side of Fig. 6). Each of the SysML models has a set of different input values w.r.t the number of slots available on the hardware, values of the cost and reliability, and the number of sub-components in the systems. The information in these SysML models is used to generate the Python script, which is then executed. It uses the python-constraint API for creating a CSP problem and the related variables, which along with the added constraints provide the set of solutions. Each of these solutions is then used to calculate the objective functions, which are represented as a point on the graph.

---

[7] https://wiki.eclipse.org/Papyrus_Software_Designer.

An output generated is shown in Fig. 7. It depicts two images with two types of variability, i.e., CV and IV. The solutions that are generated represent a quasi-optimal solution for a hardware architecture having slots for component types such as a CPU, Sensor and Transceiver. In this example, the hardware shall have three slots for each component type, i.e., to accommodate up to 3 elements of each type (3×3 matrix). For one of such experiments, the total number of discrete output solutions associated with CV is 16 and IV is 27. The X-axis of the graph represents the failure rate, while the Y-axis represents the total system cost. The engineers can use the framework to perform simulations to find the best suitable choice for the given requirement. The framework assists them by providing the Pareto Front and visual support of the threshold area based on acceptable cost or acceptable failure rates.

As seen in Fig. 7, a visual assistance is provided in the form of the Pareto diagrams and acceptance threshold (e.g. system cost $\leqslant$ 150) which support the engineers to find a (quasi) optimized system having the reliability and cost as per the requirements allocated.

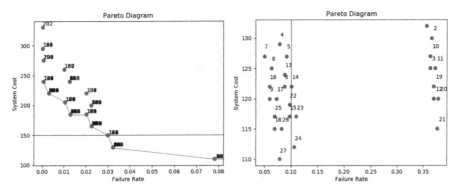

(a) Pareto Front for Component Variability   (b) Graph depicting Instance Variability

Fig. 7. Graph depicting different variability types

5   Related Work

In literature, various academic and industrial research works have addressed the problem of both safety and security for CPS [12] such as EU MERGE[8] project or the EU AQUAS[9] project. Paul [14] categorized such works into four main groups, that are: (1) *independent analysis of safety and security*, i.e., works analyzing either safety or security concerns during the SDLC without considering the other, (2) *augmenting safety engineering with security techniques*, i.e., works

---

[8] http://www.merge-project.eu.
[9] https://aquas-project.eu/.

where various processes, methods and tools in the safety engineering domain are updated with concepts and features from the security domain, (3) *augmenting security engineering with safety techniques*, and (4) *addressing safety and security co-engineering together*, i.e., approaches considering a unification of processes, methods and tools to perform both the safety and security analysis in parallel. Various works have used the strengths of MDE for conduction a co-engineering analysis. Pedroza et al. [16] proposed MDE based framework using a SysML-based environment called AVATAR that captured both safety and security related elements in SysML models. They also gave forth the importance to develop a CPS in a co-engineering manner with Safety and Security perspectives rather than in a standalone manner [15]. Our work complements these approaches by providing a way to solve architecture composition and optimization problems whereas still addressing safety and security concerns. Some existing approaches such as [18,21] assist users to perform multiple analyses, however, they are mainly focused on optimizing the component parameters, such as CPU frequency or memory management, rather than in structural features as in our approach. Meyer et al. [13] proposed an optimization technique for their microwave module but a lack of redundancy constraints is observed. Other approaches such as Design-Space Exploration (DSE) [1] and redundancy allocation problem (RAP) [5] are similar to our approach which remains nonetheless quite generic and extensible, given the few SysML profile specializations introduced and the possibility to easily integrate new user-defined objective functions.

6   Conclusion and Perspectives

In this paper, we proposed a method and framework to perform a multi-objective optimization of system hardware architecture during the design-phase of the SDLC. The approach leverages MDE techniques, especially SysML modeling and model transformation, to assist engineers in the conception of the system architecture during design-space exploration. The tooled method starts with modeling the overall system requirements, including safety and security specific requirements, and then modeling the relevant information about the hardware components to be used and optimized via block diagrams. Such information includes values related to the objective functions, architectural constraints, number of slots, number of components and type of variability. Next, these SysML BDD models are automatically transformed into Python scripts based on the underlying mathematical representation, which includes integer variables, linear constraints and objective functions. The Python scripts are executable and are solved using Python-based CSP solvers. Based, on our case study, an engineer is presented with quasi-optimal solutions, i.e., specific configuration w.r.t hardware components that maximize the objective functions such as reliability while minimizing the global costs. Since the integration of safety features (like redundancy) and security features (like encryption modules) have in common certain objective functions (like cost), the optimization of the latter can help to evaluate the impact and interplay between both. As perspectives, we intend to extend our framework by including more common or specific objective functions related to security and safety to strengthen the approach and coverage of security engineering. Furthermore, we plan to test the approach's scalability with larger and more complex case studies from different CPS domains.

References

1. Apvrille, L.: Ttool for diplodocus: an environment for design space exploration. In: NOTERE, pp. 1–4. ACM (2008)
2. Bettini, L.: Implementing domain-specific languages with Xtext and Xtend. Packt Publishing Ltd. (2016)
3. Bozdal, M., Samie, M., Jennions, I.: A survey on can bus protocol: Attacks, challenges, and potential solutions. In: ICCECE, pp. 201–205. IEEE (2018)
4. Brailsford, S.C., Potts, C.N., Smith, B.M.: Constraint satisfaction problems: algorithms and applications. Euro. J. Operat. Res. **119**(3), 557–581 (1999)
5. Coit, D.W., Smith, A.E.: Optimization approaches to the redundancy allocation problem for series-parallel systems. In: Fourth Indus. Eng. Research Conf. Proc., pp. 342–349 (1995)
6. Czarnecki, K., Helsen, S.: Feature-based survey of model transformation approaches. IBM Syst. J. **45**(3), 621–645 (2006)
7. Eigner, M., Dickopf, T., Apostolov, H., Schaefer, P., Faißt, K.G., Keßler, A.: System lifecycle management. In: IFIP International Conference on PLM, pp. 287–300. Springer (2014)
8. Fagnant, D.J., Kockelman, K.: Preparing a nation for autonomous vehicles: opportunities, barriers and policy recommendations. Transp. Res. Part A: Policy Practice **77**, 167–181 (2015)
9. Kannan, S.M., Suri, K., Cadavid, J., et al.: Towards industry 4.0: gap analysis between current automotive mes and industry standards using model-based requirement engineering. In: ICSAW 2017, pp. 29–35. IEEE (2017)
10. Lee, E.A.: Cyber physical systems: design challenges. In: ISORC, pp. 363–369. IEEE (2008)
11. Leserf, P., de Saqui-Sannes, P., Hugues, J., Chaaban, K.: Sysml modeling for embedded systems design optimization: a case study. In: MODELSWARD 2015, pp. 449–457 (2015)
12. Lisova, E., Sljivo, I., Causevic, A.: Safety and security co-analyses: a systematic literature review. IEEE Syst. J. **13**, 2189–2200 (2018)
13. Meyer, J., et al.: Process planning in microwave module production. In: 1998 AI and Manuf.: State of the Art and State of Practice (1998)
14. Paul, S., Rioux, L.: Over 20 years of research into cybersecurity and safety engineering: a short bibliography. In: Safety and Security Engineering, vol. 5, pp. 335–349. WIT Press (2015)
15. Pedroza, G.: Towards safety and security co-engineering. In: Hamid, B., Gallina, B., Shabtai, A., Elovici, Y., Garcia-Alfaro, J. (eds.) CSITS/ISSA -2018. LNCS, vol. 11552, pp. 3–16. Springer, Cham (2019). https://doi.org/10.1007/978-3-030-16874-2_1
16. Pedroza, G., Apvrille, L., Knorreck, D.: Avatar: a sysml environment for the formal verification of safety and security properties. In: NOTERE, pp. 1–10. IEEE (2011)
17. Roux, B., Gautier, M., Sentieys, O., Derrien, S.: Communication-based power modelling for heterogeneous multiprocessor architectures. In: MCSOC, pp. 209–216. IEEE (2016)

18. Spyropoulos, D., Baras, J.S.: Extending design capabilities of sysml with trade-off analysis: Electrical microgrid case study. In: CSER, pp. 108–117 (2013)
19. Suri, K., Cadavid, J., et al.: Modeling business motivation and underlying processes for rami 4.0-aligned cyber-physical production systems. In: ETFA, pp. 1–6. IEEE (2017)
20. Suri, K., Gaaloul, W., Cuccuru, A.: Configurable IoT-aware allocation in business processes. In: Ferreira, J.E., Spanoudakis, G., Ma, Y., Zhang, L.-J. (eds.) SCC 2018. LNCS, vol. 10969, pp. 119–136. Springer, Cham (2018). https://doi.org/10.1007/978-3-319-94376-3_8
21. Van Huong, P., Binh, N.N.: Embedded system architecture design and optimization at the model level. Intl. J. Comp. Comm. Eng. 1(4), 345 (2012)
22. Wolf, M., Gendrullis, T.: Design, implementation, and evaluation of a vehicular hardware security module. In: Kim, H. (ed.) Information Security and Cryptology - ICISC 2011. pp. 302–318 (2012)
23. Wolf, M., Weimerskirch, A., Wollinger, T.: State of the art: embedding security in vehicles. EURASIP J. Emb. Sys. 2007, 074706 (2007)

Towards a Model-Based Approach to Support Physical Test Process of Aircraft Hydraulic Systems

Ouissem Mesli-Kesraoui[1,2,3(✉)], Yassine Ouhammou[2], Olga Goubali[1],
Pascal Berruet[3], Patrick Girard[2], and Emmanuel Grolleau[2]

[1] SEGULA Engineering, Bouguenais, France
olga.goubali@segula.fr
[2] LIAS/ISAE-ENSMA and University of Poitiers Chasseneuil du Poitou,
Poitiers, France
{ouissem.mesli,yassine.ouhammou,grolleau}@ensma.fr,
patrick.girard@univ-poitiers.fr
[3] Lab-STICC/University of Bretagne Sud LORIENT, Lorient, France
pascal.berruet@univ-ubs.fr

Abstract. The physical integration of an aircraft consists of the assembly of several complex subsystems (including hydraulic systems) developed by different stakeholders. The cleanliness of the developed hydraulic subsystems is ensured by performing several decontamination and flushing tests. This testing phase is very tedious as it is mainly performed by SCADA (Supervisory Control and Data Acquisition) systems and depends on chemical substances. However, as the design is mainly expressed in informal textual languages and synoptic diagrams, this testing is currently done manually and is determined by the experience of the testers. This makes it error-prone and time-consuming.

In this paper, we propose to capitalize the effort for physical testing of hydraulic systems by proposing a model-based system engineering approach that allows: (i) to graphically specify the systems under test and (ii) to automatically generate the corresponding test cases. A proof of concept is proposed as well as a case study.

Keywords: DSML · MBT · Avionic test · Hydraulic system · Flushing

1 Context and Positioning

The aircraft is composed of complex subsystems called hydraulic systems such as: the hydraulic power system (ATA 29), the landing gear system (ATA 32), the flight control system (ATA 27), etc. [4]. A hydraulic system is a complex assembly of hydraulic, electronic, and mechanical components. This system uses a pressurized fluid to transfer energy from one point to another [6]. Before commissioning an avionics system, a series of tests should be performed to ensure

C. Attiogbé and S. Ben Yahia (Eds.): MEDI 2021, LNCS 12732, pp. 33–40, 2021.
https://doi.org/10.1007/978-3-030-78428-7_3

that it is free of contaminants. The presence of contaminants in a hydraulic circuit can alter its function and damage it. To reduce this damage, the hydraulic system and its components must be flushed [6] to remove contaminants that may have been created during the components fabrication, assembly and/or maintenance. Flushing is one of the best and most cost-effective solutions for removing contaminants from hydraulic circuits. For efficient flushing, the system is decompressed in loops (a portion of the circuit consisting of equipment and piping) [8].

To perform the flushing test, the testers follow three steps (Fig. 1). In the first step (Specification Translation), the testers obtain and translate the specification documentation needed to identify the test loop. These documents are essentially: (i) The physical architecture of the system under test. It is a Piping and Instrumentation Diagram (P&ID) which illustrate the connections between different the equipment and piping; (ii) The flushing process and constraints. The flushing process consists of a description of the components and their roles: the component to be flushed, the generating components, the component that stores the fluid, valves to circulate the fluid, etc. A set of constraints to be respected like the velocity, loop order, etc. are also described in constraints documents. In the second step (Identification of abstract loops), testers use the above documents to divide the circuit into sections or loops (part of the circuits on the physical architecture). This ensures that the velocity is maintained (test constraints compliance). Finally, in the third step (Translation of abstract loops to concrete loops) the testers convert the abstract loops into executable loops by adding information about the role of each component and how the components are handled (opening the generating component, operating the valves to allow the fluid to flow towards the loop, etc.). These concrete loops are then executed on the SCADA test bench according to a specific sequence to ensure that the fluid only flows through cleaning sections.

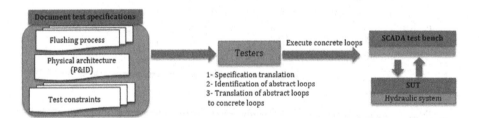

Fig. 1. Testing of the avionics hydraulic system

The manual definition of test loops after the above steps is fraught with several problems. First, the translation process (step 1) is a tedious task, since each piece of information (physical architecture, flushing process, and constraints) must be translated into a different language. In fact, nowadays, the information about the test process and test constraints is mainly scattered in a document-based manner (e.g., text and Excel files with natural languages), and the information about the hydraulic system is schematic-based (e.g., Visio and PDF files) without any domain semantics. Second, abstract and concrete loop identification are often written manually by testers. Writing tests manually is a tedious, time and resource consuming task and often prone to omissions and errors.

The problem of identifying flushing loops and their location has attracted much attention in the context of drinking water decontamination [7,8]. However, the solutions proposed for drinking water decontamination cannot be applied to hydraulic circuit flushing. Firstly, hydraulic circuits are complex and, unlike water circuits, have many different components, and secondly, the test conditions of hydraulic circuits are very different from those of water circuits.

To adapt to the increasing complexity of avionics system testing and to reduce the manual effort required for testing, MBT was developed to make it easier for testers to test avionics systems and to focus their efforts on the SUT and the features under test [10]. MBT enables the automatic generation of test cases from system models. Otherwise, while the MBT method has been used with success in automating certain tests of avionics systems [1,10], it has never been used in generating flushing loops.

In this paper, we propose a model-based approach that leads to correct-by-construction test cases (flushing loops) to reduce tester effort. This can shorten the time-to-market as designers and testers can work together on data-centric models. For this purpose, we first propose a Domain Specific Modeling Language (DSML) to play the role of a pivot language to design and collect all the information of the hydraulic system under test, as well as the flushing process and constraints. The DSML is intended to improve the reusability of hydraulic system test bench designs and to strengthen semantics. Second, we use the paradigm Model-Based Testing (MBT) to automate the generation of test cases (loops) from instances of the proposed pivot language.

The rest of the paper is as follows. In Sect. 2, we introduce our DSML and show its use for the automatic generation of flushing loops for hydraulic systems. In Sect. 3, we evaluate our approach using a case study and discuss the obtained results. Section 4 concludes this paper and discusses some future research directions.

2   Our Contribution: A Reusable and Optimized Approach for Flushing Avionics Systems

Due to the lack of space, in this paper we focus on two contributions corresponding to the first two steps of the flushing process. The first one consists in defining a new pivot language to express the P&ID of the hydraulic system, the flushing process and the flushing constraints in one language. The second contribution is the automatic generation of the abstract loops for the flushing test.

2.1   Towards a Pivot Language

A DSML is a language tailored to the needs of a particular domain. It allows to create models of complex systems that are closer to reality and use domain specific vocabularies to facilitate the understanding of different designers [9]. The use of domain concepts and terms allows experts to understand, validate, modify, or even create a DSML.

The abstract syntax of our pivot language is modelled by a metamodel augmented by several OCL (Object Constraint Language [3]) constraints that capture domain rules.

Core-Elements of the Proposed Abstract Syntax. Our pivot language (Fig. 2) allows modeling the physical architecture of a hydraulic system, the functions that ensure the flushing process, and the test constraints.

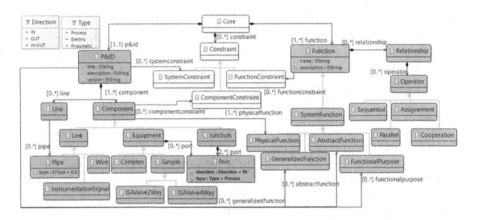

Fig. 2. Core elements of the proposed abstract syntax

The Physical Architecture Facet. The root class in the physical architecture metamodel is `P&ID` (left part in Fig. 2), identified by a title, description, and version. The P&ID consists of a set of components that may be `Equipment`, `junction` or `link`. An equipment can be `Simple` or `Complex`, which consists of other simple equipment. A junction can connect equipment by their ports (In, Out, etc.) to a link. A `Pipe` link is identified by a diameter (dash).

We also added several OCL constraints to capture domain rules. For example, the following constraint limits the number of process ports to 2 for components of type `ISAValve2ways`.

```
1    Context ISAValve2Way inv:
2    self.port->count(MM::Type::Process) = 2
```

The Function (Test Process) Facet. Each `Function` (right part in Fig. 2) is identified by a name and a description. `SystemFunction` are functions performed by the system. Functions are divided into (`FunctionnalPurpose`, `Abstract- Function`, `GeneralizedFunction` and `PhysicalFunction`). The first one describes the high level functions of the flushing process, this function is always the purpose of the design of the test bench (e.g. cleaning the circuits) which is adopted by the whole SUT. The `AbstractFunction` is used to describe the function that is actually to be realized by the system (flushing). The `Genralized- fucntion` is adopted by a subsystem of the SUT (e.g. Isolate the loop). And the low-level function is the `PhysicalFunction` inherent to the elementary components (Supply hydraulic power). Functions are connected by operators that allow the execution order of functions. There are several operators: `sequential` (functions are executed sequentially) and `parallel` (functions are executed in parallel). To assign a `PhysicalFunction` to a component, the `Assignment` operator is used. A `GeneralizedFunction` is guaranteed by more than one component, for which the `Cooperation` operator is provided.

The Constraint Facet. A `Constraint` may refer to a function, a system, or a component (Fig. 2). For example, the `FunctionalConstraint` describes constraints that refer to system functions.

Testers Designing Test Model. We develop a graphically concrete syntax to our pivot language that allows information to be represented in a familiar form to testers. In the first phase, testers used the proposed tool to design P&ID of the hydraulic system under test by dragging and dropping components and connections stored in a proposed library. During the creation of the P&ID, testers must specify properties for the different components (diameter, length, supported pressure, etc.). Once the P&ID is created, the testers can specify the different functions that describe the flushing process and the flushing constraints.

2.2   Graph and Abstract Loop Generation

We propose an MBT-based approach to identify the different flushing loops in the hydraulic system (Fig. 3). This approach consists of 3 steps. In the first step, the modelling of the P&ID, the flushing process and the test conditions is performed by our proposed tool (graphical concrete syntax). The resulting instance is called *test model*. The second step consists in transforming the test model into an oriented graph. For this purpose, each P&ID component (valve, cylinder, pump, etc.) is translated into a node and the piping into edges between nodes. In our case, loop identification remains to identify a path in the generated graph. To identify the loops, the components of P&ID are divided into: *source*, *well*, *waypoint*, and *goal*. The **source** component is the component that supplies hydraulic energy to the loop. The **goal** component is the component that needs to be flushed. The component **waypoint** is a component through which the fluid must necessarily pass, and the path ends in the component **well** that recovers the flushed fluid. The generated graph is used in the third step to derive the abstract loops of the flush. For this purpose, a search algorithm (Dijkstra algorithm [5]) is used to find a path from the source component to the target component and from the target component to the well component. Specifically, this step calculates the different loops needed to flush the hydraulic system.

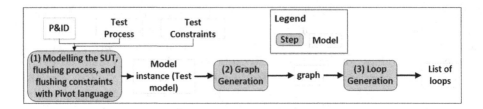

Fig. 3. MBT-based flushing loops generation

3   Proof of Concept

To emphasise our proposal and show its importance, we use the landing gear avionics system [2]. Figure 4 shows the P&ID of the hydraulic circuit that controls the landing gear and associated doors [2]. It contains 3 sets of landing gear (front, right, left). Each set consisting of a door, the landing gear and the hydraulic cylinders. This system is powered by the general Electro-valve which is connected to four Electro-valves. The first valve controls the closing of the different doors (front, right, left) and the second is used to open these doors. The third and fourth valves are used to extend and retract the three landing gears (front, right, left). The opening/closing of the doors and the retraction/extension of the landing gear are done by cylinders.

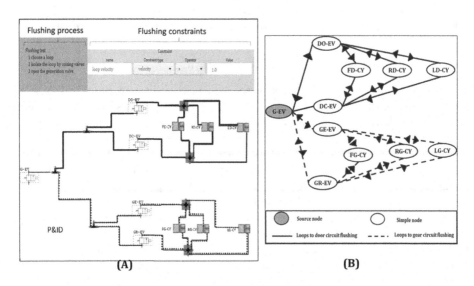

Fig. 4. (A): The modelling of the three facets of the Landing gear system with our tool. (B): Graph generated from the test model

3.1 Method

On this system, we applied our MBT-based approach to generate the flushing loops. First, we used the proposed modelling tool to capture all the required information about the flushing of the hydraulic system. We design P&ID in the new pivot language using the component library included in the modelling tool (Fig. 4). Then we specify the flushing process (functions), and the flushing constraints to obtain the Model Instance (test model). After the modelling step, we generate an oriented graph from the pivot language specification. The obtained graph of the landing gear system is shown in Fig. 4. The various steps of Dijikstra's algorithm are applied to this graph. The purpose of flushing in this system is to flush the different cylinders (LD-CY, RD-CY, DF-CY, LG-CY, RG-CY and FG-CY) corresponding to left-door cylinder, right-door cylinder, front-door cylinder, left-gear cylinder, right-gear cylinder and front-gear cylinder respectively. These cylinders correspond to the goal component. The loop used to flush a target component consists of two paths. A pressure path starts at the source component and ends at the goal component. A return path starts from the goal component to the well component.

3.2 Results and Discussion

The pivot language proved to be very useful as it successfully separated what the tester was doing (specifying) and how the test was implemented (generating loops). For the landing gear system example, our approach generates 4 loops. These loops are represented by full and dashed lines in Fig. 4. Each loop is chosen to be isolated to be flushed separately. Two full line loops allow the door circuit (front door cylinder, right door cylinder, left door cylinder, and various piping between these components) to be flushed by alternately operating the closed and open Electro-valve. In fact, if the close Electro-valve is used to circulate fluid from the General Electro-valve to supply the front door cylinder, right door cylinder, left door cylinder and various piping components, then the open Electro-valve is used to return fluid from the cylinders to the General Electro-valve, and vice versa. The second loop takes the opposite path (Electro-valve is used for supply and the close Electro-valve is used for return). The dotted line loops (two) are used to flush the front gear cylinder, the left gear cylinder and the right gear cylinder by selecting the supply valve and the return valve (selecting between the extend Electro-valve and the retract Electro-valve).

We then compared these generated loops with the manual loops created by testers and they are correct. This result proves that our pivot language contains all the necessary information to generate all the required loops for the hydraulic system. Nevertheless, it will be useful to improve the calculation process. For this purpose, some constraints must be considered in the calculation process: (i) consider the flushing constraints (velocity, flow, and pressure conditions) on the optimization algorithm; (ii) add a success constraints that is used to evaluate the cleanliness of a flushed component. If a loop is unsuccessful, that loop must be re-executed.

4   Conclusion

Testing is the most important step that cannot be neglected in the design of aircraft hydraulic systems. This step is always preceded by the design of test loops. The latter can be time consuming and tedious, depending on a large number of components and constraints that need to be considered. Our goal is to reduce this design effort by providing an approach that allows testers to easily define test models (P&ID, test constraints, test process) and generate the required loops. We first proposed a pivot language to unify the specification of all facets in one language. This language is supported by a proposed tool that is closer to the domain. Using this tool, experts from different domains can easily create specification models without learning a new language. Thus, the work of experts is focused on the specification. To generate loops from specifications, we proposed an MBT-based approach to determine the constitution of a loop.

Acknowledgement. We would like to thank Mr. Mikaël LE ROUX expert from SEG-ULA Montoir De Bretagne-France, for his collaboration and his involvement.

References

1. Arefin, S.S.: Model-based testing of safety-critical avionics systems (2017)
2. Boniol, F., Wiels, V., Aït-Ameur, Y., Schewe, K.D.: The landing gear case study: challenges and experiments. Int. J. Softw. Tools Technol. Transfer **19**(2), 133–140 (2016). https://doi.org/10.1007/s10009-016-0431-4, https://hal.archives-ouvertes.fr/hal-01851720
3. Cabot, J., Gogolla, M.: Object Constraint Language (OCL): a definitive guide. In: Bernardo, M., Cortellessa, V., Pierantonio, A. (eds.) SFM 2012. LNCS, vol. 7320, pp. 58–90. Springer, Heidelberg (2012). https://doi.org/10.1007/978-3-642-30982-3_3
4. Collinson, R.P.: Introduction to Avionics, vol. 11. Springer Science & Business Media (2012)
5. Dijkstra, E.W., et al.: A note on two problems in connexion with graphs. Numer. Math. **1**(1), 269–271 (1959)
6. ISO 16431: Hydraulic fluid power – system clean-up procedures and verification of cleanliness of assembled systems. Standard ISO 16431:2012(E), International Organization for Standardization (2012)
7. Poulin, A.: Élaboration de procédures d'intervention en réponse aux contaminations se produisant en réseaux d'eau potable. Ph.D. thesis (2008)
8. Poulin, A., Mailhot, A., Periche, N., Delorme, L., Villeneuve, J.P.: Planning unidirectional flushing operations as a response to drinking water distribution system contamination. J. Water Resour. Plan. Manag. **136**(6), 647–657 (2010)
9. Sloane, A.M.: Post-design domain-specific language embedding: a case study in the software engineering domain. In: Proceedings of the 35th Annual Hawaii International Conference on System Sciences, pp. 3647–3655. IEEE (2002)
10. Yang, S., Liu, B., Wang, S., Lu, M.: Model-based robustness testing for avionics-embedded software. Chin. J. Aeronaut. **26**(3), 730–740 (2013)

Multi-facets Contract for Modeling and Verifying Heterogeneous Systems

A. Abdelkader Khouass[1,2]([envelope]) [ORCID], J. Christian Attiogbé[1] [ORCID],
and Mohamed Messabihi[2]

[1] University of Nantes, LS2N CNRS UMR 6004, Nantes, France
`christian.attiogbe@univ-nantes.fr`
[2] University of Tlemcen, LRIT, Tlemcen, Algeria
{`abderrahmaneabdelkader.khouass,`
`mohamedelhabib.messabihi`}`@univ-tlemcen.dz`

Abstract. Critical and cyber-physical systems (CPS) such as nuclear power plants, railway, automotive or aeronautical industries are complex heterogeneous systems. They are perimeter-less, built by assembling various heterogeneous and interacting components which are frequently reconfigured due to evolution of requirements. The modeling and analysis of such systems are challenges in software engineering. We introduce a new method for modeling and verifying heterogeneous systems. The method consists in: equipping individual components with *generalized contracts* that integrate various *facets* related to different concerns, composing these components and verifying the resulting system with respect to the involved facets. We illustrate the use of the method by a case study. The proposed method may be extended to cover more facets, and by strengthening assistance tool through proactive aspects in modelling and property verification.

Keywords: Heterogeneous systems · Components assembly ·
Generalized contracts · Modeling and verifying · Formal analysis

1 Introduction

Critical and cyber-physical systems (CPS) that exist in large industries, such as nuclear power plants, railway, automotive or aeronautical industries are complex heterogeneous systems. They do not have a precise perimeter, they are open and often built by assembling various components. Their complexity forces one to have a wide variety of heterogeneous components, frequently reconfigurable due to requirements evolution.

With the advent of concurrent and distributed systems, Component Based Software Engineering (CBSE) [10] has known a high interest. The construction of a distributed system involves several specific components; this requires rigor, methods and tools. The involved components may deal with various *facets*. A facet is a specific concern or a property such as data, behaviour, time constraints,

© Springer Nature Switzerland AG 2021
C. Attiogbé and S. Ben Yahia (Eds.): MEDI 2021, LNCS 12732, pp. 41–49, 2021.
https://doi.org/10.1007/978-3-030-78428-7_4

security, etc.[1]. Therefore, if the integration of components into a global system is not mastered, it may generate considerable time losses and overcharges, because of inconsistency of requirements, incompatibility of meaning and properties, late detection of composition errors, etc. For these reasons, the modeling and formal analysis of such heterogeneous systems are challenging. The use of efficient methods and techniques is required to face these challenges.

We aim at studying and alleviating the difficulties of practical modelling and integration of heterogeneous components. We propose a novel approach based on contracts for modeling and verifying complex and heterogeneous systems. Our approach (named "ModelINg And veRifying heterogeneous sysTems with contractS" (Minarets)) consists in modeling and verifying a system with the concept of *generalized contracts*. The contract is generalized in the sense that it will allow one to manage the interaction with the components through *given facets*: the properties of the environment, the properties of the concerned components, the communication constraints and non-functional properties (quality of service for example). The use of contracts during the verification reduces the complexity of the analysis of heterogeneous systems; moreover, the structuring of contracts with facets and priority of properties makes it possible to decrease the difficulty of checking heterogeneous systems, to save time and to increase performance during verification.

The rest of the article is structured as follows. Section 2 introduces the modeling and verification methodology. In Sect. 3 we illustrate our approach with experimentation and assessment. Section 4 provides an overview of related work, and finally, Sect. 5 gives conclusions and future work.

2   Modeling and Verifying Using Generalized Contracts

An issue to be solved for heterogeneous systems is that, the interface of involved components should be composable. For the sake of simplicity of the composition, we adopt the well-researched concept of contract which is therefore extended for the purpose of mastering heterogeneity of interfaces. Moreover, for a given system we will assume *agreed-upon facets* such as data, functionality, time, security, etc.

In this work we chose the PSL language [1] to specify contrats. PSL is a formal language for specifying properties and behaviour of systems. It is an extension of the Linear Temporal Logic (LTL) and the Computation Tree Logic (CTL). PSL could be used as input for formal verification, formal analysis, simulation and hybrid verification tools. PSL improves communication between designers, architects and verification engineers. We use the ALDEC Active-HDL[2] tool that supports PSL.

For experimentation purpose, we use ProMeLa and SPIN [6] to model and verify components. SPIN is an automated model checker which supports

---

[1] This idea also appears as the separation of concerns in aspect-oriented programming/design.
[2] https://www.aldec.com/en/products/fpga_simulation/active-hdl.

parallel system verification of processes described with its input PROtocol MEta LAnguage (ProMeLa). We also use the model checker UPPAAL [4].

2.1   Definitions

We extend the traditional A-G contract with the purpose of mastering the modeling and verification of complex and heterogeneous systems.

Definition 1 (Generalized Contract). *A generalized contract is a multi-faceted Assume-Guarantee contract. It is an extension of contract, structured on the one hand with its assume and guarantee parts, and structured on the other hand according to different clearly identified and agreed-upon facets (data, functionality, time, security, quality, etc.) in its assume or guarantee.*

The generalized contract will be layered to facilitate properties analysis. Every layer will have a priority. Therefore, an analysis of a facet may be done prior to another facet.

Definition 2 (Well-Structured Component). *A well-structured or normalised component is a component equipped with a generalized contract, acting as its interface with other components.*

Normalising a component C_i consists in transforming C_i into a component equipped with a generalized contract. A multi-faceted A-G contract will be expressed in PSL.

2.2   Outline of the Proposed Method

The working hypothesis is that a heterogeneous system should be an assembly of *well-structured components* (see Definition 2). The method that we propose (Minarets) consists in, given a set of appropriately selected or predefined elementary components, normalizing these input components prior to their composition, building a global heterogeneous system, and finally analysing this global system with respect to the required properties.

For this purpose there are many issues to be solved:

i) Elementary components are from various languages and cover different facets, a pragmatic means of composition is required. We consider PSL as a wide purpose expressive language to describe generalized contracts. Each component will be manipulated through its *generalized contract* (see Definition 1) written in an appropriate language.

ii) Global properties are heterogeneous; they should be clearly expressed, integrated and analysed; they will be expressed with a wide purpose language such as PSL; we will decompose them according to the identified agreed-upon facets and spread them along the analysis of composed components.

iii) Composition of elementary components should preserve their local requirements and should also be weakened or strengthened with respect to global-level properties. For instance, some facets required by an elementary component could be unnecessary for a given global assembly, or some facets required at a global assembly may be strengthened at a component level.

iv) Global properties require heterogeneous formal analysis tools; this generates complexity. We choose to separate the concerns, so as to target various tools and try to ensure the global consistency.

v) Behaviours of components should be composable.

The Minarets method integrates solutions to these issues, as we will present in the sequel. We adopt a correct-by-construction approach for the assembly of components. Therefore, local compositions should preserve required properties of components. In the same way, global properties may impact the components; therefore, global properties are decomposed and propagated through the used components when necessary.

3   Case Study

We consider a case study of an automotive industry: a car painting workshop. This workshop is composed of three main components; a control station (CS) which manages a paint station (PS) and an automatic robot painter (RP). These components interact with each other to achieve the painting process correctly; that means to paint the cars with the desired color, in time, without any damages and without wasting color. More details can be found in [7].

Now we follow step by step the Minarets method to model and verify this system.

Step 1 *(Modeling M_1)*. Express the informal global requirements and global properties. They could be expressed with any desired language, even the natural one. The only goal here is to clearly state the requirements of the given system.

Global Requirements: Cars data are provided (type, color with RGB quantity, painting time); sufficient RGB colors are in the tanks; freeing time must be defined (the freeing time is the release time of the painting station).

Global Expected Properties: The system respects the correct RGB dosage; there is no loss of color; the painting time should be equal to the given time; the painting is done without damages (respect of car dimensions, it is known from the car type); freeing time must be equal to the given freeing time for each car; painting should be stopped and notified when there is no sufficient color; painting station status must be free before use, and busy when a car is inside; the painting starts after the end of configuration, and it finishes when the painting time is equal to given painting time.

Step 2 (*Modeling M_2*). Formalise the required global properties with appropriate expressive formal specification languages (PSL in our case), we obtain the following formalized global properties.

Global Requirements:

```
get_type = true; get_color = true; get_painting_time = true;
/* the painter never start working without car dimensions (get_type)*/
get_freeing_time = true;
R_tank_quantity >= R_GivenColor_quantity;
...
```

Global Properties:

```
R_color_PaintedQuantity = R_GivenColor_quantity;
painting_time = GivenPainting_time; car_type = given_type;
freeing_time = given_freeing_time;
/*each color is controlled separately*/
if (RGB_tank_quantity <= RGB_GivenColor_quantity)
Then (stop_process and warning_message);
painting_time = GivenPainting_time imply painting = finished;
...
```

Step 3 (*Modeling M_3*). Here we have to model or select the components. We use UPPAAL and ProMeLa to model the components RP, CS and PS. Figure 1 shows the models of the components CS and PS within both environments.

Step 4 (*Modeling M_4*). We decompose the global properties with respect to the facets that we considered (Data, functionality, time, security); we obtain the *generalized contract* decomposed with the facets. Figure 2 shows a part of the faceted and formalized properties (the guarantee part); this will emphasize the concern to be dealt with later on.

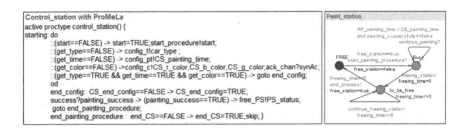

Fig. 1. CS model with ProMeLa and PS with UPPAAL

Step 5 (*Modeling M_5*). We express the structured and formalized properties with the PSL language as depicted in Fig. 2.

Step 6 (*Modeling M_6*). Normalizing the individual components (see Definition 2). We integrate the assumptions and guarantees of each individual component. Figure 3 shows the normalized individual components.

```
for RP in painting_status                    for RP in painting_status
  property p0 :always                          property p1 :always
            (get_type = true  and                      (PS in Busy_status  and
             get_color = true and                       CS in end_configuration and
             get_time = true )                          R_tank_color >= CS_R_GivenColor and
  Data : assert p0;                                     B_tank_color >= CS_B_GivenColor and
end                                                     G_tank_color >= CS_G_GivenColor)
                                               Security : assert p1;
                                             end
```

```
property p5: always  (CS_time_painting = given_time -> RP in OFF_status)
Time: assert p5;

property p6 : always (deadlock -> (PS in FREE and RP in OFF and CS in End ))
Functionality assert p6;
```

Fig. 2. A part of faceted and formalized global property with *PSL* and *Aldec Active-HDL*

Step 7 *(Modeling M_7)*. According to the result of *Step 4 (modeling M_4)*, we may add a facet of the global property to a component, or ignore some of its facets, if necessary. In the current case study it is not necessary to add or ignore a facet (see Fig. 3).

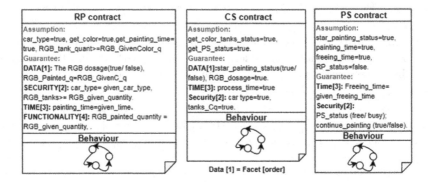

Fig. 3. Normalized components with the prioritised facets

Step 8 *(Modeling M_8)*. We attribute the following priorities to each facet (data = 1, security = 2, time = 3 functionality = 4; were 1 is the highest priority). We obtain ordered layers with respect to facets and properties as well. The order of layer is mentioned in Fig. 3. The verification by layer allows one to verify the contracts by order; from a very important layer (primary) to a less important layer (secondary). If the behaviour of our system does not satisfy a primary layer of contract, then, it is not necessary to continue the verification with the other layers.

Step 9 *(Verification V_1)*. We have to check the appropriate functioning of each normalized individual component if tools exist for that and if the required data are available. As we use ProMeLa we have an adequate tool (SPIN) but, the only component CS modelled with ProMeLa cannot be verified without composition with its environment.

Step 10 *(Modeling M_9)*. As the checking of the normalized individual component CS cannot be carried out, due to the need of composition with its environment; we translate the ProMeLa component CS to UPPAAL using the algorithm presented in our RR [7], we obtain a component CS ready for composition (see Fig. 4).

Step 11 *(Modeling M_{10})*. We compose the translated component CS with the other components PS and RP with the UPPAAL tool. As we focus on behaviours expressed with LTS, the composition results in parallel composition. We obtain the composed system depicted in Fig. 4.

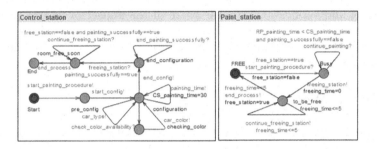

Fig. 4. CS and PS models after the component translation (in UPPAAL)

Step 12 *(Verification V_2)*. We translate the *generalized contract* (only the desired properties to be verified) of each individual component, from *PSL* to the UPPAAL language. We obtain properties ready for verification with UPPAAL. The following property is an extract of the translation from the RP component.

```
A[ ]  Robot_painter.painting imply get_type == true
and get_color == true and get_time == true³
```

Step 13 *(Verification V_3)*. We verify the properties of the same layer together; i.e. to verify each component by layer: data, security, time, functionality; also, the primary properties before the secondary. At the end of this step, we obtain the verified components. Figure 5 shows the verification status of the translated properties.

Assessment. This experimentation was conducted in order to improve our method. We have considerably detailed the steps of the method when thinking thoroughly about its applicability through the case study. Despite the success of applying in preliminary trivial exercises and on this case study, and the reproducibility of the steps, it appears that more tool assistance is needed to guide

---

³ were "A [ ] Prop" denotes the "always property".

Fig. 5. Status of properties verification with UPPAAL

the users. The experimentations even if not yet scalable to industrial cases, give the opportunity to tune the method steps and design some translators to help in modeling and verification. We are aware of the impact of treated facets on interactions between various tools; but, more lessons from various case studies will help to promote good practices through tools. Our report [7] contains more detailed experimention examples.

4   Related Work

Several works contributed to the heterogeneity issues, by proposing different methods and techniques. Interface theory [2] and theories of contracts [5] are based on the contracts of components and the pieces of information given by their interfaces; they use intensively contracts as well as the behaviours; they propose techniques and methods to promote concurrent development.

The Ptolemy project [8] proposes an approach of interaction between heterogeneous components based on models of computation (MOC); here the heterogeneity is linked to different models of computation. From our point of view, this composition method is heavy, general and constrains the use of contracts, especially when one deals with a small net of communicating processes [3]. In our Minarets method we aim to exploit more advantages of contracts in a simpler and explicit way to model and verify the assembly of heterogeneous components.

Our modeling and verification method is based on previous results [3] for heterogeneous components composition; this deals with the formal composition of heterogeneous components, and it is based on the dynamic behaviour of the components where labelled transition systems have been used as common semantic domain; the method which was proposed there is focused on the use of an algebra of operators to manage the communication mechanisms between the assembled components. This approach is supported by the aZIZA tool[4].

As a part of the B.I.P project, [9] proposes a technique with three layers: "*Behavior, Interactions, Priorities*" (B,I,P). It is a low level solution that deals with the interaction between components; it focuses on the composition of the system with the different interaction semantics; however, unlike Minarets, this approach didn't deal with the heterogeneous components modeling but it deals with the property analysis and composition of components using the B.I.P language.

---

[4] https://aziza.ls2n.fr/ [3].

5   Conclusion

We have proposed the Minarets method for complex and heterogeneous systems modeling and analysis; it is based on an extension of the traditional contracts, resulting in generalized contracts used as standard interfaces between components. Generalized contracts are structured with several facets, depending on the concerns or the properties that we are dealing with. Minarets emphasizes the stepwise composition of heterogeneous components through their generalized contracts. We have shown how one can reduce the complexity of the global modeling and the global analysis of complex and heterogeneous systems.

We illustrated our approach with an example of a painting workshop in the automotive industry domain; it involves different facets (data, functionality, security, time). We have checked the properties concerning the various facets. Concerning verification, generalized contracts are first expressed in PSL, then translated into input languages of UPPAAL/SPIN model checkers.

Future works address not only the scalability but also the study of various policies for the composition of contracts; it is important to verify the components composition as we done through the used tools but, to improve the method we are defining in a formal way the semantics of parallel composition of our normalized components. This will furthermore strengthen the foundations of the proposed method, and enable the contract management tool construction.

References

1. IEEE standard for Property Specification Language (PSL). IEEE Std 1850–2010 (Revision of IEEE Std 1850–2005), pp. 1–182 (2010). https://doi.org/10.1109/IEEESTD.2010.5446004
2. de Alfaro, L., Henzinger, T.A.: Interface theories for component-based design. In: Henzinger, T.A., Kirsch, C.M. (eds.) EMSOFT 2001. LNCS, vol. 2211, pp. 148–165. Springer, Heidelberg (2001). https://doi.org/10.1007/3-540-45449-7_11
3. Attiogbé, J.C.: Mastering heterogeneous behavioural models. In: Ouhammou, Y., Ivanovic, M., Abelló, A., Bellatreche, L. (eds.) MEDI 2017. LNCS, vol. 10563, pp. 291–299. Springer, Cham (2017). https://doi.org/10.1007/978-3-319-66854-3_22
4. Behrmann, G., David, A., Larsen, K.: A Tutorial on UPPAAL, vol. 3185, pp. 200–236 (01 2004). https://doi.org/10.1007/978-3-540-30080-9_7
5. Benveniste, A., et al.: Contracts for system design. Found. Trends Electron. Des. Autom. **12**(2–3), 124–400 (2018). https://doi.org/10.1561/1000000053
6. Holzmann, G.J.: The SPIN Model Checker. Addison-Wesley, Boston (2004)
7. Khouass, A., Attiogbé, C., Messabihi, M.: Multi-facets contract for modeling and verifying heterogeneous systems. CoRR abs/2012.13671 (2020). https://arxiv.org/abs/2012.13671
8. Lee, E.A.: Disciplined heterogeneous modeling. In: Petriu, D.C., Rouquette, N., Haugen, Ø. (eds.) MODELS 2010. LNCS, vol. 6395, pp. 273–287. Springer, Heidelberg (2010). https://doi.org/10.1007/978-3-642-16129-2_20
9. Sifakis, J.: Rigorous System Design, pp. 292–292 (2014). https://doi.org/10.1145/2611462.2611517
10. Tiwari, U.K., Kumar, S.: Component-Based Software Engineering: Methods and Metrics. CRC Press, Boca Raton (2020)

Multiperspective Modelling

An Approach to Detect Cyberbullying
on Social Media

Fatemeh Sajadi Ansari[1,2], Mahmoud Barhamgi[2], Aymen Khelifi[1],
and Djamal Benslimane[2(✉)]

[1] Kaisens Data, 9/11 Allee de l'Arche, 92400 Courbevoie, France
[2] Claude Bernard Lyon 1 University, Lyon, France
`djamal.benslimane@univ-lyon1.fr`

Abstract. Detecting Cyberbullying is still an important issue. Existing
approaches often rely on advanced techniques including machine learning
and Natural Language Processing algorithms. In this paper, we propose
an ontology and classifiers-based approach to detect cyberbullying cases
in the context of social media. We propose a cyberbullying ontology in
terms of cyberbullying categories and representative terms vocabulary.
This ontology is used to build and annotate the toxicity of our training dataset extracted from different data sources. Various unit classifiers
are used including messages toxicity detection, gender classifier, age estimation, and personality estimation. Outputs of these classifiers can be
combined to intercept contents that could be cyberbullying cases.

Keywords: Cyberbullying · Social media · Ontology · Classifiers

1 Introduction

With the prevalence of mobile and Internet technologies such as smart phones
and social media the cyberbullying phenomenon has reached unprecedented rates
over the course of only a few years. Recent surveys, such as [1] (conducted in
2018), showed that three out of four teenagers in Singapore have been victims to
cyberbullying. An important increase of cyberbullying incidents is also noticed
in the United States where 48% of teenagers surveyed in 2018 [2] reported they
have been victim to cyberbullying, as opposed to 25% in 2016 [3].

Several definitions to Cyberbullying have been provided in the literature.
In this work, we adopt the following one: "*willful and repeated harm inflicted
through the use of computers, cell phones, and other electronic devices*" [4]. Based
on these definitions, cyberbullying can be characterized as a **deliberate act**,
carried out by the perpetrator in a **repeated fashion** through the **use of digital
means** with the objective to **inflict harm** to the victim.

In this paper we present an approach for cyberbullying detection. The contributions of this paper are summarized as follows:

1. An ontology for the cyberbullying domain. Our ontology defines the various
   categories of cyberbullying and their interrelations.

© Springer Nature Switzerland AG 2021
C. Attiogbé and S. Ben Yahia (Eds.): MEDI 2021, LNCS 12732, pp. 53–66, 2021.
https://doi.org/10.1007/978-3-030-78428-7_5

2. A semi automatic and ontology-driven method to construct datasets. Terms that are assigned to categories of our proposed ontology are used to extract messages from different data providers. The extracted messages are automatically annotated by assigning them ontology categories that qualify their toxicity. To take into consideration the fact that language vocabulary used on the social network evolve constantly over time, the categories term sets are automatically enriched to include new terms used by users.
3. A combination of different unit classifiers. We think that messages content is not necessary sufficient by itself to detect cyberbullying cases. It should be assisted by analysing the potential cyberbullying case from other perspectives. In this work, four unit classifiers are combined: toxicity detection of exchanged messages, gender prediction, age estimation, and big five model-based personality estimation.

The rest of this paper is organized as follows. Section 2 describes related work. Section 3 presents an overview of the proposed approach, and describes its different components. Section 4 gives a general view of a prototype, presents experiments and discusses some results. Section 5 concludes the paper.

2   Related Work

A considerable work has been devoted to cyberbullying detection over the last few years (good surveys are [5,6]). Existing solutions have been roughly categorized by [6] into four categories including: *Machine learning based solutions*, *Lexicon-based solutions*, *Rules-based solutions* and *hybrid solutions*. In the following, we review the most proponent solutions in each category then compare them to our solution.

2.1   Machine Learning Based Solutions

Existing solutions in this category explored the usage of various Machine Learning algorithms, models and data features to detect cyberbullying.

Dinakar et al. in [7] established that classifiers achieve better classification results when the training dataset is clustered by cyberbullying topics and classifiers are applied to individual clusters. Recognizing the classification performance gain introduced by data clustering, the authors of [8] have employed the Kernel-based Fuzzy C-Means clustering algorithm to identify the natural cyberbullying topics that pertain to a training dataset.

The authors in [9,10] introduced the use of the victim behavior as an indicator of cyberbullying. The immediate reactions of victims such as their posts and statuses on social networks could constitute a warning signal of their emotional and psychological state when cyberbullied. Along the same lines, the authors of [11] demonstrated the benefits of exploiting the personal features of users such as their demographics and social dynamics, in addition to the content of their communications, to detect cyber bullies with a Decision Tree classifier.

A line of research works e.g. [12] explored the temporal properties of cyberbullying on social media including the number, frequency, and timing of posts. Dadvar et al. in [13] explored the use of deep learning models to cyberbullying detection. A compassion study showed that deep learning based models outperform the conventional machine learning models when both are applied to the same dataset.

2.2 Lexicon-Based Solutions

Solutions in this category (e.g., [14,15]) carry out lexical analyses of exchanged text messages in search for specific words or linguistic patterns that could be employed in cyberbullying conversations.

SafeChat [14] is a software tool that allows the parents to monitor the communications of their children over the Internet. Specifically, it analyses exchanged messages and automatically filters out explicit and offensive words. SafeChat uses its own offensive words dictionary that is built on top of well-maintained online repositories such as the WordNet. Along the same lines, the authors in [15] have analyzed the cyberbullying conversations on a social networking site Formspring.me and identified the commonly used cyberbullying terms.

2.3 Rules-Based Solutions

Solutions in this category, e.g. [16–18], detect cyberbullying cases by verifying the satisfiability of certain rules (or conditions) on exchanged messages. Rules could be expressed relative to the content of messages (e.g., words frequencies and combinations, the grammatical structure of sentences, etc.) and users' behavior (e.g. mobile usage patterns).

The authors of [16] proposed a proactive cyberbullying protection system that can assess the risk of being subject to cyberbullying at children and teenagers by combing their profiles (e.g., age) and their mobile usage patterns. In [17], a solution is proposed to detect offensive content and identify potential offensive users in social media. The solution combines features such as the user's writing style, the sentence structure and the use of pejoratives, profanities and obscenities wording to calculate an offensiveness score for content and users.

2.4 Hybrid Solutions

Solutions in this category combine various techniques, algorithms and features from previous solutions to achieve better detection results. Dadvar et al. in [13] explored the use of deep learning models to cyberbullying detection. A compassion study showed that deep learning based models outperform the conventional machine learning models when both are applied to the same dataset. In [19], the authors proposed an expert solution to detect bully users on social platforms. The solution combines the expertise of cyberbullying (human) experts and the results from supervised machine learning models to improve the detection precision of bully users. The experiments showcased the superiority of this expert system relative supervised machine learning models.

3   A Semantic Approach for Cyberbullying Detection

3.1   Approach Overview

We present in Fig. 1 an overview of our approach. Our approach exploits data mining and machine learning models to detect cyberbullying. In our approach, we distinguish three inter-related processes: "*Construction and enrichment of training datasets*", "*Feature calculation and content classification*" and "*Cyberbullying detection*", they are separated by dashed lines.

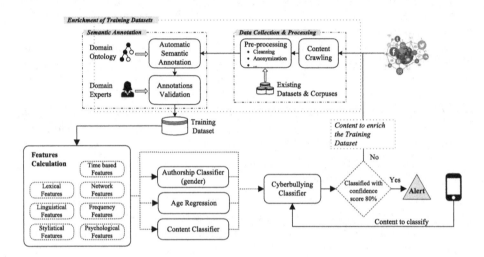

Fig. 1. An overview of the proposed approach for cyberbullying detection

The objective of the "*Construction and enrichment of training datasets*" process is to build a relevant dataset that can be used to train and sharpen our different classifiers. To build our training dataset, as shown in the *Data Collection and Processing* module in Fig. 1, we avail ourselves of existing datasets and corpuses that are used by the scientific community[1,2], but also collect real data from social medias including Twitter and Facebook. All collected data (Tweets and Facebook posts)äare cleaned and anonymized to remove personal identifying information in compliance with existing data protection regulations (e.g., GDPR). Specifically, the account names (i.e. message senders or receivers) were replaced with random numbers with the help of a mapping table that was deleted by the end of the anonymization process (e.g., @John.doe was replaced by @914812). Person names in collected messages have also been replaced with fictive names (e.g., John Doe was replaced by @Christopher Yu), ages were replaced, when available, by their categories: Child, Teenager, Adult.

---

[1] Toxic comment challenges: https://www.kaggle.com/c/jigsaw-toxic-comment-classification-challenge/data.

[2] https://www.chatcoder.com.

All collected data are annotated using a domain ontology before being fed into our training dataset. For this purpose, we have built an ontology defining the different knowledge related to cyberbullying including (i) the different categories of cyberbullying, (ii) the keywords and terminology used in cyberbullying messages, and (iii) the content (i.e. semantics) of cyberbullying messages. The semantic annotation is carried out automatically based on a set of annotation rules defined by domain experts. The semantic annotations are validated by domain experts.

The objective of the *Feature calculation and content classification* process is to calculate the different indicators of cyberbullying. Our approach involves various categories of indicators from the literature including linguistical indicators, lexical indicators, stylistical features, time-based indicators, network related indicators, psychological indicators, etc. We calculate also a set of indicators using a set of classifiers including the approximate age of the offender, their gender (i.e. man or woman) and the content type of message (e.g. sexual content, insults, negative content, etc.).

Finally, we exploit all of the calculated indicators in the *Cyberbullying detection* process. Specifically, our solution acts as a parental control system for underage subjects (e.g., children). When a new content (e.g. a message or a conversation) is exchanged, it gets intercepted and analyzed by the *cyberbullying classifier*. Its features are compared with those of the training dataset. If it can be classified as a cyberbullying case with a good confidence, then an alert is sent to competent authority (e.g., parents), otherwise the new content will be exploited to sharpen further the training dataset for handling future similar contents.

In subsequent sections, we detail those different processes.

3.2   Semantic Annotation

To train classifiers, datasets are needed. We present our collected data in the next section. Collected data needs also to be annotated.

Domain Ontology: The domain ontology plays an important role in our approach. It is firstly exploited to extract useful data from data providers to build dataset. It is also exploited to annotate the dataset in order to qualify their toxicity. We model our CyberBullying Ontology CBO as a terminology as depicted in Fig. 2. This terminology is composed of the different categories of Cyberbullying, e.g., "Harassment", "Flaming", "Trickery", etc. For each category, the ontology defines its sub categories as well as associated vocabulary composed of representative terms that might be employed by cyber-bullies in their cyberbullying messages. More formally, $CBO = (cat_1, cat_2, ..., cat_n)$ is an ontology with n categories. The function $subcat(cat_i)$ will return the subcategories of a given category. The function $Vocab(cat_i)$ will return the representative terms of any given category cat_i.

For instance, Vocab(intimidation) returns words such as: {Shut up, FUCK, Die} as representative terms of the category intimidation. The representative

Fig. 2. A part of our domain ontology

terms set of a category is not static. It evolves over time and is enriched by our cyberbullying approach. New terms that could appear in a toxic message/-conversation are automatically added to the representative terms of the more appropriate category. The Algorithm that describes how a new term is added to the current categories is not included in this paper for lack of space. It is mainly based on measuring semantic similarities between words.

Our proposed ontology differs from the one defined in [20] in mainly two aspects. They firstly differ in their objectives. Our proposed ontology considers the cyberbullying in its general meaning when the ontology proposed in [20] is more specific and deals with the exclusion cyberbullying which is a specific category of harassment. They also differ in how terms are associated to the defined categories. These terms are defined in a static way in [20] when they are dynamically computed in our ontology.

Annotation Process: We annotate our training dataset with the cyberbullying categories of our ontology. The objective of the annotation is twofold. First, we would like to build a ground-truth dataset that can be used by the community to train and calibrate cyberbullying detection algorithms. Confirmed cyberbullying cases in that dataset are annotated with their corresponding categories. Second, we exploit our annotations in computing some of cyberbullying indicators, as we detail in subsequent sections.

Annotation is carried out semi-automatically. First, we apply a set of syntaxical and linguistic rules to detect the messages with negative content, e.g., insulting, obscene, profane, aggressive, threatening, intimidating content. Then these messages are marked with the corresponding category. Then, the semantic annotations are validated manually by domain experts. In the following, we give some examples of rules and annotations.

Examples: We present in Fig. 3 some of our annotation rules. The `Preconditions` part specifies the conditions that must apply in order to annotate the message with the ontological concepts specified in the `Annotations` part.

`Rule-1` specifies that the messages that contain occurrences of the following elements: 1) an indicative verb with a negative meaning, 2) second person pronoun, 3) a racial offense and 4) a proper noun are annotated with the semantic category `Racial-Insult`. The message *"F K YOU NIGGER OBOAMA!"*, has those occurrences as follows: *"F K"* (imperative verb with a negative meaning), *"YOU"*, (second person pronoun), *"NIGGER"* (racial offense), *"OBAMA"* (proper noun).

`Rule-2` specifies that the messages that contain occurrences of the following elements: 1) second or third person pronoun, 2) state verb, 2) derogatory content are annotated with the semantic category `Mockery/Appearance`. The message[3] *"It ain't my fault you're ugly, sista"*, has those occurrences are follows: *"you"* (second person pronoun), *"are"* (state verb), *"ugly"* (derogatory content).

3.3   Unit Classifiers

Our approach relies on the use of different unit classifiers. Besides message contents, categorizing individuals according to some demographic information including gender and age could significantly help in cyberbullying detection. Four main unit classifiers are introduced: content classifier, gender classifier, age estimation, and personality estimation.

Content Classification: The content classifier is configured according to the cyberbullying semantic domain which may be that of harassment, sexual predators, insulting, racist or sexist contents. To this end, specific dictionaries and lexicons combined with a set of textual or extra-textual indicators (Sect. 4.2) are calculated and used to detect textual toxic content within the exchanges between two or more individuals. This classification is based on annotated datasets according to an ontology of cyberbullying and it characterizes a finer level of

---

[3] Extracted   from   http://www.kaggle.com/c/jigsaw-toxic-comment-classification-challenge/data.

Rule-1	
Preconditions:	Occurrences of: [*imperative/indicative verb with negative meaning*], [second person], [racial offense][proper noun] {0,1}
Annotations:	Insult. Racial
Examples:	*"Niggers and their liberal friends steal everything not tied down, just like the presidency here with acorn with its liberal defenders, FUCK YOU NIGGER OBOAMA!"*

Rule-2	
Preconditions:	Occurrences of:[second person/third person pronoun] [*state verb*], [body organ]{0,1}, [derogatory content]
Annotations :	mockery/appearance
Examples:	"Wikipedia is not the proper place for you to abuse your powers just because you're unsatisfied in life. It ain't my fault you're ugly, sista"

Fig. 3. Sample of our semantic annotation rules

toxicity that is the category of the harassment. This fine categorization of toxicity will allow us to attribute a more precise level of risk to a detected toxic content.

Gender Classification: It aims to assign male or female to the set of information given as input. A such task is known to be easy and trivial for human beings but is challenging for machines. Gender information can be determined from different sources including facial image characteristics, voices, clinical measurements (electroencephalograph, ...), and social information (handwriting, blog, ...). In our approach, automatic gender classification is performed by using information extracted from the different social networks. Face-based gender classification is firstly used when face images are available, otherwise text-based gender classification is used. Sentiment analysis is not included but could be added in the future.

Automatic Age Estimation: It can be viewed either as a classification problem where each age corresponds to a class label or as a regression problem when each age is used as a regression value. In both cases, the age estimation can be based on text, names, and can combine shape and texture of faces.

Personality Estimation: It aims to measure some basic personal characteristics. In this work, we use the big five model [21], which is one of the most used modles in the literature. Five traits are identified: (1) Openness to experience (Op) which measures to what extent a person is open to new ideas, (2) Consciousness (Co) which represents the ability to control his/her impulses and to establish long-term plans, (3) Agreeableness (Ag) to measure how a person is helpful, (4) Extroversion (Ex) to see how a person tends to communicate with others, (5) Neuroticism (Ne) represents the negative feelings of a person and their non ability to make right decisions under difficult circumstances including under stress.

4   Prototype, Experiments and Results

We built our framework on Pytorch. We conducted some experiments to evaluate the aforementioned models.

4.1   Datasets

Different messages were collected from Twitter and Facebook at different times. This extraction is ontology-driven, and representative terms of our cyberbullying categories are used. In total, 15472 messages were extracted from data providers as depicted in Table 1a. All these texts are in French. We firstly extracted 5158 message from Twitter at time t_1 then 9433 messages at time t_2, and finally 881 messages from Facebook at time t_3. We automatically annotated the extracted messages by assigning them the category labels defined at the main level (subcategories are ignored). This annotation is also manually checked by an expert. We also refined this annotation result by assigning the subcategories labels to the same set of messages as depicted in Table 1b. Most of extracted messages are positive and are related to the cyberbullying topic. Some messages are negative and are either with no connection with cyberbullying categories at all or related to these categories but from advertising perspective. Other datasets are also used to train our gender, age, and personality classification models.

Our annotation model accuracy gives us a rate of 0.75 which means that over 100 inputs (true and false together), 75 are correctly predicted as cyberbullying or non cyberbullying cases.

Our approach relies also on some others existing datasets including the Toxic Comment Classification Challenge[4]. It contains 159571 comments labeled by human raters for toxic behavior. We used this dataset to train our classifier. We also created our own datasets of images for the training needs of age and personality classifiers.

4.2   Features Calculation

To classify a textual content C_i, our toxic content classifier relies on a set of textual and extra-textual indicators to analyze the content (a set of documents) exchanged between two individuals. These indicators are:

---

[4] https://www.kaggle.com/c/jigsaw-toxic-comment-classification-challenge/data.

Table 1. Dataset characteristics

Corpus \ Labels	T_{t1}	T_{t2}	F_{t3}	T_{t1t2}	All
insult.sexual	152	601	0	753	753
insult.physical/apperance	39	95	2	134	136
insult.personal	431	1704	83	2135	2218
insult.racist	34	77	0	111	111
insult.sexist	1848	5254	1	7102	7103
insult.homophobic	176	72	0	248	248
mockery.sexual	13	50	0	63	63
mockery.physical/apperance	16	50	3	66	69
mockery.personal	387	1275	25	1662	1687
mockery.racist	12	20	0	32	32
mockery.sexist	8	18	0	26	26
mockery.homophobic	0	9	0	9	9
threat.sexual	15	29	0	44	44
threat.physical	109	327	5	436	441
threat.extortion	15	24	0	39	39
threat.psychological	80	182	3	262	265
sexual content	68	50	5	118	123
defamation	113	306	7	419	426

Corpus \ Labels	T_{t1}	T_{t2}	F_{t3}	All
insult	2252	6041	84	8377
mockery	429	1380	28	1837
threat	215	552	8	775
sexual content	68	50	5	123
defamation	113	306	7	426
advertising	479	54	19	552
No Bullying	1521	1010	464	2995
nan (other languages)	81	40	266	387

(a) General labeling of the dataset

(b) Dataset labels details

- Textual surface indicators: They result from surface syntactic analyses including syntactic, morpho-syntactic (PoS or part of Speech) and grammatical analyses. Figure 3 gives some of the rules exploited in the grammatical analysis.
- Conversational indicators involving features such as the number of questions asked by each individual U_i.
- Lexical indicators: they include n-grams of words (unigram, bigram and trigram), n-grams of characters to override the large number of spelling errors in our tweet dataset. Furthermore, these n-grams can be a representative indicator of the writing style. A lexicon of coarse words was also prepared and used and different fields are defined for its entry: the ontology category, the phraseology, the degree of severity, as well as variants and synonyms
- LIWC indicators (Linguistic inquiry and word count) obtained through a rich dictionary that maps words into 80 psychological categories obtained based on studies and psychological observations. The presence of LIWC words in the text is considered as an indicator for training of the model.
- Frequency indicators such as the number of negative words as well as the number of insults, threats, racist, obscene and toxic comments.

4.3   Implemented Classifiers and Conducted Experiments

Six classifiers were implemented as depicted in Table 2. Some of them are multi classifiers (toxic comments detection, and age detection) when others are binary classifiers (toxic tweets detection, and gender prediction). Most of our text classification are supported by SVM and Bert methods. The gender prediction from name classifier is the only one which is based on LSTM (Long Short-Term Memory) along with the recurrent neural network. All classifiers work on english texts except the CamemBERT one which works on french text.

Our dataset is vectorized by using either TF-IDF (Term Frequency-Inverse Document Frequency) which give us the importance of a word in the dataset, or word embedding technique via the bidirectional BERT model.

Table 2. Main implemented classifiers

Natural language	Model objective	Class labels	ML models
English	Toxic comments detection	Toxic, severe toxic, insult, obscene, identity hate	Bert
English	Gender prediction from text	Male, female	SVM, Bert
English	Gender prediction from name	Male, female	LSTM, CNN
English	Age prediction	Adolescent, young adult, adult	SVM, Bert
English	Personality analysis	Big 5 labels	SVM, Bert
French	Toxic tweets classification	Toxic, non toxic	CamemBERT
French	Toxic tweets classification	Cyberbullying ontology categories	CamemBERT

Table 3a illustrates the personality analysis obtained from a given text. The performance, in terms of accuracy, precision, recall, and F-score of our gender classifier from first name and using Bidirectional Character LSTM technique are shown in Table 3d. As classes of the gender classifier support are imbalanced (18270 female and 10597 male as illustrated in Table 3c), we then computed balanced accuracy which is more appropriate than the regular accuracy. The computed balanced accuracy is of 84%. The accuracy of our Camembert classifier-based toxicity detection when all categories of our cyberbullying ontology are used is 83% with an F-score of 81%. The accuracy and F-score become better when we only use the first level of categories (sub-categories are ignored): 83% of accuracy and 85% of F-score. The evolution of accuracy and loss functions are illustrated in Fig. 4.

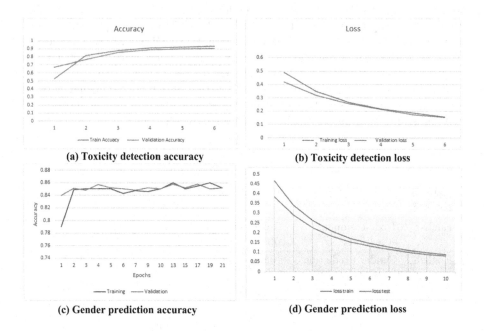

Fig. 4. Accuracy and loss

Table 3. Some performance indicators

Big 5 / Value	Op	Co	Ag	Ex	Ne
boolean value	true	true	true	false	false
probability	0.65	0.56	0.66	0.56	0.29

(a) Personality analysis from text

precision	recall	F score
88.1%	82.4%	85.1%

(b) Toxicity detection

		Actual	
		Female	Male
Predicted	Female	16189	2029
	Male	2081	8568

(c) Confusion Matrix of gender classifier

	precision	recall	F score
Female	0.89	89	0.89
Male	0.81	0.8	0.81

(d) Gender classifier

5   Conclusion

This work aimed to contribute to the cyberbullying detection problem. Our proposed approach is based on domain ontology used to collect data from different data providers and to train classifiers. Our approach is also based on the combination of different classifiers, multi and binary labels. These classifiers consider the cyberbullying detection problem from different perspectives including text exchanged between people, age, gender and personality of eventual bullyers. These different perspectives can help to profile bullyers and hence to improve the early cyberbullying detection. For best results and to not miss some cyber bullying cases, we expect to study the effect of social network features on our models. Such features could be extracted from the graph representation of users interactions. We think that using advanced technique of graph representation learning (Graph2vec, GraphSAGE, etc.) could lead to better understand the contents exchanged between users.

References

1. Mediacorp. 3 in 4 youngsters say they have been bullied online (2018)
2. AT&T, Developing Safe and Successful Mobile Device and Online Media Habits: A Survey of New York City Area Families and Recommendations for Parents, Caregivers, Communities and Companies. Tyler Clementi Foundation (2018)
3. Morel, S., Pozza, V.D., Di Pietro, A., Psaila, E.: Cyberbullying among young people. European Parliament (2016)
4. Sabella, R.A., Patchin, J.W., Hinduja, S.: Cyberbullying myths and realities. Comput. Hum. Behav. **29**(6), 2703–2711 (2013)
5. Mladenović, M., Ošmjanski, V., Stanković, S.V.: Cyber-aggression, cyberbullying, and cyber-grooming: a survey and research challenges. ACM Comput. Surv. **54**(1), 42 (2021)
6. Salawu, S., He, Y., Lumsden, J.: Approaches to automated detection of cyberbullying: a survey. IEEE Trans. Affect. Comput. **11**(1), 3–24 (2020)
7. Dinakar, K., Picard, R.W., Lieberman, H.: Common sense reasoning for detection, prevention, and mitigation of cyberbullying. In: The Twenty-Fourth International Joint Conference on Artificial Intelligence, pp. 4168–4172 (2015)
8. Nahar, V., Al-Maskari, S., Li, X., Pang, C.: Semi-supervised learning for cyberbullying detection in social networks. In: Wang, H., Sharaf, M.A. (eds.) ADC 2014. LNCS, vol. 8506, pp. 160–171. Springer, Cham (2014). https://doi.org/10.1007/978-3-319-08608-8_14
9. Choudhury, M.D., Counts, S., Horvitz, E.: Social media as a measurement tool of depression in populations. In: The Web Science Conference, pp. 47–56 (2013)
10. Dadvar, M., Ordelman, R., de Jong, F., Trieschnigg, D.: Towards user modelling in the combat against cyberbullying. In: Bouma, G., Ittoo, A., Métais, E., Wortmann, H. (eds.) NLDB 2012. LNCS, vol. 7337, pp. 277–283. Springer, Heidelberg (2012). https://doi.org/10.1007/978-3-642-31178-9_34
11. Squicciarini, A.C., Rajtmajer, S.M., Liu, Y., Griffin, C.: Identification and characterization of cyberbullying dynamics in an online social network. In: The IEEE/ACM International Conference on Advances in Social Networks Analysis and Mining, ASONAM 2015, pp. 280–285 (2015)

12. Gupta, A., Yang, W., Sivakumar, D., Silva, Y.N., Hall, D.L., Barioni, M.C.N.: Temporal properties of cyberbullying on instagram. In: Companion of the 2020 Web Conference 2020, pp. 576–583 (2020)

13. Dadvar, M., Eckert, K.: Cyberbullying detection in social networks using deep learning based models. In: Song, M., Song, I.-Y., Kotsis, G., Tjoa, A.M., Khalil, I. (eds.) DaWaK 2020. LNCS, vol. 12393, pp. 245–255. Springer, Cham (2020). https://doi.org/10.1007/978-3-030-59065-9_20

14. Fahrnberger, G., Nayak, D., Martha, V.S., Ramaswamy, S.: SafeChat: a tool to shield children's communication from explicit messages. In: 14th International Conference on Innovations for Community Services, I4CS, pp. 80–86 (2014)

15. Kontostathis, A., Reynolds, K., Garron, A., Edwards, L.: Detecting cyberbullying: query terms and techniques. In: Web Science 2013 (co-located with ECRC), WebSci 2013, Paris, France, 2–4 May 2013, pp. 195–204 (2013)

16. Serra, S., Venter, H.S.: Mobile cyber-bullying: a proposal for a pre-emptive approach to risk mitigation by employing digital forensic readiness. In: Information Security South Africa Conference ISSA 2011 (2011)

17. Chen, Y., Zhou, Y., Zhu, S., Xu, H.: Detecting offensive language in social media to protect adolescent online safety. In: IEEE International Conference on Social Computing (SocialCom), pp. 71–80 (2012)

18. Bretschneider, U., Wöhner, T., Peters, R.: Detecting online harassment in social networks. In: The International Conference on Information Systems (2014)

19. Dadvar, M., Trieschnigg, R., de Jong, F.: Experts and machines against bullies: a hybrid approach to detect cyberbullies. In: 27th Canadian Conference on Artificial Intelligence, pp. 275–281, May 2014

20. Hang, O.C., Dahlan, H.M.: Cyberbullying lexicon for social media. In: 6th International Conference on Research and Innovation in Information Systems (ICRIIS), pp. 1–6 (2019)

21. The big five trait taxonomy: History, measurement, and theoretical perspectives. In: Handbook of Personality: Theory and Research, vol. 2, pp. 102–138 (1999)

An Ontology Engineering Case Study
for Advanced Digital Forensic Analysis

Pavel Chikul$^{(\boxtimes)}$, Hayretdin Bahsi, and Olaf Maennel

Department of Software Science, Tallinn University of Technology, Tallinn, Estonia
{pavel.tsikul,hayretdin.bahsi,olaf.maennel}@taltech.ee

Abstract. Digital forensics faces some serious challenges at present. Those challenges include ever-increasing processed data volumes, heterogeneous nature of evidentiary artifacts, multiple data sources incompatible with each other, and more. Most of the commonly used forensic tools do not provide an intuitive and convenient way of accessing the data. At the same time, storage types such as relational databases cannot fully satisfy the need to store heterogeneous objects and efficiently provide access to specific properties. In this paper, we present an ontology-based approach to processing digital evidence and handling the course of digital investigation. The proposed system, named ForensicFlow, provides means of automatic artifact extraction from different origin sources, namely volatile and non-volatile memory, and reconstruction of event-artifact graphs in order to assist forensic experts in quickly and efficiently outlining the scope of an incident, and conducting an investigation.

Keywords: Ontology · Semantic web · Digital forensics · Ransomware · Event reconstruction

1 Introduction

A digital forensic investigation is a highly complex and labor-intensive task requiring identification of the data sources, extraction of the data artifacts from devices having the varied OS, applications, and hardware properties, analyze and report the relevant findings [13]. A forensic process requires the utilization of different tools and techniques as there is no unique toolkit that can handle all steps. An investigator should have a substantial amount of knowledge and experience about digital forensic procedures, various technical subjects, and tools. Despite various shortcomings, the existing forensic tools usually provide a working solution for collecting and presenting the data artifacts individually to the investigator. However, the analysis and reporting phases require deducing sophisticated correlations between these artifacts, which is mostly not fulfilled

This work in the project "ICT programme" was partly supported by the European Union through European Social Fund.

C. Attiogbé and S. Ben Yahia (Eds.): MEDI 2021, LNCS 12732, pp. 67–74, 2021.
https://doi.org/10.1007/978-3-030-78428-7_6

in the current forensic technology. Lack of this data analytic perspective in the forensic tools causes a huge cognitive burden on investigators, prolongs the investigation duration, and creates a huge dependence on the technical capabilities of investigators in an area that mostly lacks the necessary talents. All these implications reflect the need for effective knowledge management in digital forensics [5].

A body of digital forensic research under the topic of event reconstruction has addressed this analytical gap by using timeline analysis. This analysis mainly filters the system and network events based on the timing criteria and creates chronological order for the investigator [7]. Although time is a very significant parameter in revealing relations, it does not grasp all aspects implicit in the sequences of artifacts that reflect the user or system behavior in question or is not sufficient to discriminate the relevant artifacts from the enormous size of artifact pool that can be observed in any evidence. Complex cases such as malware investigations necessitate linking the user activity with some system activities by using semantic relations in addition to the time variable. Another limitation of the studies utilizing timeline analysis is that they only concentrate on the artifacts obtained from non-volatile memory sources.

Ontologies allow the logical representation of the knowledge of a certain domain by utilizing semantic relationships between class individuals. They act as a common language between experts but can also instrument transferring the created knowledge to machine-readable formats and establish a baseline for reasoning. Therefore, digital forensics can significantly benefit from ontologies and semantic frameworks as this domain mainly revolve around the application of expert knowledge. Although a body of research has explored the utilization of ontologies in the digital forensics domain [14], still, an important applied research gap exists that can be addressed by forming a more advanced reasoning system for better investigation capability in complex cases or security incidents.

In this study, we created a semantic web framework consisting of an ontology, graph model, and relevant semantic queries for the analysis of digital forensic cases. In order to populate the ontology model with forensic data obtained from volatile and non-volatile memory sources, we complemented the framework with data extraction tools. We run the forensic investigation cycle for a simulated ransomware case to demonstrate the feasibility of our solution. The main contribution of our study is the development of a comprehensive reasoning framework that utilizes a variety of resources for resolving advanced forensics cases.

2 Background Information About Digital Forensic Processes

Major principles of digital forensics evidence handling aim for adequate preservation and analysis of evidence so that it is admissible in the court. Several process models have been developed, but the majority of them share the same steps:

1. Identification and Seizure. This is a process of identifying potential sources of evidentiary data that relate to a crime scene and seizing these sources.
2. Acquisition or Preservation. This stage involves copying evidence sources in such a way that they may be later accounted for in court.
3. Analysis. After evidence sources have been acquired they are carefully analyzed by a forensic expert. The primary objective of the analysis is to obtain evidence to support or disprove a hypothesis carried out by an investigation. During this process, an investigator usually applies a wide set of different methods, tools, and techniques due to the heterogeneous nature of artifacts analyzed.
4. Presentation and Reporting. After conclusions are made a formal expert report is compiled to be later used in the court or other institutional settings.

Digital forensic processes usually have the post-mortem character, meaning that they target cases that have already fully or partially completed.

3   Literature Review

Timeline analysis studies could be classified as super, micro, or nano based on the time and data source coverage [7]. Super timeline covers the widest period enabling the investigator to form various queries [8] whereas micro or nano types address more limited timeline and source variety [3]. log2timeline is a widely-known forensic tool [9] that extracts events from different sources (e.g., images, files), creates a plaso storage file that includes the whole timeline and event content, and enables to query via different command-line tools such as pinfo and psort [1].

The high-level activity patterns (e.g., USB stick insertion) have been derived from system-level events to automatize some forensic efforts by using pre-determined rules [10]. Another study enables to do a correlation of various events retrieved from various resources including disk and memory images [4]. However, these researches do not utilize a common and understandable framework for creating the rules, limiting the extensibility and applicability of the contribution.

A review of the ontology engineering studies in the digital forensic domain is given in [14]. Various taxonomies have been developed to conceptualize the domain and classify the relevant terms in this domain. For instance, a very general taxonomy regarding the technological and professional aspects is provided in [2] whereas a more specific one categorizes the technical forensic subjects and related evidence sources [12].

A general ontology for digital investigation is proposed in [11]. This study details only the analysis of the Windows registry. In a more comprehensive study, an ontology is populated by log2timeline outputs and used for deriving inferences regarding the missing user data [6]. All these contributions usually focus on artifacts about user events extracted from non-volatile memory resources, which makes them not suitable for investigating the more complex cases requiring the comprehension of relations between the user and system-level events coming from all resources including volatile-memory.

4 Methodology

Our framework developed in this study addresses the analysis step of the digital forensic process. We formed an ontology for the forensic artifacts that can be obtained from volatile and non-volatile memory resources. This ontology constitutes the core conceptualization component of the system proposed. We utilized SPARQL to query semantic relationships. The system consists of the following major layers: data sources, extraction modules, the analyzer, knowledge base, and knowledge access interface (see Fig. 1). Extraction modules can have any implementation but the output must comply with the following rules: it should produce instances of ontology-defined classes and a mapping profile for correlations with other class instances that may or may not be found in other sources. The output of extractors is then fed to the analyzing module that performs instances correlation using object properties available from the schema. The result of the analyzer is instantiated ontology in the form of an RDF triplestore. A SPARQL query interface then allows retrieval of the stored knowledge.

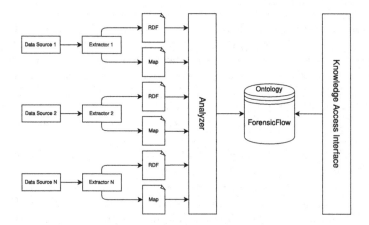

Fig. 1. Overview of the system.

ForensicFlow Ontology Design. W3C Web Ontology Language (OWL) was used for the ontology design as one of the most widely used and supported by a powerful technology stack. The ease of ontologies merging, allowing a combination of different knowledge stores, was an additional factor to decide in favor of OWL. To represent the knowledge and achieve previously discussed goals the following criteria needed to be met:

- Uniform individuals representation to allow new data sources and artifact types to be introduced, but share the same interface.
- Storing of artifact-specific data. All the type-specific knowledge needs to be carefully preserved inside the individuals.

– The ontology needs to have a pluggable architecture to allow easy adoption of any new data sources or artifact types that may appear in the future.

To achieve this goal two base classes were defined: *Event* and *ForensicArtifact*. These two major classes define a common interface by the principles of polymorphism and object-oriented design so that any sub-classes derived from them could be queried in the same manner.

The *Event* base class represents a generic event in a crime scene, e.g. an executable was run or a web page was visited. It has only one data property of its own which is event time and one generic object property that is *hasArtifact*. Those properties get inherited by all of its sub-classes. The many children of the event class may include any events that happen in the system and any data accompanying them. The exact system events are then derived from this class introducing knowledge specific to them. Object properties for Event sub-classes are generated by extending the *hasArtifact* property of the parent class. A good example of such a link can be a file download event linking to the file's metadata extracted from the MFT record.

Generic *ForensicArtifact* class includes information about the data source it was extracted from, the name of the investigator, and any particulars regarding the extraction procedure that may be relevant to those trying to replicate the process. There are several subgroups of artifacts defined to structure them better. The groups logically divide artifacts by nature, e.g. network, file system, processes, etc. Each group then defines the exact artifact type or introduces further grouping.

The described organization creates a versatile system of connection between events and artifacts involved in them with a universal interface for querying objects and related connections.

Artifacts Extraction and Ontology Instantiation. To populate the ontology with knowledge about crime scene artifacts and their mutual interconnections the expert is expected to run a set of automatic extraction tools against the evidence sources in possession. Currently, two data sources are supported: hard drive images and volatile memory snapshots. The advantage of mixed-origin data storage is that the investigator can quickly and conveniently get a different perspective of the same data or events or augment one source with evidence from another. Each tool extracts relevant pieces of information and serializes them in OWL format as ontology individuals. At the same time, it creates a mapping profile for the analyzer to digest later if artifact linking is suspected. The mapping profile is a set of links to potential individuals that could be discovered by other extractors and is used by the analyzer to create semantic relations. To address the excessive data issue mentioned earlier all extractors operate only on a subset of data that is relevant to the investigation. Currently, there are three tools developed in order to test out our hypothesis:

– Volatility plug-in operates on a volatile memory (RAM). The extracted information includes running processes information, executable path, open network connections, etc.

- Log2Timeline handler extracts valuable information from L2T reports such as prefetch events, browser activity, registry edits, etc. Data generated by this software populates the Events domain of the ontology.
- Autopsy plug-in extracts file-related data from the disk image, taking into account mapping profiles generated in previous steps to comprise a full map of forensic artifacts. At the moment the data extracted by the plug-in includes basic statistics such as file size, mime type, and location, as well as some advanced ones like entropy anomalies.

The resulting OWL individuals and mappings are then fed to the analyzer that populates the ontology with instances and their relations according to the schema. Due to the rules defined in each extractor, only relevant data is fetched from original sources that dramatically reduces the size of analyzed information.

5   Experimentation and Results

This section covers an experimental case study and covers some preliminary conclusions. The primary target here is to demonstrate the advantages of the proposed method over traditionally accepted forensic practices and outline the capabilities of the system in a life-like scenario. The simulated attack represents a system where a malicious application (ransomware) is first downloaded from a network resource and later executed following the encryption of some of the user's files. The aim of the investigator is to quickly and conveniently identify the cause, the source, and the damage caused by an attack.

A Windows 7 virtual machine acts as a target system. Some typical user files are populated in usual locations inside user space. A malicious application is uploaded to a website in advance. No antivirus software, aside from built-in Windows Defender is present in the system. The test ransomware used in the experiment when executed goes over the user directory and selectively encrypts files with AES cipher by a predefined extensions mask.

The attack begins with the user visiting a "malicious" web page and downloading the ransomware executable through the Firefox web browser to a standard download location. The malicious application is then executed and the user files get encrypted after some time. When the attack is done a ransom message is saved to the user desktop as a text file. The application stays resident and tracks any further file changes encrypting newly created files.

At this stage, the investigator moves in and performs a memory dump in order to save any volatile information that is still present in the system. The system is then shut down and a forensic disk image is taken using FTK Imager. The investigator then runs our modules against these sources in order to instantiate an ontology that is ready for analysis.

The standard forensic procedure, in this case, would include WinPrefetch events lookup. These events indicate an application execution occurrence. When suspicious execution has been identified events surrounding that execution are being manually analyzed in an attempt to figure out the context. Any traces

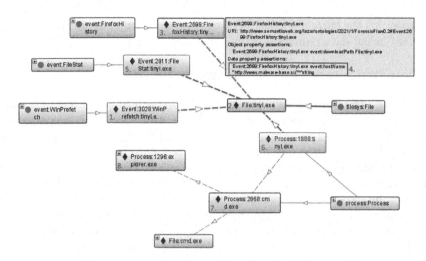

Fig. 2. The event-artifact dependency graph (cropped).

discovered during this analysis, for example, the potential location of the executable file or its origin, must be then manually traced back using different tools and data sources.

By utilizing the proposed ontology a single SPARQL query for WinPrefetch events reveals 19 instances from which only two seem suspicious since they were performed from the userspace. From the Protege graph view, these events are searched revealing the full graph of dependent artifacts (see Fig. 2). The graph reveals some crucial facts about the attack. From the suspicious WinPrefetch event (1) we discover the executable file in question (2) and all of its related events and linked artifacts. The file first appears in the Firefox web browser download event (3) that reveals the network resource of origin (4). The file is then recorded by the file system as written to disk in the user download folder (5). The next event it appears in is the execution event we started from (1). Lastly, the file is traced back to the exact process in memory that is still running (6), making it possible to extract the memory context and possibly the encryption key.

To perform a similar analysis using only standard tools the examiner would have to switch between at least 3–4 different tools performing manual searches for each artifact, noting down the correlations between them, and trying to reconstruct the flow of events. We would also like to note that currently there are no tools on the market that correlate hard disk data and memory extracted information. Thus the target goals have been met to a significant extent. An ontology for a crime scene was developed. It allows for convenient storing and querying of event-artifact-event instances making it much easier for an investigator to outline the scope of events surrounding certain actions. An extensible interface for the knowledge store population has been introduced allowing for any third party to enrich the current tool-set. The extraction modules architecture implies significant initial data reduction making it easier to analyze.

6 Conclusion and Future Work

In this paper, we proposed an ontology-based approach to accurately and effectively store digital crime scene evidence artifacts. Our ontology is accompanied by a set of evidence extraction modules to cover different evidence sources, an analyzer engine that is responsible for the mutual correlation of artifacts, and a set of SPARQL queries to fetch the data out of the ontology. The extraction modules take care of excessive data reduction by applying predefined rules that fetch only the data relevant to crime scene reconstruction. We consider several topics as parts of the future work. First of all, we want to extend the scope of data sources to include not only multiple origins for one device but systems of multiple devices connected in one way or another. This approach would allow automating investigations in a distributed environment such as the Internet of Things. Another point we would like to address is the validation with actual practitioners including law enforcement and private experts.

References

1. Plaso documentation. https://plaso.readthedocs.io/en/latest/
2. Brinson, A., Robinson, A., Rogers, M.: A cyber forensics ontology: creating a new approach to studying cyber forensics. Digit. Invest. **3**, 37–43 (2006)
3. Carvey, H. https://windowsir.blogspot.com/2015/04/micro-mini-timelines.html
4. Case, A., Cristina, A., Marziale, L., Richard, G.G., Roussev, V.: Face: automated digital evidence discovery and correlation. Digit. Invest. **5**, S65–S75 (2008)
5. Casey, E.: The chequered past and risky future of digital forensics. Aust. J. Forensic Sci. **51**(6), 649–664 (2019)
6. Chabot, Y., Bertaux, A., Nicolle, C., Kechadi, T.: An ontology-based approach for the reconstruction and analysis of digital incidents timelines. Digit. Invest. **15**, 83–100 (2015)
7. Debinski, M., Breitinger, F., Mohan, P.: Timeline2GUI: a Log2Timeline CSV parser and training scenarios. Digit. Invest. **28**, 34–43 (2019)
8. Esposito, S., Peterson, G.: Creating super timelines in windows investigations. In: Peterson, G., Shenoi, S. (eds.) DigitalForensics 2013. IAICT, vol. 410, pp. 135–144. Springer, Heidelberg (2013). https://doi.org/10.1007/978-3-642-41148-9_9
9. Gujónsson, K.: Mastering the super timeline with log2timeline. SANS Institute (2010)
10. Hargreaves, C., Patterson, J.: An automated timeline reconstruction approach for digital forensic investigations. Digit. Invest. **9**, S69–S79 (2012)
11. Kahvedžić, D., Kechadi, T.: Dialog: a framework for modeling, analysis and reuse of digital forensic knowledge. Digit. Invest. **6**, S23–S33 (2009)
12. Karie, N.M., Venter, H.S.: Toward a general ontology for digital forensic disciplines. J. Forensic Sci. **59**(5), 1231–1241 (2014)
13. Kent, K., Chevalier, S., Grance, T., Dang, H.: Guide to integrating forensic techniques into incident response. NIST Special Publ. **10**(14), 800–86 (2006)
14. Sikos, L.F.: AI in digital forensics: ontology engineering for cybercrime investigations. Wiley Interdiscip. Rev. Forensic Sci. **3**, e1394 (2021)

More Automation in Model Driven Development

Pascal André[1]([✉])[ID] and Mohammed El Amin Tebib[2]

[1] LS2N CNRS UMR 6004, University of Nantes, Nantes, France
`pascal.andre@ls2n.fr`
[2] LCIS - Grenoble Alpes University, Valence, France
`Mohammed-El-Amin.Tebib@univ-grenoble-alpes.fr`

Abstract. Model Driven Development (MDD) earned a leading role in Software Engineering but still does not play its role in practice. In MDD, models are abstractions of implementations and development is a refinement process with as much automation as possible *e.g.* code generation. In practice, the distance between logical models and implementations prevents automatic refinements and code generation. We revisit the MDD schema and we propose a method based on both forward engineering and reverse-engineering activities leading to Model Driven Engineering. Forward engineering is structured by layered macrotransformations parametrised by the abstractions of platforms obtained by reverse-engineering. We illustrate the method with a small distributed physical system case study. This work helps practitioners to automate MDD.

Keywords: Model driven development · Refinement ·
Reverse-engineering · Model transformation · Model mapping

1 Introduction

Software developers need methods and tools to design and program software applications from requirements and analysis specifications. Model Based Development (MBD) emerged with structured design in the 1970' and earned a leading role in Software Engineering with the advent of Model Driven Approach (MDA) by standards and tools for modelling and model transformations. In this context, Model Driven Development (MDD) shortens the development cycle: describe abstract models, verify and transform them to get running applications. In software engineering research, MDD lost some attention since 2010 and became one topic of the Model Driven Engineering (MDE) [1], which is now the main field of contribution; MDE includes model evolution, reverse-engineering, language engineering, etc. [2].

Years later, the goal has not been reached despite numerous contributions on techniques and tools. In current software engineering practices, developers use models as a documentation support for analysis and design (MBD style) with

© Springer Nature Switzerland AG 2021
C. Attiogbé and S. Ben Yahia (Eds.): MEDI 2021, LNCS 12732, pp. 75–83, 2021.
https://doi.org/10.1007/978-3-030-78428-7_7

guidelines given in methods like RUP [3]. To the best of our knowledge, there still miss generic proposals of MDD as a process of model transformations [4], except in specific domains such as AUTOSAR in the automotive domain or database CRUD web applications. The main question is *how to implement a generic transformation process from a logical model?* We did not found (yet) a answer, even in [5] that introduces conceptual models compilation. Reasons are: (i) The distance between logical models and implementation is really large, it includes all the software design activities. (ii) Adding technical elements (platform) at a design level is engineering activity. (iii) Tool support for transformation process is missing and tools focus on small-scale transformations. We need means to manage the abstraction gap between logical models and design models in order to enable automation i.e. the *engineering practice* challenges of [6].

To automate MDD and reduce the distance between models and code we propose: (i) a systematic approach to design by a structured MDD process with macro-transformations (being systematic enables to define the activities in order to automate them), (ii) a top-down (refinement) and bottom-up (abstraction) process to align models and programs by model composition, (iii) systematic transformations definitions instead of verbose guidelines in order to automate them and (iv) several model transformations to assist the practitioner by tooling facilities. Our long-term goal is to provide methodological assistance to MDD developers with process, techniques and tools.

The paper is organised as follows. Section 2 reports experiences on MDD practice to draw principles for automating MDD in Sect. 3. A two-track MDD process is then presented with both top-down and bottom-up entries. Section 4 focuses on implementation issues with systematic definitions of transformations and tooling. All along the paper we will refer to a case-study, which is available in a web-appendix[1] for sake of space. Section 5 discusses about this work and provides future directions.

2   Background

MDD, using MDA lexicon, is a *transformation process* from a *Platform Independent Model (PIM)* to a (more) concrete *Platform Specific Model (PSM)* injecting elements of the *Platform Description Model (PDM)*. Let PIM be the logical model produced during the analysis activity. It includes structural aspects (*e.g.* UML class), dynamic aspects (*e.g.* UML statecharts) and functional aspects (*e.g.* UML activity). Let PSM be the software system (source code, libraries) that implement the application to be deployed over devices. To answer to the above mentioned question *"how to implement a generic transformation process from a logical model (PIM)?"*, we compared in [7] three approaches with different automation degree (forward engineering, code generation, stepwise transformations) on a case study (see footnote 1), a simple automatic system implemented with Lego EV3 bricks driven by Android apps. Beyond its simplicity, this is a distributed system with networks communications that are more complex than

---

[1] https://aelos.ls2n.fr/at-medi2021app/.

simple method calls. We summarise here the experimentation feedbacks given in [7]:

1. In forward engineering we observed the student groups practice on the case study. The application code was always a free interpretation of the PIM specification, which was not considered as an abstraction to refine and to conform to. The design concerns (persistence, GUI, concurrency, quality) were not prioritized by the students, despite they have mutual influence.
2. UML code generators usually do not exploit the full model information (UML or SysML, OCL, Action Semantics). Commercial tools provide more powerful solutions. Executable UML tools, especially those including fUML and Alf, are interesting for simulation or animation purpose (platform specific) but not MDD purpose.
3. In the stepwise transformations approach, we focuses on the refinement of UML state-machines and message in Java by ATL transformations. In practice the students failed to build a rigorous and automated process due to the complexity of both the individual transformations and the transformation process.

Moreover cross cutting concerns are not easy to weave. For example, the message middleware and the state machines engine have mutual influence. This makes model driven code generation [8] very tubular solutions vs model transformations. In [7] we proposed a general transformation process in four steps. In the remaining of the paper we revisit this process and provide details about the stepwise transformation design process.

3   A Revisited MDD Transformation Process

To automate the MDD process, we lay a set of principles derived from our observations:

1. Model transformation can infer new elements but cannot create them from scratch. The quality of the inputs influences the quality of the outputs. A preliminary step is to verify the PIM quality and properties such as consistency and completeness.
2. Software design handles cross cutting aspects, such as persistence, GUI, concurrency on which the logical model must be "woven". The semantic distance between PIM and PSM requires a true transformation process not only a code generator.
3. Design, as an engineering activity, is linked to the designers' experience. A process can be automated only if all the activities are precisely known. We need a systematic definition of the design and implementation activities as transformations.
4. In practice, model transformations are effective when the source and target models are semantically close (*e.g.* class diagram ↔ relational model). Small transformations are easier to verify, to compose and to reuse. Small step transformations leave the complexity to the composition process, not to the atomic transformations.

5. A composite transformation is a process hierarchically composed of other (simpler) transformations. The atomic model transformations are preferably model compositions such as weaving or mapping [9] to enforce traceability (elements & decisions).
6. Modelling and reverse-engineering are abstraction activities, design by model transformations are refinements activities[2]. Carefully note that we do not reverse engineer applications (PSM) but platforms (PDM). A balanced design process mixes abstractions for PDM and refinements from PIM to PSM.

Based on these principles, we propose a two track process: (bottom-up) abstraction of PDM (right part of Fig. 1) and (top-down) refinement by mapping transformations (left part of Fig. 1). The (intermediate) PSM are bound to PDM models by model mapping.

An Abstraction Process. Implementation frameworks contain code archive but no PDMs. One can fill this miss by extracting a PDM from the framework source code by Model Driven Reverse Engineering (MDRE). In MDE, writing model transformations is not very simple but the source and target models are usually known. Finding abstraction in MDRE is an even more difficult problem because there miss guidelines and traceability information [10]. The source information also differs and may include binary code, source code, configuration files, tests programs or scenarios models... Such a diversity make the RE activity difficult to apply. *Fortunately, we do not target full applications but technical frameworks.*

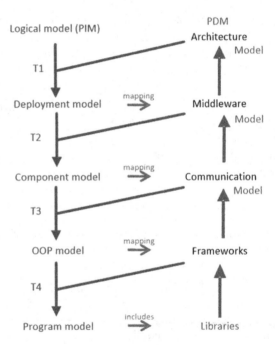

Fig. 1. Macro-transformation process

MDRE targets different levels of abstraction, from AST to high level application architectures or business processes. MDRE is itself a transformation process with *abstraction layers*, like the OSI model or the SOA Reference Architecture. In our case (*cf.* Fig. 1) the abstraction layers are: program, OOP, component and deployment models. For sake of simplicity, Fig. 1 shows one couple of models (PIM-PDM) per level but recall that the more you progress to code the more

---

[2] We do not consider *round-trip engineering* as an automated approach for MDD but as an additional facility to align model with code.

you have domains in parallel. In each layer, models are written with various modelling notations. We use the same notations as in the stepwise design process because they are the inputs of that process. Reverse engineering is automated at low level but raising in abstraction remains a difficult engineering task [11]. Intermediate languages such as Knowledge Discovery Metamodel (KDM) [12] are helpful at low level because it is a model detailed abstraction of the code. We promote the composition of small step transformation rather than monolithic transformations..

A Stepwise Transformation Design Process. The top-down design process is composed of four macro-transformations (depicted in Fig. 1). For a simple presentation, the same macro-transformation pattern is applied: $PIM \oplus PDM \rightsquigarrow PSM$. This hides a real complexity: inputs can be several models and macro-transformations are themselves transformation processes which address either a design or programming issue.

T1 The deployment transformation structures the PIM in applications (subsystem) and maps them to deployment nodes of the PDM. The link between nodes are annotated by communication protocols. The subsystems are PIM structural elements such as UML class, components or packages. If the PIM includes component and deployment diagrams for a preliminary design, T1 will resume to mapping.

T2 The Message Oriented Middleware (MOM) transformation refines object communications. The UML message sent is instrumented in MOM frameworks given by the PDM. For example, a message send can be a method call in the target language (Java, C++ or C#). Remote procedure calls (RMI), wireless communications or TCP-IP communications require high level transformation patterns.

T3 The Object Oriented Programming transformation refines UML concepts into a OOP models which in general do not natively include these concepts. This thorny problem is discussed in Sect. 4. Persistence is achieved by special transformation to Data Access Object (DAO) models provided in the PDM.

T4 The program transformation pre-processes the code generation by generating code from low level models (implementation blue print) and by matching model elements to predefined libraries of the technical *frameworks*. Recall the technical framework is the implementation of the PDM.

All the configuration parameters of the transformations and all decisions must be stored to be replayed in an iterative design process. Design aspects like GUI or security are not considered here. Since the PDM is an input for the transformations, we need different (abstract) views on it. We will discuss this point in the next section.

4   Implementation and Experimentations

The process of Fig. 1 is abstract and generic. To raise the degree of automation as challenged in the introduction, we need (i) a systematic definition of the transformations in order to program them; (ii) tool support to implement them.

a) Systematic Definition of Transformations. We define a subset of systematic transformations. Again for a sake of space, details on the transformation specifications, ATL or Papyrus implementation and their application to the case study are detailed in the web appendix (see footnote 1). For instance, T4 aims at unifying model elements and implementation (source code). All model elements are not generated from scratch, some already exist, maybe in a different nature, in the framework model (*cf.* Fig. 1). We look forward *API Mapping* a feature to map model elements to predefined elements in libraries or *frameworks*. In this section, we discuss the mapping of design classes (and operations) to predefined code source classes. To simplify, we focus on classes as the model elements, but it should be extended to packages, data types, operations, etc.

In our case study, the sensors and actuators already exist in the Lejos library. For sake of simplicity we consider that a model element maps to one implementation but an implementation can map to several model elements (1-N relation). When model and implementation elements do not match, developers usually refactor the model to converge. The mapping process includes three activities:

1. *Detect* implementation candidates in libraries with if possible matching rates. Different model elements are taken into account such as class, attribute, operation... We face here two issues: (i) Abstraction level. Basically the model and implementation elements are not comparable and we need a model of the implementation framework. (ii) Pattern matching. The model elements are not independent *e.g.* operations are in classes which are grouped in packages. The way the model elements are organised influence the matching process.
2. *Select* the adequate implementation of model elements (class, attribute, operation) and bind the model elements to implementation elements by (non-intrusive) model composition [9].
3. *Adapt* to the situation. Once a mapping link is established, it usually implies to refactor the design. Adaptation is the core mechanism to bind PIMs to PDMs according to two strategies: (i) *Encapsulate and delegate*. The model classes are preserved that encapsulate the implementation classes (Adapter pattern). The advantage is to keep traceability on base APIs. The drawback is the multiplication of the classes to maintain. (ii) *Replace* the model classes by the implementation classes. The transformation must replace the type declarations but also all messages sent. The pro and cons are the opposite of encapsulation. Replacement is possible when classes have a similar structure and behaviour including the modelling primitive types. In any other cases the Adapter pattern captures multi-feature adaptations:
   - Attribute: name, type adaptation, default value, visibility...
   - References (role): name, type adaptation, default value, visibility...
   - Operation: name, parameters (order, default), type adaptation...
   - Protocol: state machine for the model class but not the implementation class.
   - Composition: a class is implemented by several implementation classes.
   - Communication refinement: MOM communications are distributed.

- API layering: classify the methods to reduce the dependency.
- Design principles: improve the quality according to SOLID, IOC...

The high-level frameworks for MOM middlewares or state machines are not concerned by these issues because they are pluggable components. We illustrate T4 by an example in page 31 of the web appendix (see footnote 1).

b) Tool Support. A rational implementation combines model transformations tools and code generation facilities. When a PDM exists, model transformation would be *model mapping* and *weaving*. During the experimentations, we use ATL[3] to write transformations because its rule-based style supports both model-to-model (M2) and model-to-text (M2T) transformations which is convenient to our macro-transformation process (*cf.* Fig. 1). Acceleo (see footnote 3) is planned to support our M2T transformations. Code generation for the M2T transformation was implemented using Papyrus (see footnote 3) for UML-OOP-SI to Java transformations and customised ATL-transformations. In MDRE we used Modisco and AgileJ as illustrated by an example in page 36 of the web appendix (see footnote 1). Knowledge Discovery Metamodel (KDM) is a standard for software system representation [12]. It is useful at low level because it is a model representation but it keeps a very detailed information which miss abstraction. Experiments show the potential of the approach but much work remains to implement the macro-transformations. Let discuss now shortly some open implementation concerns and customisation issues:

- *Input Quality* As mentioned in the beginning of Sect. 3, the input model quality influences the transformation quality. This topic is out of the scope of this article but we discussed it in [13]. Animation and simulation help to validate the input. When the technical environment is fully mastered, the transformation can weave the model with the *framework* to make it executable.
- *DSL or Profiles zoology* Each macro-transformation handles different models written with various languages. We assumed here only UML profiles for sake of simplicity but choosing the languages (standards or DSLs) is really a big challenge in the MDE community from both the theoretical and tooling points of view.
- Transformation Process Each composite transformation is a transformation process. The more you progress in the process the more you have *parallel* transformations $T1 \mapsto T2^n \mapsto T3^{n \times m} \mapsto T4^{n \times m \times p}$. However the order in which are processed the atomic transformation will influence the final result. We discussed this tricky problem in the context of model verification in [14].
- Traceability Since the models are obtained by transformation, we implicitly assume traceability links to be inserted to trace the model element origin from PIM and/or PDM *e.g.* by the means of model annotations. An original feature is to trace from PDM, which is not usual in MDD.
- Iteration By essence this process is generative but the transformations are not fully automated. Design choice have to be made by the software designer

---

[3] https://www.eclipse.org/atl--/acceleo--/papyrus/.

and user interaction are required for those transformations that cannot be automated. Recall that the GUI part is considered to be developed separately. *Round trip* will help here to store user information when iterating on the transformations process.

– **Transformation support** Experimentations showed that no transformation tool was a panacea especially because various styles of transformation are involved *e.g.* synthesis, extraction, mapping, refactoring [15]. [16] surveys transformation tools.

5   Discussion

There exist techniques and tools for modelling and model transformations that can apply to specific situations but not to software development as a process, which remains an engineering (highly expensive) activity. We are convinced that software companies should invest more in software automation than software development. In order to add more automation in MDD and to assist the practitioner, we propose (i) a systematic approach to design by a structured MDD process with macro-transformations; (ii) a top-down and bottom-up process to align logical models with implementation framework models; (iii) systematic transformations definitions for atomic transformations and (iv) the implementation of some of them. The current contribution is a first step toward the challenging question set in the introduction.

Many validity threats remain in this empirical research mainly due to the complexity of the problem: (i) Showing the applicability requires a full abstraction of PDMs and a more complete definition of the macro-transformations which is complex, since we embrace many design and programming activities. (ii) The step from transformation definition to transformation automation is large for composite transformation leading to feasibility issues, especially for the high level transformations. We showed the applicability of low level transformations only. (iii) Asserting the genericity requires to apply it to various types of application (web, mobile...) which is not yet the case. In summary we draw here a vision, not a designer manual nor a CASE tool, that can provide research tracks to follow. Besides, from a teaching perspective, we found it useful to explain the software design to master students in a rational way.

There remain much work to obtain a MDD toolbox. A collaborative research project would be an efficient answer to the challenge. We started implementing a communication layer to refine message send in Bluetooth and Wifi communications. Future work will start from low level transformations (*e.g.* T4). Indeed the process of Fig. 1 is organised as a stack such that the low-level transformations can work independently from the high-level ones. A short-term perspective is to continue to write individual transformations (small steps) that will be composed hierarchically until reaching macro-transformations. A particular issue is to combine cross-cutting concerns during the transformations as well as multi-view models by integrating techniques of aspect-oriented transformations. A long-term perspective is to provide more systematic design guidelines for step T2 and step T1 based on Architectural Design Languages.

References

1. Selic, B.: From model-driven development to model-driven engineering. In: Proceedings of the 19th Euromicro Conference on Real-Time Systems, ECRTS 2007, USA. IEEE Computer Society (2007)
2. Brambilla, M., Cabot, J., Wimmer, M.: Model-Driven Software Engineering in Practice: Second Edition. Morgan & Claypool Publishers, San Rafael (2017)
3. Rumbaugh, J., Jacobson, I., Booch, G.: The Unified Software Development Process. Object-Oriented Series. Addison-Wesley (1999) ISBN 0-201-57169-2
4. Di Ruscio, D., Eramo, R., Pierantonio, A.: Model transformations. In: Bernardo, M., Cortellessa, V., Pierantonio, A. (eds.) SFM 2012. LNCS, vol. 7320, pp. 91–136. Springer, Heidelberg (2012). https://doi.org/10.1007/978-3-642-30982-3_4
5. Pastor, O., Molina, J.C.: Model-Driven Architecture in Practice - A Software Production Environment Based on Conceptual Modeling. Springer, Heidelberg (2007). https://doi.org/10.1007/978-3-540-71868-0
6. Paige, R.F., Matragkas, N., Rose, L.M.: Evolving models in model-driven engineering: state-of-the-art and future challenges. J. Syst. Softw. **111**, 272–280 (2016)
7. André, P., Tebib, M.E.A.: Refining automation system control with MDE. In Hammoudi, S., Pires, L.F., Selic, B. (eds.) Proceedings of MODELSWARD 2020, Valletta, Malta, 25–27 February 2020, pp. 425–432. SCITEPRESS (2020)
8. Mehmood, A., Jawawi, D.N.A.: Aspect-oriented model-driven code generation: a systematic mapping study. Inf. Softw. Technol. **55**(2), 395–411 (2013)
9. Clavreul, M.: Model and metamodel composition: separation of mapping and interpretation for unifying existing model composition techniques. Ph.D. thesis, Université Rennes 1, December 2011
10. Raibulet, C., Fontana, F.A., Zanoni, M.: Model-driven reverse engineering approaches: a systematic literature review. IEEE Access **5**, 14516–14542 (2017)
11. André, P.: Case studies in model-driven reverse engineering. In: Proceedings of the 7th International Conference on Model-Driven Engineering and Software Development, MODELSWARD 2019, Prague, Czech Republic, 20–22 February 2019, pp. 256–263 (2019)
12. Brunelière, H., Cabot, J., Dupé, G., Madiot, F.: MoDisco: a model driven reverse engineering framework. Inf. Softw. Technol. **56**(8), 1012–1032 (2014)
13. André, P., Attiogbé, C., Mottu, J.M.: Combining techniques to verify service-based components. In: Proceedings of the International Workshop, AMARETTO@MODELSWARD 2017, Porto, Portugal, February 2017
14. André, P., Ardourel, G.: Domain based verification for UML models. In: Kuzniarz, L., Reggio, G., Sourrouille, J.L., Staron, M. (eds.) Workshop on Consistency in Model Driven Engineering C@Mode 2005, pp. 47–62, November 2005
15. Karsai, G., Taentzer, G., Mens, T., Gorp, P.V.: A taxonomy of model transformation. Electron. Notes Theor. Comput. Sci. **152**, 125–142 (2006). Proceedings of the International Workshop on Graph and Model Transformation (GraMoT 2005)
16. Kahani, N., Bagherzadeh, M., Cordy, J.R., Dingel, J., Varró, D.: Survey and classification of model transformation tools. Softw. Syst. Model. **18**(4), 2361–2397 (2018). https://doi.org/10.1007/s10270-018-0665-6

Modelling and Interactions

Realisability of Control-State Choreographies

Klaus-Dieter Schewe[1(✉)], Yamine Aït-Ameur[2], and Sarah Benyagoub[2]

[1] UIUC Institute, Zhejiang University, Haining, China
kd.schewe@intl.zju.edu.cn, kdschewe@acm.org
[2] IRIT/INPT-ENSEEIHT, Université de Toulouse, Toulouse, France
{yamine,sarah.benyagoub}@enseeiht.fr

Abstract. Choreographies prescribe the rendez-vous synchronisation of messages in a system of communicating finite state machines. Such a system is called realisable, if the traces of the prescribed communication coincide with those of the asynchronous system of peers, where the communication channels either use FIFO queues or multiset mailboxes. In this article we generalise choreographies to control-state choreographies, which enable parallelism. We redefine P2P systems on grounds of control-state machines and show that a control-state choreography is equivalent to the rendez-vous compositions of its peers and that language-synchronisability coincides with synchronisability. These results are used to characterise realisability of control-state choreographies, for which we prove two necessary conditions: a sequence condition and a choice condition. Then we also show that these two conditions together are sufficient for the realisability of control-state choreographies.

Keywords: Communicating system · Control-state machine · Parallelism · Choreography · Synchronisability · Peer-to-peer system · Realisability

1 Introduction

A peer-to-peer (P2P) system is an asynchronous system of independent peers communicating through messages. Using communicating FSMs the semantics is defined by the traces of messages sent. Such a trace semantics can be defined in various ways, e.g. using a separate channel organised as a FIFO queue for each ordered pair of distinct peers [7,10]. In particular, messages on the same channel are received in the same order as they have been sent and no message is lost. Alternatives are the use of such FIFO queues with only a single channel for each receiver (as e.g. in [1]) or the organisation of the channels as multisets (see e.g. [9]), which corresponds to mailboxes, from which messages can be received in arbitrary order. Naturally, one may also consider the possibility of messages being lost (see e.g. [8]).

© Springer Nature Switzerland AG 2021
C. Attiogbé and S. Ben Yahia (Eds.): MEDI 2021, LNCS 12732, pp. 87–100, 2021.
https://doi.org/10.1007/978-3-030-78428-7_8

A common question investigated for communicating FSMs is whether the traces remain the same, if a rendez-vous (or handshake) synchronisation of (sending and receiving of) messages is considered. This *synchronisability* problem has been proven to be undecidable in general [10]. The picture changes slightly in the presence of *choreographies*. In this case the peers are projections of a choreography, and as shown in [12] the rendez-vous composition of the projected peers coincides with the choreography. Also the distinction between *language synchronisability* based only on the message traces, and *synchronisability* based in addition on the stable configurations reached becomes obsolete. The main result in [12] shows that under these restrictive circumstances realisability can be characterised by two simple necessary conditions that both together are sufficient. Actually, a hint on the sufficiency of these conditions was already given by the compositional approach to choreographies and the associated proof of realisability [3]. This compositional approach has been extended in [2] taking the new insights into account.

However, this characterisation of realisable choreographies also highlights the limitations of viewing P2P systems as systems of communicating FSMs. A choreography defined by an FSM is a purely sequential system, and as such it is far too limited to capture the needs of P2P systems. For instance, a peer may send multiple messages at the same time, different peers may operate asynchronously, or it may not be required that all sent messages are received. It is necessary to provide a more sophisticated notion of choreography capturing these possible cases.

In this article we extend the work in [12] on necessary and sufficient conditions for the realisability of choreographies. We replace choreographies modelled by FSMs by *control-state choreographies* (CSCs) modelled by control-state machines [6, pp. 44ff.]. The key difference is that whenever a (control) state is reached, the sending and receiving of several messages in parallel is enabled. We further distinguish between *triggering* and *enabling* control states depending on whether the messages must be sent or may be sent. The parallelism of sending/receiving of several messages will exploit the asynchronous semantics adopted from concurrent ASMs [4].

With communicating control-state machines we can also define a generalisation of P2P systems, for which we define different notions of compositions. Using either multisets or FIFO queues as mailboxes we obtain three different types of asynchronous compositions, which we continue to call *peer-to-peer*, *mailbox* and *queue* composition, respectively. Using handshake communication we obtain a synchronous *rendez-vous* composition. In doing so we generalise the notions of *language synchronisability* and *synchronisability* to P2P systems based on control-state machines. Analogous to our previous work we show again that a CSC is equivalent to the rendez-vous composition of its projected peers. Furthermore, language synchronisability and synchronisability coincide also for P2P systems defined by CSCs. In this way our theory extension is conservative, as it preserves key features of choreography-defined P2P systems.

Finally, we investigate the generalisation of realisability for CSCs. As for the case of FSM-based choreographies we prove the necessity of two conditions: a sequence condition and a choice condition. Then we also show that these two conditions together are sufficient for the realisability of control-state choreographies. So we also obtain a generalisation of the realisability characterisation from [12] to the more general control-state choreographies. Due to space limitations more examples and all proofs are available in an extended report [13].

Section 2 is dedicated to preliminaries, i.e. we introduce all the notions that are relevant for the work: control-state machines, P2P systems, rendez-vous, p2p, queue and mailbox semantics, and synchronisability. In Sect. 3 we introduce control-state choreographies and choreography-defined P2P systems, for which we show that synchronisability coincides with language-synchronisability. In Sect. 4 we address sufficient and necessary conditions for realisability of control-state choreographies, which gives our main result. Section 5 contains a brief summary and outlook.

2   P2P Communication Systems and Control-State Choreographies

In a P2P system we need at least peers and messages to be exchanged between them. Therefore, let M and P be finite, disjoint sets, elements of which are called *messages* and *peers*, respectively. Each message $m \in M$ has a unique *sender* $s(m) \in P$ and a unique *receiver* $r(m) \in P$ with $s(m) \neq r(m)$. We use the notation $i \xrightarrow{m} j$ for a message m with $s(m) = i$ and $r(m) = j$. We also use the notation $!m^{i \to j}$ and $?m^{i \to j}$ for the *event* of sending or receiving the message m, respectively. Write M_p^s and M_p^r for the sets of messages, for which the sender or the receiver is p, respectively.

2.1   Control-State Machines and P2P Systems

Let $s(M)$ and $r(M)$ denote the sets of send and receive events defined by a set M of messages. In [12] we defined a P2P system over M and P as a family $\{\mathcal{P}_p\}_{p \in P}$ of finite state machines[1] (FSMs) $\mathcal{P}_p$ over an alphabet $\Sigma_p = s(M_p^s) \cup r(M_p^r)$, and by abuse of terminology $\mathcal{P}_p$ was also called a *peer*.

In this article we use a more general notion of P2P system based on the notion of control-state machine. This notion is derived from *control-state abstract state machines* [6], but differs in that we only consider sending and receiving of messages instead of arbitrary ASM rules.

Definition 1. *A* control-state machine $\mathcal{M}$ *comprises*

- *a set Q of* control states *containing an* initial control state $q_0 \in Q$,
- *an* alphabet Σ,
- *a finite set δ of* transition relations $\tau \subseteq Q \times \Sigma \times Q$.

---

[1] Note that the FSM $\mathcal{P}_p$ may be deterministic or non-deterministic.

We write $\mathcal{M} = (Q, q_0, \Sigma, \delta)$.

We use control state machines to define P2P systems. As in [12] we need a set M of messages and a set P of associated peers, i.e. those peers p that appear as sender or receiver of the messages $m \in M$ as defined above.

Definition 2. *A* peer-to-peer system *(P2P system) over a set M of messages and an associated set P of peers is a family $\{\mathcal{P}_p\}_{p \in P}$ of control-state machines $\mathcal{P}_p = (Q_p, q_{p,0}, \Sigma_p, \delta_p)$ with alphabets Σ_p containing the sending and receiving events of messages in M with sender or receiver p, respectively, i.e. $\Sigma_p = s(M_p^s) \cup r(M_p^r)$.*

By abuse of terminology we also call $\mathcal{P}_p$ a *peer*. The rationale behind Definition 2 is that a P2P system is composed of several autonomous peers[2] that interact via messages.

For the semantics of a P2P system we have to define how the different peers interact. Informally, all peers are supposed to operate autonomously on their local states, while their interaction defines the sequences of global states. We will therefore first define the notion of state.

Definition 3. *A* state *of a P2P system $\{\mathcal{P}_p\}_{p \in P}$ consists of*

- *a control tuple ctl of control states q_p for $p \in P$, i.e. a function ctl $: P \to \bigcup_{p \in P} Q_p$ with $ctl(p) \in Q_p$,*
- *a message pool B, which is defined using one of the following four alternatives:*

rendez-vous. $B = \emptyset$;

p2p. *B is a family of FIFO queues $c_{i,j}$ indexed by pairs (i, j) with $i, j \in P, i \neq j$, such that all entries in $c_{i,j}$ are messages $i \xrightarrow{m} j$ from sender i to receiver j;*

queue. *B is a family of FIFO queues c_j indexed by peers $j \in P$, such that all entries in c_j are messages $i \xrightarrow{m} j$ with receiver j;*

mailbox. *B is a family of multisets b_j indexed by peers $j \in P$, such that all entries in b_j are messages $i \xrightarrow{m} j$ with receiver j.*

An initial state *is a state with $ctl(p) = q_{p,0}$ for all $p \in P$ and an empty message pool B, i.e. all queues $c_{i,j}$, c_j or multisets b_j for $i, j \in P$ are empty.*

Depending on the chosen alternative for the message pool we say that the P2P system has a p2p, queue, mailbox or rendez-vous semantics[3].

---

[2] In a more general context of concurrent systems in [4] peers are called *agents*. The special case, where messaging is added to shared locations is handled in [5].

[3] Note that it would also be possible to use a separate mailbox for each channel defined by a pair (i, j) with $i, j \in P, i \neq j$. However, as it is always possible to access all elements in a multiset, this will not make any difference. For the case of FSM-based P2P systems this was formally shown in [12, Prop. 2].

The rationale behind Definition 3 is to understand the semantics of a P2P system as an asynchronous concurrent system that progresses by state transitions. The p2p, queue and mailbox semantics capture how systems really run; they only differ in the way messages are handled by the peers. Clearly, mailboxes provide the largest flexibility, whereas the queue semantics is the most restrictive one. However, as examples in [10] show there can be P2P systems that behave oddly and hardly reflect behaviour desired from an application point of view. For this purpose we also foresee the restrictive rendez-vous semantics. With FSMs instead of control-state machines (as in [10,12]) the rendez-vous semantics enforces a strictly sequential behaviour. With the use of control-state machines this behaviour is much more relaxed, and permits parallelism.

A *location* of a state is given by a function symbol f and a n-tuple $\bar{a}$ of values, where n is the arity of f. For the states defined in Definition 3 the function symbols are ctl of arity 1, and $channel$ of arity 2 in p2p semantics, $queue$ of arity 1 in queue semantics, or $mailbox$ of arity 1 in mailbox semantics—for rendez-vous semantics no further function symbol is needed. We write $val_S(\ell)$ to denote the value of the location ℓ in state S, i.e. $val_S((ctl, p)) = ctl(p)$ is defined for peers $p \in P$. For the other locations we have:

- In p2p semantics $val_S((channel, (i, j))) = c_{i,j}$ is defined for $i, j \in P$ with $i \neq j$.
- In queue semantics $val_S((queue, j)) = c_j$ is defined for $j \in P$.
- In mailbox semantics $val_S((mailbox, j)) = b_j$ is defined for $j \in P$.

For any state $S = (ctl, B)$ and any peer $p \in P$ we obtain a *projection* $S_p = (ctl(p), B_p)$, where B_p is the family of queues $\{c_{i,p} \mid i \in P, i \neq p\}$ in p2p semantics, the queue c_p in queue semantics, the multiset b_p in mailbox semantics, and $\emptyset$ in rendez-vous semantics. A projected state S_p is also referred to as *local state* of the peer p.

Definition 4. *The set $E_p(S)$ of enabled transitions of a peer $p \in P$ in a state $S = (ctl, B)$ is defined by the following cases:*

- *In p2p, queue or mailbox semantics a sending transition $t = (q_t, !m^{p \to j}, q_t') \in \tau \in \delta_p$ of a peer $\mathcal{P}_p$ is enabled in S iff $ctl(p) = q_t$ holds.*
- *In p2p semantics a receiving transition $t = (q_t, ?m^{i \to p}, q_t') \in \tau \in \delta_p$ of a peer $\mathcal{P}_p$ is enabled in S iff $ctl(p) = q_t$ holds, and the front element in the queue $c_{i,p}$ is the message $i \xrightarrow{m} p$.*
- *In queue semantics a receiving transition $t = (q_t, ?m^{i \to p}, q_t') \in \tau \in \delta_p$ of a peer $\mathcal{P}_p$ is enabled in S iff $ctl(p) = q_t$ holds, and the front element in the queue c_p is the message $i \xrightarrow{m} p$.*
- *In mailbox semantics a receiving transition $t = (q_t, ?m^{i \to p}, q_t') \in \tau \in \delta_p$ of a peer $\mathcal{P}_p$ is enabled in S iff $ctl(p) = q_t$ holds, and the message $i \xrightarrow{m} p$ appears in the multiset b_p.*
- *In rendez-vous semantics a sending transition $t_i = (q_i, !m^{i \to j}, q_i') \in \tau_i \in \delta_i$ and a receiving transition $t_j = (q_j, ?m^{i \to j}, q_j') \in \tau_j \in \delta_j$ of a peer $\mathcal{P}_j$ are enabled simultaneously in S iff $ctl(i) = q_i$ and $ctl(j) = q_j$ hold.*

Each enabled transition $t \in E_p(S)$ yields an update set of the state S. In general, an *update* is a pair (ℓ, v) comprising a location ℓ and a value v, and an *update set* is a set of such updates.

Definition 5. *For an enabled transition* $t \in E_p(S)$ *of a peer* p *in state* S *the yielded update set* $\Delta_t(S)$ *is defined as follows:*

- *If* $t = (q_t, !m^{p \to j}, q_t')$ *is a sending transition or* $t = (q_t, ?m^{i \to p}, q_t')$ *is a receiving transition, then* $((ctl, p), q_t') \in \Delta_t(S)$.
- *If* $t = (q_t, !m^{p \to j}, q_t')$ *is a sending transition, then in p2p semantics (using* $\oplus$ *to denote concatenation)*

$$((channel, (p, j)), val_S((channel, (p, j))) \oplus [p \overset{m}{\to} j]) \in \Delta_t(S) ,$$

in queue semantics (using again $\oplus$ *for concatenation)*

$$((queue, j), val_S((queue, j)) \oplus [p \overset{m}{\to} j]) \in \Delta_t(S) ,$$

and in mailbox semantics (using $\uplus$ *to denote multiset union)*

$$((mailbox, j), val_S((mailbox, j)) \uplus [p \overset{m}{\to} j]) \in \Delta_t(S) .$$

- *If* $t = (q_t, ?m^{i \to p}, q_t')$ *is a receiving transition, then in p2p semantics*

$$((channel, (i, p)), c_{i,p}) \in \Delta_t(S) \; for \; val_S((channel, (i, p))) = [i \overset{m}{\to} p] \oplus c_{i,p} ,$$

in queue semantics $((queue, p), c_p) \in \Delta_t(S)$ *for* $val_S((queue, p)) = [i \overset{m}{\to} p] \oplus c_p$, *and in mailbox semantics* $((mailbox, p), val_S((mailbox, p)) - \langle i \overset{m}{\to} p \rangle) \in \Delta_t(S)$.
- *No other updates are in* $\Delta_t(S)$.

This notion of yielded update set generalises naturally to sets of transitions and to peers in a P2P system.

Definition 6. *Let* S *be a state of a P2P system* $\{\mathcal{P}_p\}_{p \in P}$. *For each set* $\tau \in \delta_p$ *of transitions the* update set yielded by τ in S *is* $\Delta_\tau(S) = \bigcup_{t \in \tau, t \in E_p(S)} \Delta_t(S)$. *The* set of update sets *of the peer* $p \in P$ *in state* S *is* $\boldsymbol{\Delta}_p(S) = \{\Delta_\tau(S) \mid \tau \in \delta_p\}$.

The definition highlights the nature of the sets of transition sets defining peers. Transitions t in the same set $\tau \in \delta_p$ yield updates that are to be applied in parallel, whereas the different transition sets τ provide choices. As we build update sets by unions, we have to handle updates to the same location with function symbol *channel*, *queue* or *mailbox*, respectively. For receiving transitions in p2p or queue semantics such updates request a particular front element in a FIFO queue, so two such updates can only occur in parallel, if they require the same front element. For receiving transitions in mailbox semantics the messages are removed according to their multiplicity. For sending transitions in p2p and queue semantics different messages are appended to the queues in arbitrary order, and they are cumulatively added to the multisets in mailbox semantics. In doing so, we actually realise *partial updates* as defined in [11].

Furthermore, in p2p, queue and mailbox semantics update set yielded by different peers are handled independently from each other, whereas in rendez-vous semantics we always have to consider pairs of a sending and a receiving transition. We will take care of this dependency in the definition of runs of P2P systems.

2.2   Runs of P2P Systems

We can use update sets to define successor states. For this we require consistent update sets.

Definition 7. *An update set Δ is consistent if and only if for any updates $(\ell, v_1), (\ell, v_2) \in \Delta$ of the same location we have $v_1 = v_2$.*

Now assume that S is a state and Δ is an arbitrary update set such the updates in Δ correspond to the semantics underlying S. If Δ is consistent, we define a successor state $S' = S + \Delta$ by

$$val_{S'}(\ell) = \begin{cases} v & \text{if } (\ell, v) \in \Delta \\ val_S(\ell) & \text{else} \end{cases}.$$

For inconsistent Δ we extend the definition to $S + \Delta = S$.

Using update sets of the peers yielded in states we can now express the semantics of P2P systems by the runs they permit. As peers are supposed to operate concurrently, we adopt the definition of concurrent run from [4]. That is, peers yielding an update set in some state contribute with these updates in building a new later state, but not necessarily the immediate successor state.

Definition 8. *A run of a P2P system $\{\mathcal{P}_p\}_{p \in P}$ in p2p, queue or mailbox semantics is a sequence $S_0, S_1, S_2, \ldots$ of states such that S_0 is an initial state, and $S_{i+1} = S_i + \Delta_i$ holds for all $i \geq 0$, where the update sets are defined as $\Delta_i = \bigcup_{p \in P_i} \Delta_{a(i,p)}$ with update sets $\Delta_{a(i,p)} \in \Delta_p(S_{a(i,p)})$ formed in previous states, i.e. $a(i,p) \leq i$, for subsets $P_i \subseteq P$ of peers such that $p \in P_i$ implies $p \notin P_j$ for all $a(i,p) \leq j < i$.*

According to our remark at the end of the previous subsection Definition 8 is sufficient to define the runs for P2P systems with p2p, queue or mailbox semantics. For rendez-vous semantics the update sets Δ_i in the definition require further restrictions.

Definition 9. *A run of a P2P system $\{\mathcal{P}_p\}_{p \in P}$ in rendez-vous semantics is a sequence $S_0, S_1, S_2, \ldots$ of states as in Definition 8, where in addition the update sets $\Delta_{a(i,p)}$ satisfy the following condition: if $\Delta_{a(i,p)}$ contains an update defined by a sending transition t with the sending event $!m^{p \to j}$, then $a(i,j) = a(i,p)$ and $\Delta_{a(i,j)} \in \Delta_j(S_{a(i,j)})$ contains the update defined by the corresponding receiving transition t' with the receiving event $?m^{p \to j}$ and vice versa.*

Extending our previous work on the basis FSMs we can now define message languages for our generalised P2P systems. For a state transition from S_i to S_{i+1} in a run consider the multiset M_i containing those messages $m \in M$ that appear in an update in Δ_i defined by a sending transition. If m appears in several such updates, it appears in M_i with the corresponding multiplicity. Then let $\hat{M}_i$ be the set of ordered sequences containing all the elements of M_i.

Definition 10. *For a run $R = S_0, S_1, \ldots$ let $\mathcal{L}(R)$ be the set of all sequences with elements in M that result from concatenation of $\hat{M}_0, \hat{M}_1, \ldots$. Let $\mathcal{L}$ be the language of all finite sequences in $\bigcup_R \mathcal{L}(R)$, where the union is built over all runs of the P2P systems. We write $\mathcal{L}_0$, $\mathcal{L}_{p2p}$, $\mathcal{L}_q$ and $\mathcal{L}_m$ for the languages in rendez-vous, p2p, queue and mailbox semantics, respectively, and these languages the trace languages of the P2P system in rendez-vous, p2p, queue and mailbox semantics, respectively.*

Two P2P systems are called equivalent *in rendez-vous, p2p, queue and mailbox semantics, respectively, if and only if they define the same trace language.*

Note that finite sequences correspond to finite runs. So we can also define sublanguages $\hat{\mathcal{L}}_{p2p}$, $\hat{\mathcal{L}}_q$ and $\hat{\mathcal{L}}_m$ containing only those sequences of message that result from runs $S_0, S_1, \ldots, S_k$, where all channels or mailboxes in the state S_k are empty. These languages are called the *stable trace languages* of the P2P system in the different semantics.

We use these generalised trace and stable trace languages associated with a P2P system to generalise the notions of synchronisability and language-synchronisability from [12] and [10].

Definition 11. *A P2P system $\{\mathcal{P}_p\}_{p \in P}$ is* language-synchronisable *in p2p, queue or mailbox semantics, respectively, if and only if $\mathcal{L}_0 = \mathcal{L}_{p2p}$, $\mathcal{L}_0 = \mathcal{L}_q$ or $\mathcal{L}_0 = \mathcal{L}_m$ holds, respectively.*

A P2P system $\{\mathcal{P}_p\}_{p \in P}$ is synchronisable *in p2p, queue or mailbox semantics, respectively, if and only if in addition $\hat{\mathcal{L}}_{p2p} = \mathcal{L}_{p2p}$, $\hat{\mathcal{L}}_q = \mathcal{L}_q$ or $\hat{\mathcal{L}}_m = \mathcal{L}_m$ holds, respectively.*

Following our previous remarks about the rationale behind P2P systems defined as families of control-state machines the notion of language-synchronisability emphasises a desirable property of P2P systems. On an abstract level we may view the system via its rendez-vous semantics, in which messaging is handled in an atomic way, whereas in p2p, queue and mailbox semantics we emphasise the reification by autonomous peers with the only difference concerning how messages are kept and processed. The stricter notion of synchronisability further emphasises the desire that messages should not be allowed to be ignored by the receiver peer.

Clearly, synchronisability implies language-synchronisability. These two properties are known to be undecidable in general even for the restricted case, where the peers are defined by FSMs [10].

3   Choreography-Defined P2P Systems

While synchronisability and language-synchronisability are undecidable in general, it was shown in [12] that decidability results, if we assume P2P systems that are choreography-defined.

3.1   Control-State Choreographies

We first define a generalised notion of choreography based on control-state machines, which we call control-state choreographies.

Definition 12. *Let M be a set of messages with peers P. A control-state choreography (CSC) over M and P is a control-state machine $\mathcal{C} = (Q, q_0, M, \delta)$.*

We define the semantics of CSCs analogous to the rendez-vous semantics of a P2P system with a single peer. The fact that we now only consider the messages instead of the associated sending and receiving events eases the formulation of runs.

Definition 13. *A* state *of a CSC $\mathcal{C} = (Q, q_0, M, \delta)$ is a control state $ct \in Q$. An* initial state *is $ct = q_0$.*

A transition $t = (q, i \xrightarrow{m} j, q') \in \tau \in \delta$ is enabled in state $S = ct$ iff $ct = q$ holds; we write $t \in E_{\mathcal{C}}(S)$. Then the update set $\Delta_t(S)$ *contains exactly the update (ct, q').*

For each set $\tau \in \delta$ the update set *yielded by τ in S is $\Delta_\tau(S) = \bigcup_{t \in \tau, t \in E_{\mathcal{C}}(S)} \Delta_t(S)$, and the* set of update sets *in state S is $\boldsymbol{\Delta}_{\mathcal{C}}(S) = \{\Delta_\tau(S) \mid \tau \in \delta\}$.*

All remarks in the preceding section we made in connection with the partial updates apply analogously to the update sets defined by CSCs. We now obtain the notion of run of a CSC.

Definition 14. *A* run *of a CSC $\mathcal{C} = (Q, q_0, M, \delta)$ is a sequence $S_0, S_1, S_2, \dots$ of states such that S_0 is an initial state, and $S_{i+1} = S_i + \Delta_i$ holds for an update set $\Delta_i \in \boldsymbol{\Delta}_{\mathcal{C}}(S_i)$.*

For a state transition from S_i to S_{i+1} in a run consider the multiset M_i containing those messages $m \in M$ that appear in the update in Δ_i. If m appears in several such updates, it appears in M_i with the corresponding multiplicity. Then let $\hat{M}_i$ be the set of ordered sequences containing all the elements of M_i.

For a run $R = S_0, S_1, \dots$ let $\mathcal{L}(R)$ be the set of all sequences with elements in M that result from concatenation of $\hat{M}_0, \hat{M}_1, \dots$. We define the *trace language* $\mathcal{L}(\mathcal{C})$ of the CSC as the language of all finite sequences in $\bigcup_R \mathcal{L}(R)$, where the union is built over all runs R of the CSC.

3.2   Choreographies Defined by P2P Systems

Now consider a P2P system $\{\mathcal{P}_p\}_{p \in P}$. For each peer $p \in P$ and each transition set $\tau \in \delta_p$ we may assume without loss of generality that transitions $t, t' \in \tau$ have the same initial and final control states q_τ and q'_τ, respectively. According to Definition 4 equal initial control states are necessary for the transitions to be enabled simultaneously, and according to Definition 5 equal final control states are necessary to ensure that consistent update sets are yielded. With this assumption we can select a transition set $\tau_p \in \delta_p$ for each peer $p \in P$ are choose $\tau_p = \emptyset$.

Definition 15. *A selection* $\{\tau_p \mid p \in P\}$ *such that either* $\tau_p \in \delta_p$ *or* $\tau_p = \emptyset$ *holds is* closed under pairing *iff whenever* τ_p *contains a sending transition* $t = (q_{\tau_p}, !m^{p \to p'}, q'_{\tau_p})$, *then* $\tau_{p'}$ *contains the corresponding receiving transition* $t' = (q_{\tau_{p'}}, ?m^{p \to p'}, q'_{\tau_{p'}})$ *and vice versa.*

If a selection $\{\tau_p \mid p \in P\}$ is closed under pairing, we can define tuples $q, q' : P \to \bigcup_{p \in P} Q_p$ with $q(p) = q_{\tau_p} \in Q_p$ and $q'(p) = q'_{\tau_p} \in Q_p$ for all $p \in P$; in case $\tau_p = \emptyset$ we extend q and q' chosing $q(p) = q'(p) \in Q_p$ arbitrarily. Then the set

$$\bigcup_{p,p' \in P} \{(q, p \xrightarrow{m} p', q') \mid (q_{\tau_p}, !m^{p \to p'}, q'_{\tau_p}) \in \tau_p, (q_{\tau_{p'}}, ?m^{p \to p'}, q'_{\tau_{p'}}) \in \tau_{p'}\}$$

is the *set of choreography transitions* of the selection.

Definition 16. *The* choreography *defined by a P2P system* $\{\mathcal{P}_p\}_{p \in P}$ *is* $\mathcal{C} = (Q, q_0, M, \delta)$, *where the set* Q *of control states is the set of all tuples* $q : P \to \bigcup_{p \in P} Q_p$ *with* $q(p) \in Q_p$, *the initial control state is* q_0 *with* $q_0(p) = q_{p,0}$ *for all* $p \in P$ *and* δ *contains exactly the non-empty sets of choreography transitions defined by all selections that are closed under pairing.*

Proposition 1. *The trace language* $\mathcal{L}(\mathcal{C})$ *of the choreography* $\mathcal{C}$ *defined by a P2P system* $\{\mathcal{P}_p\}_{p \in P}$ *is equal to the trace language* $\mathcal{L}_0$ *defined by the rendez-vous semantics of the P2P system.*

3.3   P2P Systems Defined by Choreographies

We now investigate the inverse of Proposition 1, i.e. we will define a P2P system associated with a given choreography and show that the trace language $\mathcal{L}_0$ defined by its rendez-vous semantics is the same as the trace language of the choreography, which generalises [12, Prop. 4]. For this we will introduce projections to peers, which will be straightforward except for a little subtlety associated with ϵ-transitions, in which a control state is changed without a sending or receiving event. Analogous to [12, Prop. 2] such ϵ-transitions can be eliminated; they are neither needed nor particularly useful.

So let $\mathcal{C} = (Q, q_0, M, \delta)$ be a CSC over M and P. For each $p \in P$ we define the projection $\mathcal{C}_p$ of $\mathcal{C}$ in two steps:

Step 1. First define $\mathcal{C}_p' = (Q, q_0, \Sigma_p, \delta_p')$ with alphabet $\Sigma_p = s(M_p^s) \cup r(M_p^r)$ and set of transition sets $\delta_p' = \{\tau_p \mid \tau \in \delta\}$, where $\tau_p = \{(q, \pi_p(m), q') \mid (q, m, q') \in \tau\}$ with

$$\pi_p(i \xrightarrow{m} j) = \begin{cases} !m^{p \to j} & \text{if } i = p \\ ?m^{i \to p} & \text{if } j = p \ . \\ \epsilon & \text{else} \end{cases}$$

Here, $\mathcal{C}_p'$ is not a control-state machine as defined in Definition 1, because it admits transitions, in which only the control state is modified, but no message is sent nor received. Nonetheless, for such ϵ-*extended machines* we can still define enabled transitions as in Definition 4 and update sets as in Definition 6—the only difference concerns Definition 5 in that ϵ-transitions do not yield updates to locations with function symbols *channel*, *queue* or *mailbox* in p2p, queue and mailbox semantics, respectively. Therefore, we can treat $\{\mathcal{C}_p'\}_{p \in P}$ also as a P2P system.

Step 2. We can eliminate the ϵ-transitions and define a control state machine $\mathcal{C}_p = (Q, q_0, \Sigma_p, \delta_p)$. As before we can assume without loss of generality that for each $\tau \in \delta_p'$ we have unique states q_τ and q_τ' such that the transitions $t \in \tau$ all have the form $(q_\tau, \tilde{m}_i, q_\tau')$; otherwise we could split τ.

Then take transitions $t = (q, \epsilon, q') \in \tau$ and $t' = (q', \tilde{m}, q'') \in \tau'$. We may combine t and t' to the new transition $t' \circ t = (p, \tilde{m}, q'')$.

Then we can replace τ by $\tau' \circ \tau$ containing all those composed transition $t' \circ t$ with $t \in \tau$ and $t' \in \tau'$. This replacement can be iterated until there are no more transition sets with ϵ-transitions, which defines δ_p.

This construction suggests the following terminology of a choreography-defined P2P system.

Definition 17. *A P2P system $\{\mathcal{P}_p\}_{p \in P}$ is* choreography-defined *iff there exists a CSC $\mathcal{C}$ such that the projected P2P system $\{\mathcal{C}_p\}_{p \in P}$ is equivalent to $\{\mathcal{P}_p\}_{p \in P}$.*

Proposition 2. *The trace language $\mathcal{L}(\mathcal{C})$ of the choreography $\mathcal{C}$ is equal to the trace language $\mathcal{L}_0$ defined by the rendez-vous semantics of the P2P system $\{\mathcal{C}_p\}_{p \in P}$ and the P2P system $\{\mathcal{C}_p'\}_{p \in P}$ with ϵ-transitions.*

Proposition 2 shows that a choreography-defined P2P systems and choreographies mutually define each other, and the language defined by a choreography is the rendez-vous language of the corresponding P2P system. This justifies the following notion of realisability.

Definition 18. *A CSC $\mathcal{C}$ is called* realisable *if and only if its projected P2P system $\{\mathcal{C}_p\}_{p \in P}$ is synchronisable.*

Clearly, the notion of realisability depends on p2p, queue or mailbox semantics.

3.4 Synchronisability for Choreography-Defined P2P Systems

Definition 11 also defines the weaker notion of language synchronisability, so we could analogously define "language-realisability" for CSCs. However, for the restricted case of choreographies defined by FSMs we showed in [12, Prop. 5] that language synchronisability and synchronisability coincide for choreography-defined P2P systems. This result also holds for our generalised CSCs.

Proposition 3. *A choreography-defined P2P system* $\mathcal{P} = \{\mathcal{P}_p\}_{p \in P}$ *is synchronisable with respect to p2p, queue or mailbox semantics, respectively, if and only if it is language synchronisable with respect to the same semantics.*

4 Characterisation of Realisability

We now investigate realisability of CSCs. From Proposition 2 we know that CSCs are equivalent to their projected P2P systems concerning rendez-vous semantics in the sense of defining the same trace language, so a CSC prescribes the behaviour of a P2P system capturing communication in a synchronous way, i.e. sending and receiving of messages are always synchronised. We have to show the synchronisability of these P2P systems, which means to show that the semantics expressed by the trace languages is preserved, when a more realistic asynchronous behaviour using channels, queues or mailboxes is taken into account. As a consequence of Proposition 3 it suffices to concentrate on language synchronisability.

Not all P2P systems are choreography-defined, and not all choreographies will be realisable. We will now derive necessary and sufficient conditions for CSCs that guarantee realisability. For this we will generalise the sequence and choice conditions from [12].

4.1 The Sequence Condition

In [12, Prop. 6] we proved that if a choreography defined by an FSM is realisable, then any two messages that follow each other in a sequence must either be independent, i.e. their order can be swapped, or the sender of the second message must be the same as the sender or receiver of the first message. The first alternative condition was a workaround for parallelism, and the second condition expressed that messages in a sequence must either have the same sender or the receiver of a message must be the sender of the next one. We will now show that in a slightly generalised way this condition must also be satisfied for CSCs in order to obtain realisability.

Definition 19. *A CSC* $\mathcal{C} = (Q, q_0, M, \delta)$ *is said to satisfy the* sequence condition *if and only if for any two transitions* $(q_1, i \overset{m_1}{\twoheadrightarrow} j, q_2) \in \tau \in \delta$ *and* $(q_2, k \overset{m_2}{\twoheadrightarrow} \ell, q_3) \in \tau' \in \delta$ *we have that* $(q_1, k \overset{m_2}{\twoheadrightarrow} \ell, q_2) \in \tau \in \delta$ *and* $(q_2, i \overset{m_1}{\twoheadrightarrow} j, q_3) \in \tau' \in \delta$ *hold or* $k \in \{i, j\}$ *holds.*

Proposition 4. *If* $\mathcal{C} = (Q, q_0, M, \delta)$ *is a realisable CSC with respect to p2p, queue or mailbox semantics, then* $\mathcal{C}$ *satisfies the sequence condition.*

4.2   The Choice Condition

In [12, Prop. 7] we further proved that if a choreography defined by an FSM is realisable, then any two messages subject to a choice must be independent or have the same sender. The second condition expresses that messages in a choice can only refer to the same sender. We will now show that in a slightly generalised way this condition must also be satisfied for CSCs in order to obtain realisability.

Definition 20. *A CSC $C = (Q, q_0, M, \delta)$ is said to satisfy the choice condition if and only if for any two transitions $(q_1, i \xrightarrow{m_1} j, q_2) \in \tau \in \delta$ and $(q_1, k \xrightarrow{m_2} \ell, q_3) \in \tau' \in \delta$ we have that $(q_1, k \xrightarrow{m_2} \ell, q_2) \in \tau \in \delta$ and $(q_1, i \xrightarrow{m_1} j, q_3) \in \tau' \in \delta$ hold or $k = i$ holds.*

Proposition 5. *If $C = (Q, q_0, M, \delta)$ is a realisable CSC with respect to p2p, queue or mailbox semantics, then C satisfies the choice condition.*

4.3   Sufficient Conditions for Realisability

In [12, Thm. 1] we also proved that the two necessary conditions for realisability, the sequence and the choice conditions, together are also sufficient. We will now generalise this result for CSCs, which gives the main result of this article.

Theorem 1. *A CSC $C = (Q, q_0, M, \delta)$ is realisable with respect to p2p, queue or mailbox semantics, if and only if it satisfies the sequence and the choice conditions.*

5   Conclusion

Choreographies prescribe the rendez-vous synchronisation of messages in P2P systems. If defined by communicating finite state machines, the synchronisability problem, i.e. to decide whether a reification by asynchronous peers is equivalent to the rendez-vous composition of the peers, is undecidable in general [10]. However, for choreography-defined P2P systems it becomes decidable, as for these synchronisability coincides with language synchronisability [12]. On these grounds it was possible to discover two necessary conditions for realisability, the sequence and the choice condition, which together are also sufficient.

Nonetheless, despite its solid theory we felt that the value of this result is limited, as the notion of choreography defined by FSMs is rather weak; it enforces a strictly sequential behaviour of the P2P system. Therefore, we now introduced control state choreographies replacing FSMs by control-state machines and thus introducing parallelism. On these grounds the results from [12] could be generalised to CSCs using slightly generalised definitions for the sequence and choice condition.

Still the parallelism and the choice in choreographies and P2P systems is bounded. So a natural continuation of the research would be to look at unbounded choice and unbounded parallelism. However, this requires to consider not just messages, but also other state changes. In fact, it seems reasonable to focus on realisable choreographies only on high levels of abstraction, whereas in a thorough refinement process weaker synchronisation properties are well acceptable.

References

1. Basu, S., Bultan, T.: On deciding synchronizability for asynchronously communicating systems. Theor. Comput. Sci. **656**, 60–75 (2016). https://doi.org/10.1016/j.tcs.2016.09.023
2. Benyagoub, S., Aït-Ameur, Y., Schewe, K.-D.: Event-B-supported choreography-defined communicating systems. In: Raschke, A., Méry, D., Houdek, F. (eds.) ABZ 2020. LNCS, vol. 12071, pp. 155–168. Springer, Cham (2020). https://doi.org/10.1007/978-3-030-48077-6_11
3. Benyagoub, S., Ouederni, M., Aït-Ameur, Y., Mashkoor, A.: Incremental construction of realizable choreographies. In: Dutle, A., Muñoz, C., Narkawicz, A. (eds.) NFM 2018. LNCS, vol. 10811, pp. 1–19. Springer, Cham (2018). https://doi.org/10.1007/978-3-319-77935-5_1
4. Börger, E., Schewe, K.-D.: Concurrent abstract state machines. Acta Informatica **53**(5), 469–492 (2015). https://doi.org/10.1007/s00236-015-0249-7
5. Börger, E., Schewe, K.D.: Communication in abstract state machines. J. UCS **23**(2), 129–145 (2017). http://www.jucs.org/jucs_23_2/communication_in_abstract_state
6. Börger, E., Stärk, R.F.: Abstract State Machines. A Method for High-Level System Design and Analysis. Springer, Berlin (2003). https://doi.org/10.1007/978-3-642-18216-7
7. Brand, D., Zafiropulo, P.: On communicating finite-state machines. J. ACM **30**(2), 323–342 (1983). https://doi.org/10.1145/322374.322380
8. Chambart, P., Schnoebelen, P.: Mixing lossy and perfect fifo channels. In: van Breugel, F., Chechik, M. (eds.) CONCUR 2008. LNCS, vol. 5201, pp. 340–355. Springer, Heidelberg (2008). https://doi.org/10.1007/978-3-540-85361-9_28
9. Clemente, L., Herbreteau, F., Sutre, G.: Decidable topologies for communicating automata with FIFO and bag channels. In: Baldan, P., Gorla, D. (eds.) CONCUR 2014. LNCS, vol. 8704, pp. 281–296. Springer, Heidelberg (2014). https://doi.org/10.1007/978-3-662-44584-6_20
10. Finkel, A., Lozes, É.: Synchronizability of communicating finite state machines is not decidable. In: Chatzigiannakis, I., et al. (eds.) 44th International Colloquium on Automata, Languages, and Programming (ICALP 2017). LIPIcs, vol. 80, pp. 122:1–122:14. Schloss Dagstuhl - Leibniz-Zentrum für Informatik (2017). https://doi.org/10.4230/LIPIcs.ICALP.2017.122
11. Schewe, K.D., Wang, Q.: Partial updates in complex-value databases. In: Heimbürger, A., et al. (eds.) Information and Knowledge Bases XXII, Frontiers in Artificial Intelligence and Applications, vol. 225, pp. 37–56. IOS Press (2011). https://doi.org/10.3233/978-1-60750-690-4-37
12. Schewe, K.-D., Aït-Ameur, Y., Benyagoub, S.: Realisability of choreographies. In: Herzig, A., Kontinen, J. (eds.) FoIKS 2020. LNCS, vol. 12012, pp. 263–280. Springer, Cham (2020). https://doi.org/10.1007/978-3-030-39951-1_16
13. Schewe, K.D., Ameur, Y.A., Benyagoub, S.: Realisability of control-state choreographies. CoRR abs/2009.03623 (2020). https://arxiv.org/abs/2009.03623

A Low-Cost Authentication Protocol Using Arbiter-PUF

Fahem Zerrouki[1] , Samir Ouchani[2(✉)] , and Hafida Bouarfa[1]

[1] University of Blida, Blida, Algeria
[2] Lineact CESI, Aix-en-Provence, France
`souchani@cesi.fr`

Abstract. Integrated Circuits (ICs) and electronic devices become part of human daily life (mobile, home, car, etc.) and for the safety of the information transported to and from these devices, some specifics security measures are proposed. Some ICs are a source of high randomness due to the manufacturing variation of these ICs and exploited as physically unclonable functions (PUFs), called Silicon PUFs (SPUFs). Many propositions have shown that SPUFs could guarantee the three pillars of security; confidentiality, authenticity, and privacy of the information used by these devices. However, to extract strong cryptographic keys from them, error correction and hashing code techniques should be considered. Unfortunately, generating many times the same response for the same input is not possible due to many factors especially the environment variables such as noise, temperature, supply voltage, and aging of an IC. To overcome this issue and recover the original response from the noisy one, a secure sketch can be used. In this paper, we propose a low-cost authentication protocol based on SPUFs by using Arbiter PUF that is classified as a strong PUF.

Keywords: Physical Unclonable Function · Arbiter PUF · Fuzzy extractor · Cryptographic key generation · Authentication

1 Introduction

The minimal common requirements for a good cryptosystem are i) a secure key generation, to ensure that the generated keys have to be random, unique, unpredictable, and reproducible, and ii) a secure key-storage, this due to the need access to the key by cryptographic systems several times. Of course, the key-storage systems' have to protect the key's information from unauthorized parties [5]. However, the generation process used by traditional cryptography systems generally relies on uniform distribution and random string of the generated key such as random passwords and tokens. Besides, this random string must also consist of high entropy which makes it not easy to create and memorize [6]. In addition, the secret key information is generally stored or embedded on nonvolatile memory which makes it vulnerable to physical attacks such as invasive, semi-invasive, or side-channel attacks [7].

© Springer Nature Switzerland AG 2021
C. Attiogbé and S. Ben Yahia (Eds.): MEDI 2021, LNCS 12732, pp. 101–116, 2021.
https://doi.org/10.1007/978-3-030-78428-7_9

Due to the manufacturing variation process of an integrated circuit (IC), the latter is considered a good source of high entropy. Silicon physical unclonable functions (SPUFs) [7] are a primitive cryptosystem that exploits the intrinsic randomness found in ICs to generate a uniform string and use it as a cryptographic key. By challenging SPUFs with different inputs, they generate responses, and the set of challenges and responses represent a unique behaviour, which can be used as a fingerprint of an IC or the device where it is embedded in. Consequently, SPUFs can generate cryptographic keys to be used in the cryptographic application without storing them on the device, which avoids the need for key-storage systems. Unfortunately due to aging and environmental variations such as temperature or voltage variations, when regenerating the same response several times, some error bits appeared and it impacts directly the security of the system deploying the IC. Therefore, in order to use this output as a secret key, their errors must be corrected. Fuzzy extractor (FE) and Secure sketch (SS) [8] are two algorithms used respectively to extract safely a uniform string from a non-uniform one and to generate a stable key from a noisy unstable output of PUFs using error correction techniques [9].

With the expansion of the deployed IoT (Internet of Things) and smart devices [18], a low-cost authentication is required as a process between a prover and a secure trusted server. Thus, a dedicated PUF authentication protocol that orchestrates the communication between the network entities should exploit the deployed PUF challenges and responses [23]. This work exploits the randomness of the Arbiter PUF to propose a low-cost authentication protocol that can be deployed on low energy and low memory devices to be applied in many areas: autonomous vehicles and smart transportation, smart cities, etc. In this contribution, we detailing the needed background to understand PUFs, in general. Then, we discuss how PUFs can generate a good response which can be used as a cryptographic key, and how can be corrected using a fuzzy extractor. Then, we present how PUFs can be used to authenticate individual ICs without costly cryptographic primitives. The proposed protocol details and shows the different phases and algorithms that can be used in the authentication steps to generate a stable uniform response each time.

2   Related Work

PUFs based Protocols are often dependent on the type of PUF (weak or strong), in this section, we mainly emphasize protocols that have been proposed mainly for the smart devices.

Aman et al. [14] used PUF's response as part of the authentication process by proposing two mutual authentication schemes: server-IoT and IoT-IoT devices. The protocol guaranteed the integrity of data without saving any secret information on the IoT device. Then, Chatterjee *et al.* [20] used the PUF's response and timestamp of an authentication session to provide the identity of the IoT device. Also, Xie *et al.* [16] proposed a lightweight mutual authentication mechanism for body area network (BAN) sensors with PUFs, and the sensor's identity is used as the PUF challenge to generate the needed response.

Chatterjee *et al.* [17] used the combination of encryption (IBE), PUFs, and hash functions to develop authentication and key exchange protocol for IoT devices where the PUF's response is used as a session key. In addition to a trusted server, this protocol uses two other entities: the security credential generator and the security association provider. Gope *et al.* [18] used PUFs as authentication factors that allow an anonymously to communicate between the IoT device and the trusted server. Baruah *et al.* [19] proposed a PUF-based authentication protocol where a captcha image is used as a challenge to generate a PUF's response.

Yanambaka *et al.* [13] proposed a suitable device authentication scheme for Internet-of-Medical-Things (IoMT) using a database as a third party in the authentication process. They store the device information server where the response generated by the server is used as the challenge of the IoMT device in the authentication process. Chatterjee *et al.* [20] have demonstrated several vulnerabilities such as Denial-of-Service (DoS) and replay attacks, synchronization problems, token/server impersonation, modeling attack, lack of integrity checking of the helper data, and limited local authentication.

Since all PUF based protocols are based on the challenge-response behavior where the response plays an important role in the authentication process as a secret cryptography key. It is first generated in the enrollment phase to be stored in the trusted server or another database for the authentication phase. But when generating the same response several times in order to compare it with the initial and the stored one, the response differs from the first one in some error bit which makes it a noisy response compared with the initial one and this due to the environmental factors. Compared to the discussed initiatives, it is necessary to consider the impact of noise in the authentication.

3   Physical Unclonable Functions

Instead of storing the secret information used in identification, authentification, and other security applications domain, a PUF is a physical function that extracts this secret information from the physical characteristics of the physical object. This is due to the physical disorder phenomena of the subject. The latter has to respond to two criteria inter-distance and intra-distance. The integrated circuit is a good example of the physical disorder, where the first criteria are

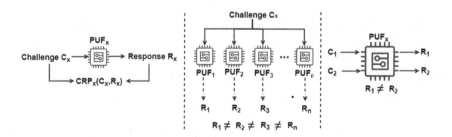

Fig. 1. The challenge response behavior [12].

guaranteed by the manufacturing process and the second one by the environment variables. But the latter causes errors in the regeneration of the secret key, which needs an error correction technique to repair the error in the generated information. Challenge-response pair (CRP) is a fingerprint of the physical object, where the challenge is the input given to the PUF as an input and the response is its output. The process or the way that the challenge followed to generate the response is called Challenge-response behavior. Figure 1 shows the PUF's challenge-response behavior. Given the same challenge c_x as input to many PUF_i instances, so each one should outputs a unique and different response R_i than others. Also, inputting two different challenges C_1 and C_2 to the same PUF_x, that latter should generate different responses $R_{i=1}$ and R_2.

Figure 1 describes the Physical Unclonnable Function (PUF) as a parameterized random function denoted by $\Pi_{S \in \mathcal{S}}$ ($\Pi_{S \in \mathcal{S}} \in \mathcal{P}$) where the structural parameters ($S \in \mathcal{S}$) are random variables describing the PUF structure.

Definition 1. *A PUF is a function* $\Pi_{S \in \mathcal{S}} : \mathcal{C} \to \mathcal{R}$, *where:*

- $\mathcal{S}$: *set of parameters that characterize the structure of the physical system.*
- $\mathcal{C}$: *set of challenges.*
- $\mathcal{R}$: *set of responses.*

3.1 PUF's Performances

To perform the quality analysis of PUF, we propose to evaluate its performance metrics, mainly: *uniqueness, uniformity, randomness,* and *steadiness*. We notice that the first requirement is an inter property whereas others are of type intra properties.

Definition 2 evaluates the uniqueness of PUFs by estimating the distance between two responses produced by two diverse PUF instances, for a similar challenge. For $R_i, R_j \in \mathcal{R}$, Δ is the distance (i.e. measure the difference between R_i and R_j), and i and j abbreviate the PUFs p_i and p_j, respectively.

Definition 2 (Uniqueness). *The uniqueness of a class of PUF, for a challenge $c \in \mathcal{C}$ is given by:*

$$\alpha_{\mathcal{S}}^{C} = \frac{2}{|\mathcal{S}|(|\mathcal{S}| - 1)|R|} \sum_{i=1}^{|\mathcal{S}|-1} \sum_{j=i+1}^{j=|\mathcal{S}|} \Delta(R_i, R_j).$$

We define the uniformity of n-bit PUF identifier as the percentage Hamming Weight (HW) of the n-bit. There should be a uniform distribution of 0s and 1s in a given response r for PUF instance i.

Definition 3 (Uniformity). *The uniformity of a response $r \in \mathcal{R}$ for a PUF p_i and a challenge c is evaluated by:*

$$\vartheta_i^c = \frac{1}{|r|} \sum_{j=1}^{|r|} r_j$$

Definition 4 relies on the entropy to measure the randomness β_i of responses $\mathcal{R}$ of a PUF p_i with a uniformity ϑ_i.

Definition 4 (Responses' Randomness). *The randomness of the set of responses of a PUF p_i is measured by: $\beta_i = -\sum_{j=1}^{|\nabla|} \vartheta_j \log \vartheta_j$.*

In the case when the structure parameters are known, the randomness of p_i is evaluated with the conditional entropy on $\mathcal{S}$ as follows.

Definition 5 (PUFs' Randomness). *The randomness of a set of PUFs, 'S' with respect to a set of responses ∇, is defined by: $\beta_P(\mathcal{S}|\nabla) = -\sum_{i=1}^{|\nabla|} \rho(r_i)\beta_P(P|r_i)$.*

Definition 6 evaluates the PUF reliability using the min-entropy, where a less reliability means more changes and instability.

Definition 6 (Reliability). *The reliability of a PUF p_i is estimated by:*

$$\gamma_i = 1 + \frac{1}{|\mathcal{R}|(|\mathcal{R}|-1)|\mathcal{C}|} \sum_{i=1}^{|\mathcal{R}|-1} \sum_{j=i+1}^{j=|\mathcal{R}|} \log_2 \max(\vartheta_i, 1-\vartheta_i)$$

3.2 Silicon PUF

Silicon PUFs [22] refer to PUFs that are based on conventional integrated circuit. By exploiting the inherent manufacturing variations of wires and transistors that differ from an IC to another, SPUF is classified as intrinsic PUF as it does not require any modification in the manufacturing process to be used. According to the different sources of variation, Silicon PUF is mainly divided into three major categories:

- Delay-based PUFs: The response generated by the delay PUFs depends on the propagation delay between the different delay paths of the PUF's circuits. This response can be affected by the temperature changes of the circuit [1]. This type of PUF mainly includes Arbiter PUF [24] and Ring Oscillator (RO) PUFs [25].
- Memory-based PUFs: The response generated by the memory-based PUFs depends on the initial state of the memory structures. By power-up, these structures are set in an unstable state, and the response corresponds to the stable state of the structures caused by an external data signal input [1]. This type of PUF family mainly include SRAM PUF [28], Bistable Ring (BR-PUF) [29], Latch PUFs and Flip-flop [1].
- Analog electronic PUFs: The response generated by the Analog electronic PUFs depends on the analog movements of the electric or electronic components such as resistance, and capacitance. This type of PUF mainly includes ICID or Threshold Voltage (TV) PUF [30] and Power distribution or Power distribution PUF [31].

3.3   Arbiter PUF

Arbiter PUF were first introduced by Lee *et al.* [24]. As shows the Fig. 2, two parallel 64 order multiplexer chains are used in its first implementation. It takes a challenge of 64 bits to generate a one-bit response. Its first evaluation was focused on two metrics: security and reliability. To build a unique secret key of an integrated circuit, Lee *et al.* [24] propose to exploit the variations of wires and transistors of an IC to generate a one-bit response from a challenge of 64 bits with a good factor of reliability. APUF could be used in the identification and authentication of ICs without demonstrating their security against many types of attacks such as modeling attacks. APUF is classified as a strong PUF, which makes it suitable for the authentication protocol.

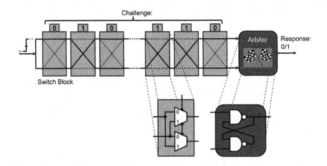

Fig. 2. First structure of Arbiter PUF [24].

4   Fuzzy Extractors

Biometrics like fingerprint, iris, face and voice [6], PUFs [22] and quantum information [21] are sources of non-uniform random data with high entropy. Due to environmental characteristics and aging, they provide similar but not identical reading at each enrollment [6]. To exploit these sources as cryptographic key generation process, two problems should be solved: 1) clean the noise or errors found in data when reading multiple times (known as *information-reconciliation*) and, 2) uniforming data and this process is known as *privacy amplification* [8]. Fuzzy Extractor (FE) [8] is a solution used to extract uniformly random string from noisy and not uniformly random data. This string can be used as a secure key in cryptographic applications. To perform information reconciliation and privacy amplification, most of fuzzy extractors are based on the combination of two other processes, *secure sketch* and *randomness extractor* [8].

As shown by Fig. 3, a fuzzy extractor process consists of a pair of efficient algorithms (*Gen, Rep*), and it works as follows. The generation algorithm *Gen* takes as input the initial reading w from a noisy source and produces a uniformly random string R, which is used as a cryptographic key, and a non-secret string P (Public helper data). The second algorithm *Rep* takes two inputs: the public helper data P and w' which is a noisy version of w, *Rep* reproduces R if w and w' are close enough.

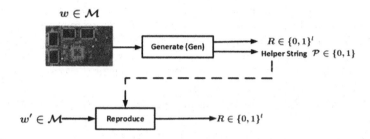

Fig. 3. Fuzzy extractor.

4.1 Generation Phase

This part is based on a secure sketch and a randomness extractor. The uniformly random string "R" is extracted from the initial response w, using the extractor and a random value "x". The helper data "P" is constructed by the combination of the output of the sketching procedure where the sketch "s" is extracted from the initial response "w" and the random value "x". Figure 4 shows the steps to generate helper data and the keys proper to each response w.

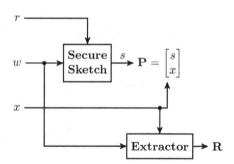

Fig. 4. Generation phase.

The information reconciliation is granted by the secure sketch algorithm [8], which takes as input noisy data w', then outputs the initial non-noisy data w.

Secure Sketch. The secure sketch (SS) [8] takes as input the initial response w, and outputs a public sketch $s \in \{0,1\}^*$ that does not reveal too much information of w and it can be made publicly available. SS is secure as the publicly available Helper Data reveals little to no information about w. Figure 6 summarizes the description of a secure sketch. It allows the construction of a fuzzy extractor by using error correction techniques. There are two prevalent ECC used in secure sketch constructions [10]:

1. Code-offset construction: For an initial input w, it selects a uniformly random code word c, and SS(w) be the shift needed to get from c to w: $ss(w) = s = w - c$. To recover w from a noisy input w', $Rec(w', s)$ is computed by subtracting the shift s from w' to get $c' = w' - s$. Then, decoding c' to get c and compute the initial input w by shifting back to get $w = c + s$. This is only possible if w' is close to w [8,11].
2. Syndrome construction: For the initial input w, set $SS(w) = s = syn(w)$. To recover the initial input w from a noisy input w', a unique vector e is chosen such that $syn(e) = syn(w') - s$ and output $Rec(w', s) = w = w' - e$ [8,11].

Randomness Extractor. In order to use low entropy random source as a cryptographic key. Randomness extractors [5] is required to output near-uniform or high-quality randomness from a weakly random entropy source. This method can be constructed from a universal hash function.

4.2   Reproduction Phase

In the reproduction phase, the secure sketch is used to recover the original input w from the noisy input by using the helper data "P" and the random extractor used in the generation phase. Figure 5 shows the reproduction phase using the secure sketch and randomness extractors.

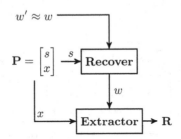

Fig. 5. Reproduction phase.

The ability to recover w from w' is dependent on the distance between the initial and noisy input. If this distance is too large, it may not be possible to recover w from w'.

Recover. In order to recover the original input w from the noisy one w', the second procedure Rec of secure sketch algorithm is used [8], on input a noisy response w' and a public sketch s, Rec compute w from w', this if w' is close to w.

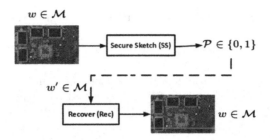

Fig. 6. Secure sketch.

5   A Low-Cost Authentication Protocol

Since PUFs are classified as a cryptosystem, then they can guarantee the three main goals of any good cryptosystem (IAC): the integrity to ensure that messages have not tampered with during the communication process, the authentication to prove that the device is trusted or a part of a trusted group, and the confidentiality which protects the communication between two devices.

5.1   Overview

Several PUFs based authentication protocols have been proposed to ensure that an adversary cannot obtain the PUF output for misusing authentication. In the case of a weak PUF, the protocol requires cryptographic methods such as hashing [3]. However, any PUF based authentication protocol should respect the following requirements:

- A protocol should be clear in its function with graphical and textual representation.
- The use of a particular PUF should be suggested.
- Ability to cope with unstable PUFs (error correction).
- PUF authentication protocol must be robust against machine learning attacks.
- Take charge low-cost and resource-constrained of the device.
- The capability to resist specific protocol attacks.
- Do not store any secrets on the device.
- Compatible with most possible PUF architecture.
- Do not use any cryptography or secure memory technique.

Figure 7 shows an overview of the process of a PUF based authentications protocol that proceeds as follows.

1. The server obtains access to the PUF with a random challenge to generate the response. Then, it repeats this action till generating enough tables of CRPs.
2. When the client wants to authenticate, he submits a request to the trusted server. Then, he receives a challenge from a database server.

3. The client runs the challenge on the PUF and returns the response to the server.
4. Server checks his database to see that the received response is matched or not.
5. If the response matches, the device is authentic and the client authenticates, else the authentication will be refused. In the first case, the server mark that the previous challenge is used, and this challenge will be never reused.

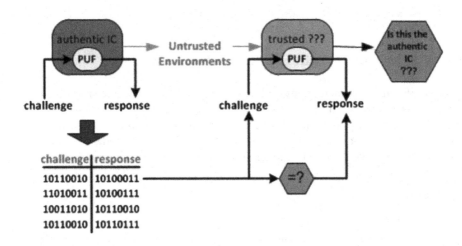

Fig. 7. A PUF-based authentications protocol overview.

5.2   Proposed PUF Based Authentication Protocol

In this section, we present our proposed authentication protocol between a device with an arbiter PUF and a server, the device as a prover uses the response generated by the SPUF to prove its identity in the network, and this, through a trusted entity which is a trusted server. The used symbols and cryptographic function are defined in Table 1.

Enrolment Phase. When a new device needs to be added as a member of the trusted network, the device and the trusted server go first through the enrollment phase. First, the prover device sends its identity in plain text to the server with a registration request. Then, the server stores the device identity and the server sends a set of k random challenges to the device. Each challenge is passed through the PUF mechanism of the electronic device in order to obtain the corresponding response. For each received response, the server sends it with the id of the device to the prover in order to extract a uniform stable key from the response using the fuzzy extractor process. A hash of the extracted key and the helper data is computed, where the latter and the device identity are sent to the database,

Table 1. Authentication protocol's symbols.

Symbol	Definition
ID_i	Device Identity i
Reg_{req}	Registration request
C_j	Challenge j
H(.)	One-way hash function
PUF_i	Physically unclonable functions of device ID_i
R_j	A response of C_j
R'_j	A noisy response of C_j
Gen(R_j)	Generation procedure of Fuzzy Extractor
Rep(R'_j, P_j)	Reproduction procedure of Fuzzy Extractor
K_j	Extracted key from R_j
P_j	Helper data of R_j

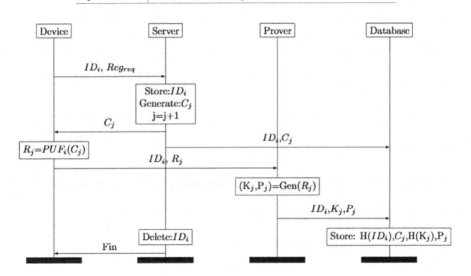

Fig. 8. Enrollment phase.

which is responsible for the secure storage of the challenge-response data for each device. In the end, the server informs the device about the end of the registration process. The enrollment phase is executed in a trusted and secure environment through steps shown in Fig. 8.

Authentication Phase. As shown in Fig. 9, the device first sends the authentication request and the hash of its id, $H(ID_i)$ to the server. Then, the server sends the received $H(ID_i)$ to the database to search if $H(ID_i)$ exists, and the authentication fails if the match does not found. Otherwise, the challenge C_j) stored against $H(ID_i)$ is selected and communicated to the device ID_i, and the latter generates its corresponding response. Thus, the device sends $H(ID_i),C_j$, R'_j to the prover. In order to extract the cryptography key, the prover stores the received noisy data R'_j and recovers from the database the corresponding helper data P_j, which is used with the noisy response as an input in the reproduction step to regenerate the secret key K_j using fuzzy extractor and secure sketch algorithms Using its hash values, the prover compares the old key within the new one. If the operation of comparison fail, the authentication fails, else the device is authenticated and it receives a $welcome's$ message. Finally, the used challenge and all stored data against it in the database as well as, the noisy response stored by the prover will be deleted.

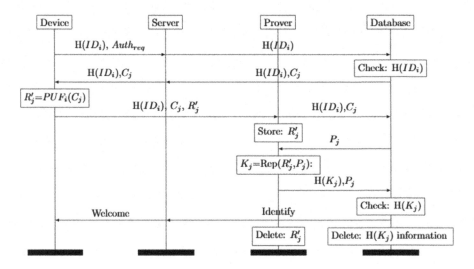

Fig. 9. Authentication phase.

6   Implementation and Results

This section validates the proposed protocol through an experimental evaluation of the prover step The experimental setup uses an arbiter PUF as a module for authentication, in which, the noise of the new response was eliminated and the secret key and the helper data were generated.

The device is equipped with an arbiter PUF, of 16 switching components, which means it accepts a challenge of 16 bits. Given this challenge as input, the PUF outputs 32 bit as a response. To check the environmental factors, the response of the same challenge was generated at three different temperatures: $-40C$, $25C$, and $85C$. The result of the simulation process of this arbiter PUF is used as a dataset[1]. For the prover, we implemented the fuzzy extractor and the error correction that is based on the BCH error correction code.

---

[1] https://github.com/salaheddinhetalani/PUF.

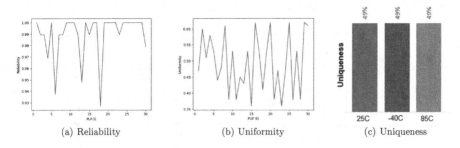

(a) Reliability   (b) Uniformity   (c) Uniqueness

Fig. 10. The performance evaluation of the deployed Arbiter PUF.

First, we evaluate the performance of the used 16-arbiter PUF before evaluating the fuzzy extractor efficiency on the implemented protocol. The obtained results showed that it is strong since it has good reliability, uniqueness, and uniformity. As shown in Fig. 10, the uniformity of the generated response is around 50% (the ideal value is 50%), which means the same ratio of ones and zeros in each 32-bit response. Its uniqueness is 49.58% (the ideal value is 50%), which means giving the same challenge to two different 16-arbiter PUF, each instance will generate its unique response. Finally, the reliability of the used PUF is 98.96% (the ideal value is 100%), and this means that our used PUF generates stable responses at different temperatures 25C, −40C, and 85C.

Secondly, we evaluate the correctness of the prover by extracting the uniform key from the response and recover the original one from the noisy response. From the used dataset, we used the response of each PUF_i instance generated at 25C, as the initial PUF_i's response. Then, we used a fuzzy extractor and secure sketch based on the BCH error correction algorithm to correct two other responses of each PUF_i generated at two different temperatures −40C and 85C. A shown in Table 2, the reliability and the stability of the used arbiter PUF, where there are not a lot of error bits, compared to the uniformity of the response and the extracted key. We showed that our prover extracts a string with high uniformity, using the hash function $sha1$, that is used as a cryptography key. For the reproduction steps of the prover, the results show that all the noisy responses with error bits (0,5) are corrected and recover the initial response.

Table 2. Fuzzy extractor statistics

PUF_i	$R_{Reliability}$ (%)	$R_{Uniformity}$ (%)	$Error_{-40C}$	$Error_{85C}$	$K_{Uniformity}$ (%)	Correction
1	100	0.47	0	0	48	✓
2	99	0.6	1	0	45	✓
3	99	0.51	1	0	50	✓
4	97	0.58	1	2	54	✓
5	100	0.53	0	0	51	✓
6	94	0.44	1	5	50	✓
7	99	0.48	0	1	48	✓
8	99	0.61	0	1	48	✓
9	100	0.38	0	0	51	✓
10	100	0.53	0	0	50	✓
11	100	0.38	0	0	53	✓
12	99	0.45	1	0	47	✓
13	95	0.43	3	2	51	✓
14	100	0.53	0	0	50	✓
15	99	0.36	0	1	50	✓
16	100	0.62	0	0	51	✓
17	100	0.53	0	0	45	✓
18	93	0.41	4	3	51	✓
19	100	0.53	0	0	50	✓
20	100	0.62	0	0	50	✓
21	100	0.38	0	0	48	✓
22	100	0.47	0	0	50	✓
23	99	0.36	0	1	48	✓
24	100	0.47	0	0	50	✓
25	100	0.62	0	0	50	✓
26	100	0.38	0	0	51	✓
27	100	0.53	0	0	50	✓
28	100	0.38	0	0	50	✓
29	100	0.62	0	0	50	✓
30	98	0.61	1	1	55	✓

7   Conclusion

In this paper, we have developed a low-cost protocol that exploits the randomness of the Arbiter PUF. We have used a fuzzy extractor as a prover in the protocol with the role to identify the trusted objects and correct the keys in the case of an allowed noise or transmission errors. The experiments were run on a benchmark related to the Arbiter PUF and showed good results.

As future work, we intend to deploy the proposed protocol with a blockchain architecture that exploits a PUF based on IoT and smartphones. Further, we look to prove the correctness and security properties of the developed protocol using an automatic cryptographic protocol verifier [4]. Finally, we target to apply the protocol on real autonomous vehicles.

References

1. Adames, I.A.B., Das, J., Bhanja, S.: Survey of emerging technology based physical unclonable funtions. In: 2016 International Great Lakes Symposium on VLSI, pp. 317–322. IEEE, May 2016
2. Suh, G.E., Devadas, S.: Physical unclonable functions for device authentication and secret key generation. In: 44th, ACM/IEEE Design Automation Conference, p. 1 (2007)
3. Babaei, A., Schiele, G.: Physical unclonable functions in the Internet of Things: state of the art and open challenges. Sensors 19(14), 3208 (2019)
4. Ouchani, S.: Otmane Aït Mohamed, pp. 1–7. A formal verification framework for Bluespec System Verilog. FDL, Mourad Debbabi (2013)
5. Maes, R.: PUF-based entity identification and authentication. Physically Unclonable Functions, pp. 117–141. Springer, Heidelberg (2013). https://doi.org/10.1007/978-3-642-41395-7_5
6. Wen, Y., Liu, S.: Robustly reusable fuzzy extractor from standard assumptions. In: Peyrin, T., Galbraith, S. (eds.) ASIACRYPT 2018. LNCS, vol. 11274, pp. 459–489. Springer, Cham (2018). https://doi.org/10.1007/978-3-030-03332-3_17
7. Zerrouki, F., Ouchani, S., Bouarfa, H.: Towards an automatic evaluation of the performance of physical unclonable functions. In: International Conference in Artificial Intelligence in Renewable Energetic Systems, pp. 775–781. Springer, Cham, December 2020. https://doi.org/10.1007/978-3-030-63846-7_74
8. Dodis, Y., Ostrovsky, R., Reyzin, L., Smith, A.: Fuzzy extractors: how to generate strong keys from biometrics and other noisy data. SIAM J. Comput. 38(1), 97–139 (2008)
9. Wen, Y., Lao, Y.: Efficient fuzzy extractor implementations for PUF based authentication. In: 2017 12th International Conference on Malicious and Unwanted Software (MALWARE), pp. 119–125. IEEE, October 2017
10. Gao, Y., Su, Y., Yang, W., Chen, S., Nepal, S., Ranasinghe, D.C.: Building secure SRAM PUF key generators on resource constrained devices. In 2019 IEEE International Conference on Pervasive Computing and Communications Workshops (PerCom Workshops), pp. 912–917. IEEE, March 2019
11. Korenda, A.R., Afghah, F., Cambou, B., Philabaum, C.: A proof of concept SRAM-based physically unclonable function (PUF) key generation mechanism for IoT devices. In: 2019 16th Annual IEEE International Conference on Sensing, Communication, and Networking (SECON), pp. 1–8. IEEE, June 2019
12. Zerrouki, F., Ouchani, S., Bouarfa, H.: Quantifying security and performance of physical unclonable functions. In: 2020 7th International Conference on Internet of Things: Systems, Management and Security (IOTSMS), pp. 1–4. IEEE, December 2020
13. Yanambaka, V.P., Mohanty, S.P., Kougianos, E., Puthal, D.: PMsec: physical unclonable function-based robust and lightweight authentication in the internet of medical things. IEEE Trans. Consum. Electron. 65(3), 388–397 (2019)
14. Aman, M.N., Chua, K.C., Sikdar, B.: Mutual authentication in IoT systems using physical unclonable functions. IEEE Internet Things J. 4(5), 1327–1340 (2017)

15. Chatterjee, U., Chakraborty, R.S., Mukhopadhyay, D.: A PUF-based secure communication protocol for IoT. ACM Trans. Embed. Comput. Syst. (TECS) **16**(3), 1–25 (2017)

16. Xie, L., Wang, W., Shi, X., Qin, T.: Lightweight mutual authentication among sensors in body area networks through physical unclonable functions. In: 2017 IEEE International Conference on Communications (ICC), pp. 1–6. IEEE, May 2017

17. Chatterjee, U., et al.: Building PUF based authentication and key exchange protocol for IoT without explicit CRPs in verifier database. IEEE Trans. Dependable Secure Comput. **16**(3), 424–437 (2018)

18. Gope, P., Sikdar, B.: Lightweight and privacy-preserving two-factor authentication scheme for IoT devices. IEEE Internet Things J. **6**(1), 580–589 (2018)

19. Baruah, B., Dhal, S.: A two-factor authentication scheme against FDM attack in IFTTT based smart home system. Compu. Secur. **77**, 21–35 (2018)

20. Chatterjee, U., et al.: PUF+ IBE: blending physically unclonable functions with identity based encryption for authentication and key exchange in IoTs. IACR Cryptol. ePrint Arch. **2017**, 422 (2017)

21. Bennett, C.H., DiVincenzo, D.P.: Quantum information and computation. Nature **404**(6775), 247–255 (2000)

22. Lim, D., Lee, J.W., Gassend, B., Suh, G.E., Van Dijk, M., Devadas, S.: Extracting secret keys from integrated circuits. IEEE Trans. Very Large Scale Integr. (VLSI) Syst. **13**(10), 1200–1205 (2005)

23. Braeken, A.: PUF based authentication protocol for IoT. Symmetry **10**(8), 352 (2018)

24. Lee, J.W., Lim, D., Gassend, B., Suh, G.E., Van Dijk, M., Devadas, S.: A technique to build a secret key in integrated circuits for identification and authentication applications. In: 2004 Symposium on VLSI Circuits. Digest of Technical Papers (IEEE Cat. No. 04CH37525), pp. 176–179. IEEE, June 2004

25. Suh, G.E., Devadas, S.: Physical unclonable functions for device authentication and secret key generation. In: 2007 44th ACM/IEEE Design Automation Conference, pp. 9–14. IEEE, June 2007

26. Mills, A.: Design and evaluation of a delay-based FPGA physically unclonable function (2012)

27. Bai, C., Zou, X., Dai, K.: A novel ThyristorBased Silicon physical Unclonable function. IEEE Trans. Very Large Scale Integration (VLSI) Syst. **24**, 290–300 (2015)

28. Guajardo, J., Kumar, S.S., Schrijen, G., Tuyls, P.: FPGA intrinsic PUFs and their use for IP protection. In: Proceedings of the 9th International Workshop on Cryptographic Hardware and Embedded Systems, pp. 63–80, September 2007

29. Chen, Q., Csaba, G., Lugli, P., Schlichtmann, U., Rührmair, U.: The bistable ring PUF: a new architecture for strong physical unclonable functions. In: 2011 IEEE International Symposium on Hardware-Oriented Security and Trust, pp. 134–141. IEEE, June 2011

30. Lofstrom, K., Daasch, WR., Taylor, D.: IC identification circuit using device mismatch. In: 2000 IEEE International Solid-State Circuits Conference. Digest of Technical Papers (Cat. No. 00CH37056), pp. 372–373. IEEE, February 2000

31. Helinski, R., Acharyya, D., Plusquellic, J.: A physical unclonable function defined using power distribution system equivalent resistance variations. In: 2009 46th ACM/IEEE Design Automation Conference, pp. 676–681. IEEE, July 2009

Aspect-Oriented Model-Based Testing with UPPAAL Timed Automata

Jüri Vain$^{(\boxtimes)}$ [ID], Leonidas Tsiopoulos [ID], and Gert Kanter [ID]

Department of Software Science, Tallinn University of Technology, Ehitajate tee 5,
19086 Tallinn, Estonia
{juri.vain,leonidas.tsiopoulos,gert.kanter}@taltech.ee

Abstract. This paper presents a method for offline test derivation from formal aspect-oriented models so that the tests provide coverage in terms of aspects related metrics. A test purpose specification method in temporal logic TCTL is proposed that enables referring to the attributes of aspect models symbolically. The method is exemplified on a health monitoring system and the quantitative evidence of the advantages provided by the method are evaluated in terms of work effort put into the test development and by analytical reasoning on the complexity.

Keywords: Model-Based Testing · Aspect-Oriented Modeling · UPPAAL Timed Automata · Offline test generation · Test coverage

1 Introduction

Model-Based Testing (MBT) provides the opportunities for test automation and reducing testing effort [10]. MBT suggests the use of models for specifying the expected behavior of the System Under Test (SUT) and automatically generating tests from models. Modularization attempts have been driven by the need to improve the comprehension of test models, to address the complexity of test generation and to reduce the test cases to a manageable size, both in terms of test execution time and computational complexity. We apply aspect-orientation as modularization mechanism through Aspect-Oriented Modeling (AOM) [4] promoting the idea of separation of concerns aiming at well-targeted test goals and back traceability of root causes in terms of SUT design aspects.

The study in [7] provides a survey and assessments of AOM techniques concentrating on UML. Test generation from aspect-oriented (AO) UML state machines has been suggested in [1]. Our approach differs from it by stating explicitly AO test coverage criteria and abstracting them in temporal logic TCTL. In contrast to highly expressive UML, we target semantic unambiguity and mature verification tool support while introducing the AOM constructs. We focus on conformance testing by merging UPPAAL Timed Automata (UTA) with the concepts of AOM [9]. This allows defining the test coverage metrics relative to AO UTA model attributes. Also, mature UTA tool support enables deriving cost/time efficient test sequences providing the required coverage with optimal

© Springer Nature Switzerland AG 2021
C. Attiogbé and S. Ben Yahia (Eds.): MEDI 2021, LNCS 12732, pp. 117–124, 2021.
https://doi.org/10.1007/978-3-030-78428-7_10

resource needs. We revisit the coverage metrics of [9] and provide explicit encoding of coverage criteria in terms of AO model elements. The approach is exemplified on a Home Rehabilitation System (HRS). The quantitative evaluation of the advantages is provided by analytical reasoning on the complexity of the test generation process.

2   Aspect-Oriented Modeling and Testing with UTA

UTA [3] are defined as a closed network of extended timed automata that are called *processes*. The processes are composed by synchronous parallel composition. The nodes of the automata graph are called *locations* and directed vertices between locations are called *edges*. The *state* of an automaton consists of its current location and valuation of all variables, including clocks. Synchronization of processes is defined by *channels*. A channel *ch* synchronizing a pair of transitions in parallel processes is introduced by labeling edges with input action ch? and output action ch! on channel ch.

Formally, an UTA is given as the tuple $(L, E, V, CL, Init, Inv, T_L)$, where L is a finite set of locations, E is the set of edges defined by $E \subseteq L \times G(CL, V) \times Sync \times Act \times L$, where $G(CL, V)$ is the set of constraints in guards, $Sync$ is a set of channels and Act is a set of assignments with integer and boolean expressions and clock resets. V denotes the set of variables of boolean and integer type and arrays of those. CL denotes the set of real-valued clocks $(CL \cap V = \varnothing)$. $Init \subseteq Act$ is a set of initializing assignments to variables and clocks. $Inv : L \to I(CL, V)$ maps locations to the set of invariants over clocks CL and variables V. $T_L : L \to \{ordinary, urgent, committed\}$ defines types of locations. An example of UTA model that consists of several processes communicating via channels is depicted in Fig. 1 which represents a part of the AO model of HRS [6] which will act as the running example for the rest of the paper. HRS collects a patient's health data using sensors attached to the patient's body.

AO UTA consists of a *base model* that is a set of processes representing the core functionality of the system. Similarly, advice model being also a set of UTA processes represents a crosscutting concern. *Weaving* means composing a base model with the advice model. *Join points* are model fragments in the base model to which an aspect can be woven. A *pointcut* is the set of join points and conditions under which an advice can be woven. A *woven model* is composition of base and (possibly several) advice model(s).

Figure 1 exemplifies the weaving of HRS base model with aspects *Patients safety*, *Exercising quality* and *Exercising performance* woven by *location and/or edge superposition refinement* [8]. The base models are depicted on the left side of the figure and the advice models on the right. For instance, *Patient_exercising* is woven with advice *Patient_Exercising1* via location join point *Exercising* to introduce quality indicators such as exercising counter (*Ex_Counter*) and the duration interval [*Ex_Lb*, *Ex_Ub*] of performing an exercise. The weaving correctness is verified by model-checking conditions specified in [8].

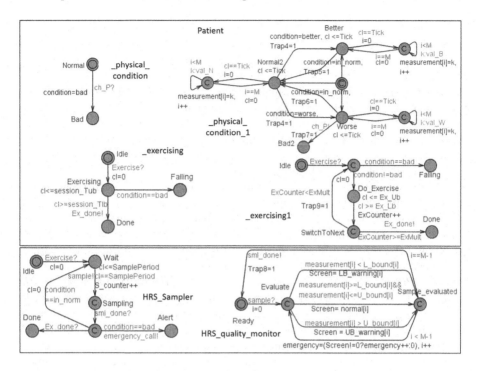

Fig. 1. Exercising quality aspect model (Base mode left, advice model right)

We presume the augmented UTA test model is deterministic and includes structures (join points, weavers, advices) that allow specifying various aspect related structural coverage criteria. The semantics and scoping of AO coverage constraints can be defined in compact form by following the hierarchy of AO model structures. For specifying the coverage criteria we use the expressions of TCTL [2] with 1st order terms on the alphabet of AO UTA model elements and call these formulae *coverage expressions* (CE). In CE the AO coverage sub-expressions are nested into higher level expressions according to the following scoping order: AC, JPC, APC, MEC, where AC stands for aspect coverage, JPC - join point coverage, APC - aspect path coverage and MEC - advice model element coverage. Having this nesting hierarchy by coverage item types, each of the CE defines the contextual scope within which the lower level nested operator is interpreted. For instance, if for JPC the AC-prefix specifies aspects A_1 and A_2 then join points in JPC are assumed to be only those of A_1 and A_2.

For AC, given an aspect model M, all of its advices (for strong coverage) or some (for weak coverage) must be covered by the tests. To implement AC we use the parameterized UTA templates where the template parameter p_i identifying an aspect ranges over indexes $[1, n]$. Let $P(i)$ be the predicate symbol in M assigned value *true* only when the i-th aspect advice model is executed. Then the traces of $M(p_i)$, assuring AC, can be generated by model-checking query:

$E\Diamond$ *forall* $(i : int[1, n])$ $P(i)$ for strong coverage and query $E\Diamond$ *exists* $(i : int[1, n])$ $P(i)$ for weak AC.

JPC requires executing join points of the aspects specified in AC-prefix of CE. *Strong Join Point Coverage* (SJPC) defines a conjunct of the form ...$forall$ $(j : int[1, m])$ $P(i) \wedge R(j)$, where j ranges over the join point indexes of the aspects referred in the AC-prefix and $R(j)$ is a Boolean variable at each join point assigned *true* whenever this join point is visited. *Weak Join Point Coverage* (WJPC) is similarly expressed with formula ...$exists$ $(j : int[1, m])$ $P(i) \wedge R(j)$. Predicate symbols P and R are also called "traps" which "close" when test run traverses model elements labeled with them.

APC requires executing all or some paths between advice entry and exit points. Assume the entry and exit transitions of an advice model are labeled with $entry(i, j, k)$ and $exit(i, j, l)$ predicates where i, j, k, l range over the set of aspects, join points, and their advice entry and exit points respectively. Then, the *Strong Aspect Path Coverage* (SAPC) is specified by the sub formula prefixed with aspect and join point constraints as follows: $forall$ $(k : int[1, K]$, $l : int[1, L])$ $P(i) \wedge R(j) \wedge [(\vee_{k=1, K} entry(k)) \wedge (\vee_{1, L} exit(l))]$. *Weak Aspect Path Coverage* (WAPC) replaces universal quantifiers with existential ones for variables k and l. More specific AO coverage constraints can be specified using type discriminating predicates on the data variables being nested into higher level constraints if more concrete specification of test coverage criteria is needed.

To exemplify the use of coverage criteria we consider the HRS system which has test interfaces with two external actors, Doctor and Patient, whose behavior defines the test cases. In order to demonstrate how the tests that satisfy AO coverage criteria are specified and generated, we present *Strong Model Element Coverage* which presumes covering model elements such as locations, edges and their attributes in advice models.

For instance, in AO HRS model the edges of *Patient_physical_condition1* which depart from sub-states of *Normal* and the edge that enters the state *Bad*, as well as the edge from *Sample_evaluated* to *Ready* in *HRS_quality_monitor* and the edge modeling the repetition of the exercise in *Patient_exercising1* are labelled with *Trap4* - *Trap9* (Fig. 1). Note that for model-checking efficiency, AO coverage (AC, JP, AP) prefixes can be discarded in the test generation queries by unique renaming of traps in the AO test model and flattening the coverage formula to form where these traps are explicitly referred. For example, query $E \Diamond Trap4 \wedge \ldots \wedge Trap9$ is the result of flattening AO formula for strong AC. It was executed for both models and the verification/elapsed time used was $0\,s/0{,}016\,s$ for the AO model and $0{,}047\,s/0{,}047\,s$ for the non-AO model, respectively. The simulation trace for this query was also shorter by 12% for the AO model compared to that of the non-AO model.

Same differences in favor of the AO models are observed when comparing the rest of their bisimilar AO models for HRS against the non-AO counterparts.

3   Analytical Validation of the Approach

3.1   Comparison of Test Purpose Specification Effort

We estimate the test purpose specification effort by the time taken for finding and labeling edges with the trap variables to define edges as test coverage items. For quantitative characterization of test purpose specification effort we derive an empirical formula that correlates with factors of human capability of finding the edge to be labelled amongst the set of all edges of the model. As the base case, we estimate the specification time on the non-AO model, at first, and then, compare it with a case of AO models splitting the same set of traps between aspects.

For the non-AO case, we assume the total number of edges in model M is $n = |E(M)|$ and the number of edges to be labeled with traps is k. We also assume that once an edge has been labeled it is memorized and does not need to be inspected when searching for the next trap. Thus, in the worst case when searching for an edge to label it with i-th trap, $n - i$ options should be inspected. Let the duration for inspection of one edge be constant d. Then the upper time bound T^{tr} of labeling k edges with traps can be calculated by formula

$$T^{tr}(M) = \frac{n^2 - (n - (k - 1))^2}{2} d \qquad (1)$$

For comparison of test purpose specification effort we assume that the non-AO model M and its AO counterpart M^{ao} are observationally (with respect to test interface) bisimilar. Also, the edges in M labeled with traps should occur in the augmented model M^{ao}. Thus, the set of trap labeled edges in M is equal to the union of trap labelled edges in the individual aspect models: $E^{tr}(M) = \cup_i E^{tr'}(M_i^{ao})$. In other words, we assume that the original trap labeled set of edges is partitioned so that: k_1 traps specify coverage in M_1^{ao}, k_2 traps in M_2^{ao}, and k_m traps specify the coverage in M_m^{ao} so that for m aspects condition $\Sigma_{i=1}^m k_i = k$ is satisfied. By second assumption the set of edges of any AO model is a strict subset of edges of the non-AO model M, i.e.,

$$\forall i : E(M_i^{ao}) \subset E(M) \Rightarrow |E(M_i^{ao})| < |E(M)|. \qquad (2)$$

When denoting the number of edges $|E(M_i^{ao})|$ with n_i and applying formula (1) we get the labeling effort upper bound for each AO model M_i^{ao}

$$T_i^{tr}(M_i^{ao}) = \frac{n_i^2 - (n_i - (k_i - 1))^2}{2} d \qquad (3)$$

and total test purpose specification time

$$T^{tr}(M^{ao}) = \sum_{i=1}^m \left(\frac{n_i^2 - (n_i - (k_i - 1))^2}{2} d \right) \qquad (4)$$

In the following we demonstrate under which conditions AO test models have advantage over non-AO test models, i.e.,

$$T^{tr}(M) > T^{tr}(M^{ao}) \qquad (5)$$

By substituting (1) and (3) in (5) and dividing by $d/2$ we get

$$n^2 - (n - (k-1))^2 > \sum_{i=1}^{m} n_i^2 - \sum_{i=1}^{m}(n_i - (k_i - 1))^2) \tag{6}$$

Since for any n, m, $a \geq 2$, if $a = \sum_{i=1}^{m} b_i$ then $a^n > \sum_{i=1}^{m} b_i^n$ entails that $T^{tr}(M) > T^{tr}(M^{ao})$.

3.2   Comparison of Test Generation Effort

The generation of test sequences in this approach is based on using timed witness traces produced by UPPAAL model checker as the result of checking queries that encode the test purposes (coverage reachability) symbolically. Thus, the test generation effort is measurable in terms of time complexity of model-checking such TCTL formulas that express AO coverage criteria.

The worst-case time complexity $\mathcal{O}$ of model checking TCTL formula ϕ over timed automaton M, with the clock constraints of ϕ and of M in ψ is [5]:

$$\mathcal{O}(|\phi| \times (n! \times 2^n \times \Pi_{x \in \psi} c_x \times |\mathsf{L}|^2)) \tag{7}$$

where n is the number of clock regions, ψ is the set of clock constraints, c_x is the maximum constant the clock x is compared with, $\mathsf{L} : \mathsf{L} \rightarrow 2^{AP}$ is a labeling function for symbolic states of M. L denotes the product of data constraints over all locations and edges defined in the UTA model and AP is the set of atomic propositions used in guard conditions and invariants.

According to (7) the time complexity of model-checking TCTL is linear in the length of the formula ϕ, exponential in the number of clocks and exponential in the maximal constants c_x with which each clock x is compared in model M and in ϕ. However, using state space reduction techniques the worst case time complexity can be reduced to being quadratic in the number of symbolic states on data variables in the model [5]. The lower bound of TCTL model-checking space complexity is known to be PSPACE-hard [2]. In practice, time and space complexity of model-checking TCTL on UTA, boils down to the size of the symbolic state space and to the number of symbolic states (including clock zones) to be explored, and respectively stored during the verification. Since by definition the number of locations $|L(M^{ao})|$ and edges $|E(M^{ao})|$, as well as the number of variables $|V(M^{ao})|$ of an AO model M^{ao} is not greater than that of a behaviorally equivalent non-AO model M, we can conclude from (7) that every reduction in the number of model elements and the number of symbolic states, provides an exponential decrease in the number of steps of the model exploration. This applies also to checking the aspect related properties for AO test generation. Due to the superposition refinement based weaving, our approach does not introduce additional interleaving between the base and advice model transitions.

Relying on the arguments above we can state the following claims on symbolic states to be processed and stored for model-checking-based test generation. Let us consider the non-AO model M and, respectively, the AO model M^{ao} that specifies the same behavior of a system. Recall that M^{ao} consists of a base model M^B with which a set of models of non-interfering aspects A_i, $i = 1, \cdots, n$ are woven. Then for any reachability property ϕ_i of any A_j, $i = 1, \cdots, n$ decidable on $M^B \oplus M^{A_j}$ the model-checking effort $\mathbf{E}$ (in terms of time or space) is equal or less than that of checking the property ϕ_i on the non-AO model M, where the semantics of composition $M^B \oplus M^{A_j}$ is a subset of semantics of M, i.e.,

$$M^B \oplus M^{A_j} \models \phi_i \wedge [\![M^B \oplus M^{A_j}]\!] \subseteq [\![M]\!] \Rightarrow$$
$$M \models \phi_i \wedge \mathbf{E}(M^B \oplus M^{A_j} \models \phi_i) \leq \mathbf{E}(M \models \phi_i). \tag{8}$$

Assuming the tests are generated using model-checking traces formula (8), under the assumptions for model-checking effort, the effort $\mathbf{E}$ (in terms of time or space) of generating the test case T^{ϕ_i} that is interpretation of property ϕ_i of aspect A_j and is bounded with AO model $M^B \oplus M^{A_j}$, is less than or equal to the effort of generating the test T^{ϕ_i} from non-AO model M:

$$M^B \oplus M^{A_j} \models \phi_i \wedge [\![M^B \oplus M^{A_j}]\!] \subseteq [\![M]\!] \Rightarrow$$
$$\mathbf{E}(M^B \oplus M^{A_j}, T^{\phi_i}) \leq \mathbf{E}(M, T^{\phi_i}). \tag{9}$$

Here notation $[\![M]\!]$ denotes the operational semantics of model M. The validity of formulas (8) and (9) stems from the fact that AO models represent subsets of the behavior of the non-AO model of the same system.

Comparison of Length of Generated Test Sequences. Generating the traces by model-checking a TCTL formula ϕ on a non-AO model has disadvantages compared to generating the traces of ϕ sub-formulas (when their conjunction is equivalent to ϕ) on AO models. Since the AO and non-AO models must be bisimilar, then by definition, their execution traces are observationally equivalent. Thus, regarding test execution effort we can conclude that the test sequences derived from AO models and non-AO models are coverage equivalent although their traces can differ due to the interleaving introduced by the other structural elements of the monolithic non-AO model.

4   Conclusions

We presented an AOT methodology for UTA targeting efficiency improvement of MBT through better comprehension of test models, better traceability of SUT quality attributes related to particular aspects, and reduction of test generation and execution effort. Tests can be generated offline automatically by running TCTL model-checking queries on AO models and applying the resulting witness traces thereafter as test sequences of AO test cases. A set of AO test coverage criteria was elaborated and formalized in TCTL language that makes possible automatic test generation by model-checking. The usability of the AOT method was demonstrated on the HRS system testing providing experimental evidence that AOT improves the efficiency of MBT compared to the methods that are based on non-AO models. The quantitative evidence of the advantages of AOT was provided with reference to complexity results of TCTL model-checking. It is shown analytically that the effort for specifying AO coverage criteria and the complexity of test generation from AO models is less than that of non-AO models. Some verification effort is required to prove the correctness of aspect weaving guaranteeing aspect non-interference which in turn enables compositional AOT. Though, proving weaving correctness properties presumes model-checking of queries on advice models locally, that is computationally more feasible than running the same queries on full augmented model.

Acknowledgement. This work has been supported by EXCITE (2014-2020.4.01.15-0018) grant.

References

1. Ali, S.L., Briand, L., Hemmati, H.: Modeling robustness behavior using aspect-oriented modeling to support robustness testing of industrial systems. Softw. Syst. Model. **11**(4), 633–670 (2012)
2. Alur, R., Courcoubetis, C., Dill, D.: Model-checking for real-time systems. In: Proceedings of Fifth Annual IEEE Symposium on Logic in Computer Science, LICS 1990, pp. 414–425. IEEE (1990)
3. Behrmann, G., David, A., Larsen, K.G.: A tutorial on UPPAAL. In: Bernardo, M., Corradini, F. (eds.) SFM-RT 2004. LNCS, vol. 3185, pp. 200–236. Springer, Heidelberg (2004). https://doi.org/10.1007/978-3-540-30080-9_7
4. France, R.B., Ray, I., Georg, G., Ghosh, S.: An aspect-oriented approach to early design modelling. IEE Proc. Softw. **151**(4), 173–185 (2004)
5. Katoen, J.P.: Concepts, Algorithms, and Tools for Model Checking. IMMD, Erlangen (1999)
6. Kuusik, A., Reilent, E., Sarna, K., Parve, M.: Home telecare and rehabilitation system with aspect oriented functional integration. Biomedical Engineering/Biomedizinische Technik **57**(SI-1 Track-N), 1004–1007 (2012). https://doi.org/10.1515/bmt-2012-4194
7. Mehmood, A., Jawawi, D.: A quantitative assessment of aspect design notations with respect to reusability and maintainability of models. In: Proceedings of 8th Malaysian Software Engineering Conference (MySEC), pp. 136–141. IEEE (2014)
8. Sarna, K., Vain, J.: Exploiting aspects in model-based testing. In: Proceedings of the Eleventh Workshop on Foundations of Aspect-Oriented Languages - FOAL 2012, pp. 45–48. ACM, New York (2012). https://doi.org/10.1145/2162010.2162023
9. Sarna, K., Vain, J.: Aspect-oriented testing of a rehabilitation system. In: Kanstren, T., Helle, P. (eds.) Proceedings of the Sixth International Conference on Advances in System Testing and Validation Lifecycle - VALID 2014, pp. 73–78. IARIA (2014)
10. Utting, M., Pretschner, A., Legeard, B.: A taxonomy of model-based testing approaches. Softw. Test. Verification Reliab. **22**(5), 297–312 (2012)

Machine Learning

Social Neural Hybrid Recommendation with Deep Representation Learning

Lamia Berkani[1,2(✉)] [iD], Dyhia Laga[2], and Abdelhak Aissat[2]

[1] Laboratory for Research in Artificial Intelligence (LRIA), USTHB University, 16111 Bab Ezzouar, Algiers, Algeria
lberkani@usthb.dz

[2] Department of Computer Science, USTHB University, 16111 Bab Ezzouar, Algeria

Abstract. Deep learning techniques are proved to be very effective in various domains, such as computer vision, pattern recognition, and natural language processing. With the rapid development of this technology, more and more deep models have been used in the recommendation system. However, only few attempts have been made in social-based recommender systems. This paper focuses on this issue and explores the use of deep neural networks for learning the interaction function from data. A novel recommendation model called SNHF (Social Neural Hybrid Filtering) is proposed. It combines collaborative and content-based filtering with social information in an architecture based on both models: (1) Generalized Matrix Factorization (GMF); and (2) Hybrid Multilayer Perceptron (HybMLP). Extensive experiments on two real-world datasets show that SNHF significantly outperforms state-of-the-art baselines and related work, especially in the cold start situation.

Keywords: Social recommendation · Neural collaborative filtering · Content based filtering · Generalized Matrix Factorization · Multi-layer perceptron

1 Introduction

Recommender systems (RSs) are used to help users find new items or services, based on information about the user and/or the recommended items [1]. They can filter information from overwhelming data based on the user's historical preference, find out what the user is really interested in, and help the user efficiently obtain the information the user wants [2]. The most commonly used algorithms are: (1) Collaborative filtering (CF) [3, 4], based on the principle that a prediction of a given item can be generated by aggregating the ratings of like-minded users; and (2) Content-based filtering (CBF) [5], based on a comparison of user profiles with item profiles to recommend items that match their preferences and tastes. To obtain better accuracy, hybrid recommender systems combine the two recommendation techniques in different ways [6] in order to overcome their drawbacks (e.g., cold start and sparsity problems). Furthermore, with the expansion of social media platforms, the performance of traditional recommender systems can be improved with the integration of social information [7].

C. Attiogbé and S. Ben Yahia (Eds.): MEDI 2021, LNCS 12732, pp. 127–140, 2021.
https://doi.org/10.1007/978-3-030-78428-7_11

On the other hand, in order to provide users with better recommendations, machine learning (ML) algorithms are being used in recommender systems. Recently, with the great success of deep learning (DL) technology in many domains such as computer vision, natural language processing, and machine translation, this emergent technology began to gain widespread interest and attention in recommender systems. DL has provided more opportunities to improve the recommendation effectiveness of traditional algorithms. The reason is that DL techniques can achieve complex and nonlinear user-item relationships among users and items, and encode feature representations into higher layers, which can lead to improvements of the accuracy [8, 9, 10]. However, the state of the art shows that only few approaches have been proposed in the social networking context, on one side, and on the other side, some neural network-based methods are proposed to learn user and item representations from text notation and comments data [11, 12]. He et al. [13] modeled the user-item assessment matrix using a multilayer feedback neural network.

Accordingly, in this paper we propose a social and neural hybrid recommendation approach, combining collaborative and content-based filtering with social information in an architecture based on both models: (1) Generalized Matrix Factorization (GMF); and (2) Hybrid Multilayer Perceptron (HybMLP).

The remainder of the paper is organized as follows: Sect. 2 presents some related works about deep learning-based recommender systems. Section 3 proposes a novel hybrid recommendation approach combining CF and CBF algorithms using social information in a multi-layered neural architecture using the linear GMF model and the non-linear HybMLP model. Section 4 presents experimental methodology and results. Finally, conclusion and future work are shown in Sect. 5.

2   Background and Related Work

2.1 Deep Learning

Machine learning (ML), one of the artificial intelligence fields, is a set of methods that automates analytical model building. It is based on the idea that systems can learn from data, identify patterns and make decisions with minimal human intervention. ML algorithms can be initially classified as supervised, semi-supervised, unsupervised, or reinforcement learning.

Recently, deep learning (DL) technology, an emerging field of ML, has achieved great success as it aims at automatically mine multi-level feature representations from data by merging low-level features to form denser high-level semantic abstractions, therefore enable encoding feature representations into higher layers [8, 9].

2.2 Deep Learning-Based Recommender Systems

Considering the latest successes in a variety of domains, DL models have been exploited recently in recommender systems. Current models mainly use deep neural networks to learn user preferences on items to make final recommendations. According to Ni et al. [2], two categories of models can be distinguished: (1) recommendation models with a single neural network structure [10], such as MLP (Multi-Layer Perception), CNN (Convolutional Neural Network), RNN (Recurrent Neural Network) and its variants LSTM (Long Short-Term Memory), and GRU (Gated Recurrent Unit); and (2) recommendation models that combine the attention mechanism with the above described neural network structures [11].

The state of the art shows that some neural network-based methods are proposed to learn user and item representations from text notation and comments data [12, 13]. He et al. [6] modeled the user-item assessment matrix using a multilayer feedback neural network. On the other hand, He et al. [14] proposed an approach based on CF and CNN. They considered that it was more interesting and semantically plausible to use a dyadic product to represent correlations between user and item embeddings, since the objective was to highlight the link between these data. Zheng et al. [15] combined all user comments and items and applied the CNN to jointly model user preferences and item characteristics. According to Liu et al. [16], these methods only use notes or comments, individually, to model users and items and ignore the intrinsic link between them. Lu et al. [13] proposed a model of mutual learning between notes and comments, given that the method of modeling notes is based on PMF (Probabilistic MF), which can only learn linear characteristics. Liu et al. [16] used a MLP network in a note-based coder to learn deeper and higher-level features from note models. To overcome the cold start problem, Berkani et al. [17] proposed an extension of the NCF model [13], considering a hybridization of CF and CBF based on GMF and HybMLP architecture. Similarly to NCF approach [13], MLP allows the approximation of the interaction function without making a product between latent factors. The GMF model has been associated to the MLP, to improve the accuracy of the algorithm. From the concatenation of the two models, a third neural matrix factorization model (NeuMF) will be created.

In this article, we will adapt this work to the social context. The ubiquitous use of social media and the massive growth of interaction between users at an unprecedented rate, generates huge amounts of information. This information could be exploited to increase the correlation between users, with the aim of improving the performance of traditional recommendation algorithms. Exploiting social information such as friendship and trust between users could significantly improve the quality of recommendation, by taking into account the interactions between them and exploiting their "trusted network" (the users they are connected with). Accordingly, we propose, a novel hybrid method called SNHF (Social Neural Hybrid Filtering), combining CF and CBF algorithms and using social information (friendship and trust between users) in the same architecture based on GMF and HybMLP (Hybrid MLP) models.

3 Social Neural Hybrid Recommendation (SNHF)

This section introduces our method SNHF in detail. The overview of the proposed approach is shown in Fig. 1. We describe below the different components of our model.

3.1 Description of the Different Layers of the SNHF Model

Input Layer. In addition to the NHybF model entries [13], i.e., user and item IDs with corresponding characteristics, we also consider the social information of each

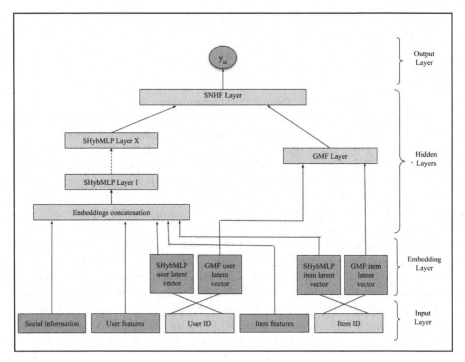

Fig. 1. Overview of the proposed SNHF method

user (friends and trusted persons). This information will be concatenated later to the Embedding. Friendship and trust are extracted as follows.

The friends list of a given user u includes all of its friends from the social network. The extraction of this degree is calculated with the following Jaccard formula:

$$Friendship(u, v) = \frac{|F_u \cap F_v|}{|F_u \cup F_v|} \tag{1}$$

where: F_u is the set of friends of u and F_v is the set of friends of v.
The friendship degree is calculated as follows:

$$D_{Friendship(u,v)} = 1 - Friendship(u, v) \tag{2}$$

Trust describes the degree of trust of a user u towards a user v. The calculation of the degree of trust is done in two steps: (1) calculation of the degree of trust directly between two users; and (2) propagation of the trust through the trust network.

The degree of trust between two directly related users is calculated according to the following Tanimoto formula:

$$D_{Trust(u,v)} = \frac{1}{deg(u) + deg(v) - 1} \tag{3}$$

where:

$deg(u)$: is the number of users that user u trusts, including v; and.

$deg(v)$: is the number of users that user v trusts, including u.

We used the MoleTrust algorithm [18], to compute the trust degree between a source user and a target user, by browsing the trust graph and propagating it along the arcs. The social degree between two users is calculated according to this weighted formula:

$$D_{social} = \alpha * D_{Friendship(u,v)} + \beta * D_{Trust(u,v)} \tag{4}$$

where: α, and β: represent the importance weights related, respectively, to friendship and trust, with: $\alpha + \beta = 1$.

Note that in our model, the hybridization of the three algorithms (collaborative, content-based and social filtering), is based on machine learning, and therefore learns to deduce from itself the best parameters (degree of involvement of each algorithm) to achieve the best prediction.

Embedding Layer. The embedding layer of a neural network allows converting and representing discrete or categorical variables into a vector of continuous numbers. A common use in natural language processing is to take a word and apply Word Embedding to make it denser. The words are encoded in a binarized sparse vector with one-hot encoding [19]. The goal is to design an algorithm capable of forming less sparse vectors that would have a logical relationship between them.

Generalized Matrix Factorisation (GMF) - Linear Model. This part represents the operation made by the GMF to calculate the predictions. The textual description of the model layers is given as follows:

- *GMF User Embedding*: User latent factor vectors.
- *Embedding of GMF items*: Vectors of latent factors of the items.
- *Multiplication layer*: ensures the element by element multiplication of the user and item embedding (factors) according to the following formula to calculate the predictions:

$$\phi^{GMF} = p_u^G \odot q_i^G \tag{5}$$

where:

p_u^G: Latent factors vector of the user u of GMF; and

q_i^G: Latent factors vector of the item i of GMF.

Social Hybrid Multilayer Perceptron (SHybMLP) - Nonlinear Model. This part deals with the learning of the interaction function which returns "1" if there is an interaction between a user u and an item i and "0" otherwise. The textual description of the different layers of the SHybMLP model is given as follows:

- *Embedding of SHybMLP users:* Vectors of user latent factors.

- *Embedding of SHybMLP items:* Vectors of items latent factors.
- *Concatenation Layer:* is the first hidden layer of the SHybMLP model. It allows the concatenation of embeddings with user and item characteristics as well as social information related to each user.

These data will be passed through other hidden layers in order to better learn the interaction between both users and items characteristics, as well as the influence that a user's circle (his friendships and trusted people) can have on their choice of items. To decide on the optimal number of nodes for each layer, we set a node for the last one, called PF (Predictive Factors). The previous layer will have the double of this number of nodes (PF * 2), and so on until the first layer (PF * X layers, where X is the number of layers). The activation function used in each hidden layer is *ReLU* (Rectified Linear Unit: ReLU(x) = max(0, x)), because it reduces the risk of overfitting and neuron saturation. Below is the formula of the model activation function [13]:

$$\phi^{\text{SHybMLP}} = a_L(W_L^T(a_{L-1}(...a_2(W_2^T\left[p_u^M f_u f_s q_i^M f_i\right] + b_2)...)) + b_L) \qquad (6)$$

where:

a_L: Layer activation function L;
b_L: Bias of the layer L (has the same role as the threshold);
W_L^T: Weight of the layer L;
p_u^M: Latent factors vector of the user u of SHybMLP;
f_u: Information vector of the user u;
q_i^M: Latent factors vector of the item i of SHybMLP; and
f_i: Information vector of the item i.

Social Neural Hybrid Matrix Factorization (SNHF). The last layer of SHybMLP is concatenated to that of GMF so that the combination of the results of these models would lead to better predictions. The training of this model can be done in two ways: with or without the pre-training of the models composing it (GMF and SHybMLP).

Output Layer. This layer takes as input the vectors of the last SHybMLP and GMF layers previously concatenated in the SNHF layer, then passes them through the Sigmoid activation function: $\sigma(x) = 1/(1 + e^x)$. The formula of the output layer activation function is given as follows [13]:

$$\hat{y}_{ui} = \sigma(h^T\begin{bmatrix}\phi^{GMF} \\ \phi^{SHybMLP}\end{bmatrix}) \qquad (7)$$

where:

h^T: Predictions vector of GMF and *SHybMLP* models; and
$\hat{y}_{ui}$: Prediction of SNHF model.

3.2 Training

Dataset Management. During our experimentation, we used two databases: Yelp[1] and FilmTrust[2]. The data extracted from these two databases must be processed and transformed before passing through the model (using a one-hot vector representation). Then, explicit data would have to be transformed into implicit data due to the lack of explicit data provided by users. We consider all user-item interactions in the dataset as positive instances (to have more data to exploit and an idea of the type of items the user is likely to be interested in). In order to guarantee better learning, it is necessary to provide the model with a diversity of data (including negative instances, considering any non-interaction as a negative instance). However, the term "negative instance" does not mean that an item i is not relevant for a given user u, but he did not have the opportunity to interact with this item although it could be interesting for him.

In order to integrate social information, we chose the method proposed by Ma et al. [20], which assumes that the taste of a user u is close to the average tastes of his friends. Although in practice, a user can have hundreds of friends, but some of them may have completely different tastes with him/her. In order to avoid biasing the results, it is necessary to filter the list of friends using the friendship degree according to a given threshold. Similarly, for trust, a degree is calculated using Tanimoto's formula MoleTrust algorithm, in order to benefit from the opinions of other users who are not directly related to u. Once the transformation of the data is done, the dataset will be divided into two parts: (1) *Training data*, which will serve as learning data for the model (experiments will help to find the right balance between positive and negative instances); and (2) *Testing data*, which will be used to evaluate the efficiency of our model. To simulate the real conditions of recommender systems, we consider that only 1% of the data is relevant for the user, i.e., for each user, we randomly select an item with which he has interacted (representing the positive instance), and add 99 items with which he has not interacted yet (representing the negative instances).

SNHF Training. Neural networks are trained using an optimization process that requires a cost function to be minimized. This function calculates the difference between the prediction made by the model and the actual value it was supposed to predict. For comparison purposes, we used the same functions as [13] for the training of our models (GMF, SHybMLP, and SNHF): the binary cross-entropy or log loss functions because it is a Logistic Regression problem with a binary classification. This function is given as follows:

$$L = -\sum\nolimits_{(u,i)\in Y\cup Y^-} y_{ui}log\hat{y}_{ui} + (1 - y_{ui})\ log\left(1 - \hat{y}_{ui}\right) \tag{8}$$

where:

y_{ui}: Value of the interaction between the user u and item i;

$\hat{y}_{ui}$: Prediction of the interaction between user u and item i;

Y: Set of positive instances; and Y^-: Set of negative instances.

---

[1] https://www.yelp.com/dataset.

[2] http://konect.cc/networks/librec-filmtrust-trust/.

Optimization. In order to update the values of the model weights, we used the Adam (Adaptive Moment Estimation) algorithm for the training of GMF and HybMLP and the SGD (Stochastic Gradient Descent) algorithm for the training of NHybSoc. The first algorithm adapts the learning rate to each parameter while the second uses a single learning rate to update the parameters.

4 Experiments

Our objective is to show the contribution of DL in the prediction of the interaction as well as the contribution of our hybrid approach, in particular, in the cold start situation.

4.1 Datasets

We conduct the experiments on two real-world datasets: Yelp and FilmTrust. Real database conditions were simulated (few items relevant to the user) considering that only one item out of one hundred could be relevant to the user. A pre-processing has been performed on these data to transform them into exploitable vectors. The statistics of these sample data sets are shown in Table 1:

- Yelp: we use a sample of the database collected and processed by Berkani et al. [21], including information on users, restaurants and ratings.
- FilmTrust: allows users to share movie ratings and explicitly specify other users as trusted neighbors.

Table 1. Statistics of the sample datasets

Dataset	# Users	#Items	#Ratings	Density (%)
Yelp	5,436	4,733	110,894	0.43
FilmTrust	1,508	2,071	35,497	1.14

4.2 Metrics

The performance of a ranked list is evaluated by Hit Ratio (HR) and Normalized Discounted Cumulative Gain (NDCG) [22]. We truncated the ranked list at 10 for both metrics. The HR measures whether the test item is present on the top-10 list. While the NDCG accounts for the position of the hit by assigning higher scores to hits at top ranks. Both metrics are calculated for each test user and the average score is reported.

4.3 Evaluation of the SNHF Model

To evaluate our model, it is important to study the evolution of the models composing it after the integration of social information, in order to determine the most optimal values of the parameters. We start first with the evaluation of the impact of social information on CF, and then its impact on the combination of CF and CBF algorithms.

Collaborative and Social Filtering. To evaluate the contribution of social information on CF, we only take into account user and item identifiers and the social information related to each user. This version of our model is called SNCF, and its sub-models are SMLP and GMF.

SMLP. We evaluated the SMLP model with Yelp and FilmTrust with different embedding sizes and a different number of hidden layers and a Predictive factor = 16. We have seen an improvement and increase in accuracy with the increasing number of layers and the size of the embedding especially when considering three or five layers. Figure 2 illustrates the evolution of prediction according to the number of layers of the two models SMLP and MLP.

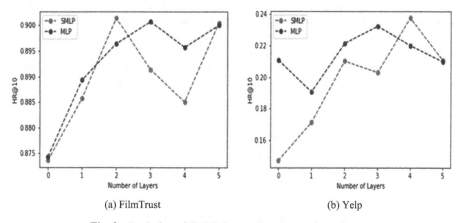

(a) FilmTrust   (b) Yelp

Fig. 2. Evolution of SMLP, by varying the number of layers

The results obtained show the importance of using deep neural networks instead of a simple network (SMLP-0) and that the best performance was obtained with the SMLP model. The optimal number of layers is equal to 2 and 4 layers respectively for FilmTrust and Yelp.

Figure 3 shows the values of the HR metric when the embedding size is varied with the SMLP, MLP and GMF models. We notice the superiority of the deep neural network models over the GMF model. This evaluation allowed us to find that the most optimal number of latent factors is 8 factors for FilmTrust and 16 factors for Yelp.

SNCF. There are two ways to train the SNCF model. The first is to train the whole model, i.e., GMF and SMLP simultaneously. While the second proposes to train SMLP

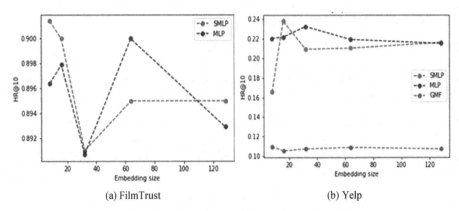

(a) FilmTrust   (b) Yelp

Fig. 3. Evolution of SMLP according to the number of layers

and GMF separately, then train the last concatenation layer after having the weights of both models. The evaluation without pre-training showed that the model fails to make good predictions (a decrease in accuracy is observed with the increase in the number of layers). This is due to the combination of two models that learn to predict the same thing at the same time. On the other hand, the results with the pre-training of the sub-models are much better than the previous one, demonstrating its effectiveness for the SNCF training.

Figure 4 shows the predictions according to the number of nodes in the last hidden layer of SMLP, using FilmTrust. The optimal number of predictive factors is 16.

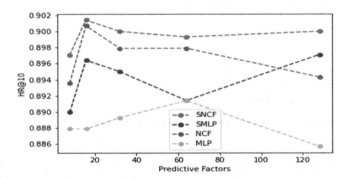

Fig. 4. Performance according to the number of predictive factors - FilmTrust.

Collaborative, Content-Based and Social Filtering. We evaluated the impact of social information on the hybridization of CF and CBF algorithms, using the Yelp dataset.

SHybMLP. We evaluated the SHybMLP model with different embedding sizes and different number of hidden layers - Predictive factors = 16. We noticed an improvement of the results with the increase of the number of layers, where the best values were obtained with 5 hidden layers.

Figure 5 illustrates the evolution of SHybMLP prediction according to the number of layers and the size of embeddings with the Yelp database. We can notice that the SHybMLP model performs better with 5 hidden layers and 8 latent factors, but that the SMLP model achieved better results.

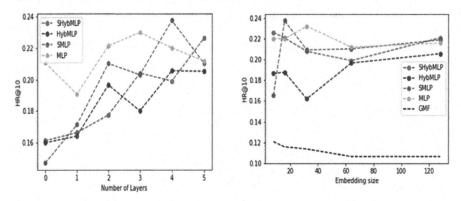

Fig. 5. Evolution of SHybMLP according to the number of layers and embedding size

SNHF. Similarly to the evaluation of the SNCF model, we evaluated the SNHF model with and without pre-training. The best results were obtained with the pre-training of its SHybMLP and GMF sub-models. Figure 6 shows the predictions according to the number of predictive factors of SHybMLP, with an increase in the number of epochs in order to see the effects on the results.

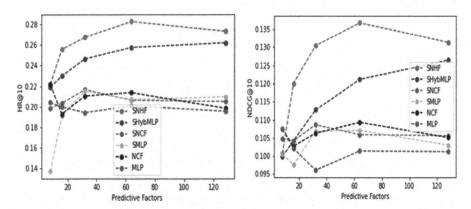

Fig. 6. SHybMLP performance according to the number of predictive factors – Yelp

We can notice the improvement of both SNHF and SHybMLP models with the increase of the number of epochs to 15 epochs for both metrics. On the other hand, no effect was observed for the other models such as SNCF. The optimal number of nodes is 64 for SNHF and 16 for SNCF with both metrics.

Dataset Scrambling. In order to demonstrate the effectiveness of the models, we proceeded to scramble the learning data, i.e., increase the number of negative instances compared to positive ones. The objective was to find the right number of negative instances to allow the model to be more accurate.

The following tests were performed on the best SNCF, SMLP, SHybMLP and SNHF models obtained from the training on the balanced dataset.

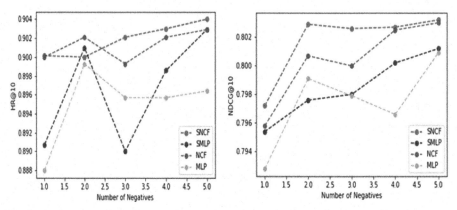

Fig. 7. Performance according to the number of negative instances per positive instance - FilmTrust

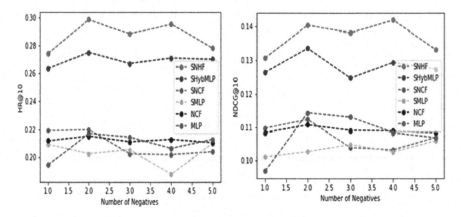

Fig. 8. Performance according to the number of negative instances per positive instance - Yelp.

As shown in Fig. 7 and Fig. 8, this evaluation shows that the SNHF model performs better when we have a dataset with two negative instances per positive instance for Yelp and five negative instances for FilmTrust. Furthermore, when using Yelp, we can notice the effectiveness of SNHF, which outperforms all the other models. While SNCF is slightly more efficient than NCF in the case the dataset has one or two negative instances. On the other hand, with FilmTrust, the SNCF model is more efficient than NCF except for the case where the dataset contains two negative instances.

5   Conclusion

This article explored neural network architectures for item recommendation in social networks. We proposed a hybrid algorithm that combines CF and CBF using social information in an architecture based on GMF and HybMLP models. Extensive experiments carried out on two wide-used datasets, Yelp and FilmTrust, demonstrated the effectiveness of the proposed SNHF hybrid algorithm which outperforms the other baselines and related work. This proves the impact and contribution of social information on the recommendation. Furthermore, the empirical evaluation shows that using deeper layers of neural networks offers better recommendation performance.

As perspectives to this work, it would be interesting to enrich the social information with other features (such as the influence and credibility of users) and consider the use of other more complex data such as user comments on items, linked open data (LOD) [23, 24] or multi-criteria ratings [25, 26] to improve the quality of the recommendation. Moreover, we plan to explore other deep learning architectures such as convolutional neural networks to learn high-order correlations among embedding dimensions.

References

1. Adomavicius, G., Tuzhilin, A.: Toward the next generation of recommender systems: a survey of the state-of-the-art and possible extensions. IEEE Trans. Knowl. Data Eng. **17**(6), 734–749 (2005)
2. Ni, J., Huang, Z., Cheng, J., Gao, S.: An effective recommendation model based on deep representation learning. Inf. Sci. **542**(1), 324–342 (2021)
3. Resnick, P., Iakovou, N., Sushak, M., Bergstrom, P., Riedl, J.: GroupLens: an open architecture for collaborative filtering of netnews. In: Proceedings of the Computer Supported Cooperative Work Conference, pp. 175–186 (1994)
4. Sarwar, B., Karypis, G., Konstan, J., Reidl, J.: Item-based collaborative filtering recommendation algorithms. In: Proceedings of the 10th International Conference on World Wide Web, pp. 285–295 (2001)
5. Peis, E., Morales-del-Castillo, J.M., Delgado-López, J.A.: Semantic recommender systems. Analysis of the state of the topic. Hipertext.net **6** (2008), http://www.hipertext.net, ISSN 1695-5498
6. Burke, R.: Hybrid recommender systems: survey and experiments. User Model. User Adap. Inter. **12**(4), 331–370 (2002)
7. Sun, Z., et al.: Recommender systems based on social networks. J. Syst. Softw. **99**, 109–119 (2015)
8. LeCun, Y., Bengio, Y., Hinton, G.: Deep learning. Nature **521**, 436–444 (2015)
9. Batmaz, Z., Yurekli, A., Bilge, A., Kaleli, C.: A review on deep learning for recommender systems: challenges and remedies. Artif. Intell. Rev. **52**(1), 1–37 (2018). https://doi.org/10.1007/s10462-018-9654-y
10. Zhang, S., Yao, L., Sun, A.: Deep learning based recommender system: a survey and new perspectives. arXiv preprint arXiv:1707.07435 (2017)
11. Hu, B., Shi, C., Zhao, W.X., Yu, P.S.: Leveraging meta-path based context for top-n recommendation with a neural co-attention model. In: Proceedings of the ACM SIGKDD, pp. 153–1540 (2018)
12. Chen, C., Zhang, M., Liu, Y., Ma, S.: Neural attentional rating regression with review-level explanations. In: Proceedings of the WWW, pp. 1583–1592 (2018)

13. Lu, Y., Dong, R., Smyth, B.: Coevolutionary recommendation model: mutual learning between ratings and reviews. In: Proceedings of the WWW, pp. 773–782 (2018)
14. He, X., Liao, L., Zhang, H., Nie, L., Hu, X., Chua, T.-S.: Neural collaborative filtering. In: 2017 International World Wide Web Conference Committee (IW3C2), WWW 2017, Perth, Australia, 3–7 April (2017). https://doi.org/10.1145/3038912.3052569
15. Zheng, L., Noroozi, V., Yu, P.S.: Joint deep modeling of users and items using reviews for recommendation. In: Proceedings of the WSDM, pp. 425–434 (2017)
16. Liu, H., et al.: Hybrid neural recommendation with joint deep representation learning of ratings and reviews. Neuro-Comput. **374**, 77–85 (2020)
17. Berkani, L., Kerboua, I., Zeghoud, S.: Recommandation Hybride basée sur l'Apprentissage Profond. In: Actes de la conférence EDA 2020, Revue des Nouvelles Technologies de l'Information, RNTI B.16, pp. 69–76 (2020). ISBN 979-10-96289-13-4
18. Avesani, P., Massa, P., Tiella, R.: Moleskiing it: a trust-aware recommender system for ski mountaineering. Int. J. Infonom. **20**(35), 1–10 (2005)
19. Harris, D., Harris, S.: Digital Design and Computer Architecture, 2nd edn., p. 129. Morgan Kaufmann, San Francisco (2012). ISBN 978-0-12-394424-5
20. Ma, H., Zhou, D., Liu, C., Lyu, M., King, I.: Recommender systems with social regularization. In: Proceedings of the 4th ACM International Conference on Web Search and Data Mining, pp. 287–296. ACM, New York (2011)
21. Berkani, L., Belkacem, S., Ouafi, M., Guessoum, A.: Recommendation of users in social networks: a semantic and social based classification approach. Exp. Syst. **38** (2020).https://doi.org/10.1111/exsy.12634
22. He, X., Chen, T., Kan, M.-Y., Chen, X.: TriRank: review-aware explainable recommendation by modeling aspects. In: CIKM, pp. 1661–1670 (2015)
23. Natarajana, S., Vairavasundarama, S., Natarajana, S., Gandomi, A.H.: Resolving data sparsity and cold start problem in collaborative filtering recommender system using Linked Open Data. Exp. Syst. Appl. **149**, 113248 (2020)
24. Sejwal, V.K., Abulaish, M., Jahiruddin: CRecSys: a context-based recommender system using collaborative filtering and LOD. IEEE Access **8**, 158432–158448 (2020).https://doi.org/10.1109/ACCESS.2020.3020005
25. Shambour, Q.: A deep learning based algorithm for multi-criteria recommender systems. Knowl. Based Syst. **211**, 106545 (2021)
26. Nassar, N., Jafar, A., Rahhal, Y.: A novel deep multi-criteria collaborative filtering model for recommendation system. Knowl. Based Syst. **187**, 104811 (2020)

Cloud Query Processing with Reinforcement Learning-Based Multi-objective Re-optimization

Chenxiao Wang[1](✉), Le Gruenwald[1], Laurent d'Orazio[2], and Eleazar Leal[3]

[1] University of Oklahoma, Norman, Oklahoma, USA
{chenxiao,ggruenwald}@ou.edu
[2] CNRS IRISA, Rennes 1, University, Lannion, France
laurent.dorazio@univ-rennes1.fr
[3] University of Minnesota Duluth, Duluth, Minnesota, USA
eleal@d.umn.edu

Abstract. Query processing on cloud database systems is a challenging problem due to the dynamic cloud environment. The configuration and utilization of the distributed hardware used to process queries change continuously. A query optimizer aims to generate query execution plans (QEPs) that are optimal meet user requirements. In order to achieve such QEPs under dynamic environments, performing query re-optimizations during query execution has been proposed in the literature. In cloud database systems, besides query execution time, users also consider the monetary cost to be paid to the cloud provider for executing queries. Thus, such query re-optimizations are multi-objective optimizations which take both time and monetary costs into consideration. However, traditional re-optimization requires accurate cost estimations, and obtaining these estimations adds overhead to the system, and thus causes negative impacts on query performance. To fill this gap, in this paper, we introduce ReOptRL, a novel query processing algorithm based on deep reinforcement learning. It bootstraps a QEP generated by an existing query optimizer and dynamically changes the QEP during the query execution. It also keeps learning from incoming queries to build a more accurate optimization model. In this algorithm, the QEP of a query is adjusted based on the recent performance of the same query so that the algorithm does not rely on cost estimations. Our experiments show that the proposed algorithm performs better than existing query optimization algorithms in terms of query execution time and query execution monetary costs.

Keywords: Query optimization · Cloud databases · Reinforcement learning · Query re-optimization

1 Introduction

In a traditional database management system (DBMS), finding the query execution plan (QEP) with the best query execution time among those QEPs

© Springer Nature Switzerland AG 2021
C. Attiogbé and S. Ben Yahia (Eds.): MEDI 2021, LNCS 12732, pp. 141–155, 2021.
https://doi.org/10.1007/978-3-030-78428-7_12

generated by a query optimizer is the key to the performance of a query. In a cloud database system, minimizing query response time is not the only goal of query optimization. As hardware usages are charged on-demand and scalability is available to users, monetary cost also needs to be considered as one of the objectives for optimizing QEPs in addition to query response time. In order to do that, the query optimizer evaluates the time and monetary costs of different QEPs in order to derive the best QEP for a query. These time and monetary costs are estimated based on the data statistics available to the query optimizer at the moment when the query optimization is performed. These statistics are often approximate, which may result in inaccurate estimates for the time and monetary costs needed to execute the query. Thus, the QEP generated before query execution may not be the best one.

To solve the problem, researchers have developed learning-based algorithms to adjust the data statistics to get more accurate cost estimations [4,13]. These methods are heuristic-based and the adjustment of QEP is not adaptable to the dynamic environment. More recently, machine learning-based algorithms are introduced [9,20]. More accurate cost estimations are made by the data statistics estimated by the machine learning models. The optimizer uses these cost estimations to adjust the QEP. Again, even those methods improved the accuracy of data statistics estimation such as cardinalities, the overall performance is not improved much. This is usually because updating data statistics for the optimizer to use is a very expensive operation by itself. This becomes the main source of negative impacts on the overall performance. To eliminate this problem, in this paper we propose an algorithm, called Re-OptRL, for query re-optimization that makes use of reinforcement learning (RL) to find the best QEP without relying on data statistics.

RL is about taking suitable actions to maximize reward in a particular situation. The decision of choosing actions follows the trial-and-error method and is evaluated by the reward. The learning model is adjusted with the reward after each action has been performed. We choose RL instead of supervised learning methods [10] because unlike supervised learning, RL does not require a labeled dataset of past actions to be available to train the learning model in advance before it can be used to predict future actions. Query optimization in the database system can be regarded as a series of actions and the best actions can be learned from evaluating the historical optimizations. In the proposed algorithm ReOptRL, the QEP of a query is adjusted during the query execution independently of data statistics. The adjustment is based on the performance of historical queries that are executed on the system recently. Evaluating these historical executions is reasonable especially on the cloud database system. As a large number of different queries are executed frequently, for the same query, the time gap between its incoming query and its historical query is short. The algorithm monitors and keeps learning from these executions while more incoming queries are executed. Some algorithms used reinforcement learning in adjusting their QEPs also, but these adjustments are only focusing on the JOIN order of queries [5,8]. Our algorithm extends these features to not only adjust the QEP itself but also to find the best allocations to execute different operators in the query, i.e., to find which machines should be used to execute which operator. In a cloud database system, each operator can be executed on a different machine, and different machines can have different hardware configurations with different usage prices. This means that for the same query operator, not only different execution times but also different monetary costs can incur depending on which machine is used to execute the operator. An operator executed on machine A may have a lower execution time but may cost more money, on the other hand, the same operator executed on machine B may have a higher execution time but may cost less money. Those differences in execution times and monetary costs for individual operators eventually affect the overall performance in accumulation for the entire query. Our goal is to fulfill both user's execution time and monetary cost requirements of the overall query performance. Due to this reason, optimal allocations of operator executions to appropriate machines on the cloud are as important as optimal QEPs themselves. Our contributions in this work are the following:

- We propose an algorithm that uses RL to perform multi-objective query processing for an end-to-end cloud database system.
- Our algorithm employs a new reward function designed specifically for query re-optimization.
- We present comprehensive experimental performance evaluations. The experimental results show that our algorithm improves both query execution time and monetary costs when comparing to existing query optimization algorithms.

The remaining of this paper is organized as follows: Sect. 2 discusses the related work; Sect. 3 presents the proposed reinforcement learning-based query re-optimization algorithm; Sect. 4 discusses the experimental performance evaluations; and finally, Sect. 6 presents the conclusions and discusses future research directions.

2   Related Work

The problem of query re-optimization has been studied in the literature. In the early days, heuristics were used to decide when to re-optimize a query or how to do the re-optimization. Usually, these heuristics were based on cost estimations which were not accurate at the time when query re-optimization takes place. Also, sometimes, a human-in-the-loop was needed to analyze and adjust these heuristics [5]. These add additional overheads to the overall performance of queries. Unfortunately, these heuristic solutions can often miss good execution plans. More importantly, traditional query optimizers rely on static strategies, and hence do not learn from previous experiences. Traditional systems plan a query, execute the query plan, and forget that they ever optimized this query. Because of the lack of feedback, a query optimizer may select the same bad plan repeatedly, never learning from its previous bad or good choices.

Machine learning techniques have been used recently in query optimizations for different purposes. In earlier works, Leo [15] learns from the feedback of executing past queries by adjusting its cardinality estimations over time, but this algorithm requires human-tuned heuristics, and still, it only adjusts the cardinality estimation model for selecting the best join order in a query. More recently, the work in [20] presents a machine learning-based approach to learn cardinality models from previous job executions and these models are then used to predict the cardinalities in future jobs. In this work, only join orders, not the entire query, are optimized. In [18], the authors examine the use of deep learning techniques in database research. Since then, reinforcement learning is also used. The work proposes a deep learning approach for cardinality estimation that is specifically designed to capture join-crossing correlations. SkinnerDB [16] is another work that uses a regret-bounded reinforcement learning algorithm to adjust the join order during query execution. None of these machine learning-based query optimization algorithms is designed for predicting the action that should follow after one query operator has been executed. Recently, there have been several exciting proposals in putting reinforcement learning (RL) inside a query optimizer. As described in [19] and shown in Fig. 1, reinforcement learning describes the interaction between an agent and an environment. The possible actions that the agent can take given a state S_t of the environment are denoted as $A_t = \{a_0, a_1, \ldots, a_n\}$. The agent performs an action from the action set A_t based on the current state S_t of the environment. For each action taken by the agent, the environment gives a reward r_t to the agent and the environment turns into a new state S_{t+1}, and the new action set is A_{t+1}. This process repeats until the terminal state is reached. These steps form an episode. The agent tries to maximize the reward and will adjust after each episode. This is known as the learning process.

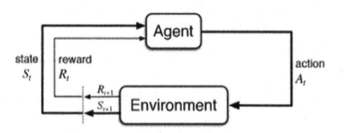

Fig. 1. General procedure of reinforcement learning [19]

Ortiz et al. [12] apply deep RL to learn a representation of queries, which can then be used in downstream query optimization tasks. Liu [6] and Kipf [5] use DNNs to learn cardinality estimates. Closer to our work is Marcus et al.'s work of a deep RL-based join optimizer, ReJOIN [8], which orders a preliminary view of the potential for deep RL in this context. The early results reported in

[8] give a 20% improvement in the query execution time of Postgres. However, they have evaluated only 10 out of the 113 JOB queries.

3 Deep Reinforcement Learning Based Query Re-optimization

In this section, we present our query re-optimization algorithm that makes use of reinforcement learning, ReOptRL. First, we discuss our query processing procedure in the cloud database system. Then we describe how the deep reinforcement learning algorithm is used in our query processing to select the best actions for the performance of queries. In this algorithm, a query will be converted into a logical plan by a traditional query parser. Then for each logical query operator, we use a deep reinforcement learning model to select what physical operator and which machines (containers) should be used to execute this logical operator so that each operator execution is optimized to gain the maximum improvement on overall performance. These selections are learned from the same operator executed previously in the system. Besides that, as in large applications, there will be a large number of queries running frequently. For a query operator, it is reasonable to utilize the performance of its previous executions of the same query to predict the performance of its current execution because the time between the two executions is short.

3.1 The Environment of Our Cloud Database System

In this section, we first briefly describe our environment for query processing. Our applications mainly focus on processing queries in a mobile cloud environment. Figure 2 shows the process flow of query processing in the mobile-cloud database system that we have developed [17]. In this architecture, the mobile device is for the user to access the database and input queries, the data owner is a server on-site that contains private data, and the cloud provider owns the cloud database system. Executing a query incurs three different costs: the monetary cost of query execution on the cloud, the overall query execution time, and the energy consumption on the mobile device where the query might be executed. These three costs constitute the multi-objectives that the query optimizer needs to minimize to choose the optimal query execution plan (QEP). Different QEPs are available due to the elasticity of the cloud which considers multiple nodes with different specifications. In this paper, we focus on query processing on the cloud provider's part. We consider both the query execution time and monetary cost, but not the energy cost on mobile devices.

Different users have different preferences for choosing a suitable QEP for their purposes. In an application scenario where many queries are executed per day, organizations may want to minimize the monetary cost spent for query execution to fit their budget. They may also want to minimize query execution time to meet customers' query response time requirements and to optimize employees' working time. In order to incorporate these user's preferences into the query optimization algorithm, we use the Normalized Weighted Sum Model we have developed in [3] to select the best plan. In this model, every possible QEP alternative is rated by a score that combines both the objectives, query response time, and monetary cost, with the weights defined by the user and the environment for each objective, and the user-defined acceptable maximum value for each objective. These weights defined by the user are called Weight Profile (wp), which is a two-dimensional vector and each dimension is a number between 0 to 1 which defines the preferences. The following function is used to compute the score of a QEP:

$$A_i^{WSM-SCORE} = \sum_{j=1}^{n} W_j \frac{a_{ij}}{m_j} \tag{1}$$

where a_{ij} is the value of QEP alternative i (QEP_i) for objective j, m_j the user-defined acceptable maximum value for objective j, and w_j the normalized composite weight of the user and environment weights for objective j, which is defined as follows:

$$w_j = \frac{uw_j * ew_j}{\sum(uw * ew)} \tag{2}$$

where uw_j and ew_j describe the user and the environmental weight for objective j, respectively. These weights are user-defined. Since the different objectives are representative of different costs, the model chooses the alternative with the lowest score to minimize the costs.

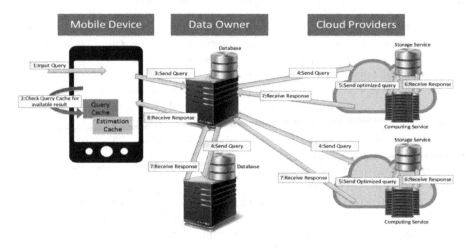

Fig. 2. Mobile cloud database environment

3.2   Overview of Reinforcement Learning Based Query Processing Algorithm (ReOptRL)

In this work, we use a policy gradient deep RL algorithm for query re-optimization. This algorithm uses a deep neural network to help the agent decide the best action to perform under each state. In this work, the agent is the query optimizer, an action is a combination of a physical operator to execute a logical operator and a machine to execute this operator, and a state is a fixed-length vector encoded from the logical query plan parsed from a traditional query optimizer.

The input of the neural network is the vector of the current state. The input is sent to the first hidden layer of the neural network whose output is then sent to the second layer, and so on until the final layer is reached, and then an action is chosen. The policy gradient is updated using a sample of the previous episodes, which is an operator execution in our case. Once an episode is completed (which means a physical operator and execution container are selected in our case), the execution performance is recorded and a reward is received where a reward is a function to evaluate the selected action. The details of the reward function will be explained in Sect. 3.3. The Weights of the neural network is updated after several episodes using existing techniques, such as back-propagation [6].

Figure 3 shows the major steps in query processing when ReOptRL is incorporated for query re-optimization. First, the optimizer receives a query and then the query is compiled into a QEP (logical plan). Secondly, the first available operator is converted into a vector representation and is sent to the RL model. The RL model will select the optimal action, which is the combination of a physical operator and a container to execute this operator. Then this operator is sent to execution and the execution time and monetary costs of this execution are recorded to update the reward function. Once the reward function is updated, the weights of this RL model are adjusted according to the updated reward function. Then the updated RL model is ready for future action selections of the same operator.

There are various kinds of reinforcement learning algorithms. Q-Learning is one of the popular value-based reinforcement learning algorithms [19]. In Q-Learning, a table (called Q-table) is used to store all the potential state-action pairs (S_n, a_n) and an evaluated Q-value associated with this pair. When the agent needs to decide which action to perform, it looks up the Q-value from the Q-table for each potential action under the current state and selects and performs the action with the highest Q-value. After the selected action is performed, a reward is given and the Q-value is updated using the Bellman equation [11].

$$Q(S_t, a_t) \leftarrow Q(S_t, a_t) + \alpha(R_t + \gamma Q(S_{t+1}, a_{t+1}) - Q(S_t, a_t)) \tag{3}$$

In Eq. (3), $Q(S_t, a_t)$ is an evaluated value (called Q-value) of executing Action a_t at State S_t. This value is used to select the best Action to perform under the current state. To keep this value updated with accurate evaluation is the key to reinforcement learning. α is the learning rate and γ is the discount rate. These two values are constant between 0 and 1. The learning rate α controls how fast

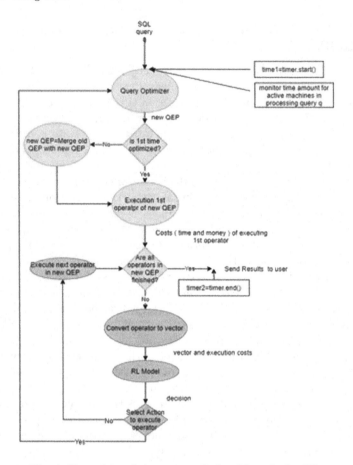

Fig. 3. Procedure of the proposed algorithm, ReOptRL

the new Q-value is updated. The discount rate γ controls the weight of future rewards. If $\gamma = 0$, the agent only cares for the first reward, and if $\gamma = 1$, the agent cares for all the rewards in the future [5]. R is the reward which is described in Sect. 3.3

Figure 4 shows the pseudo-code of the proposed algorithm. First, the optimizer receives a query and then the query is compiled into a QEP (logical plan) (Line 4). Then, the QEP is converted into a vector representation. By doing this, the current QEP represents the current state and can be used as the input of a neural network. We use a one-hot vector to present this vector and this technique is adapted from the recent work [7] (Line 5). This vector is sent to the RL model, which is a neural network as described in Sect. 3.1. The RL model will evaluate the Q-values for all the potential actions to execute the next available query operator (Line 6). Each of these actions includes two aspects, the optimal physical operator and the best container to execute the next available query operator. Then the action with the best Q-value will be selected and performed by the

DBMS (Line 7). After that, the executed operator is discarded from the QEP and the time and monetary costs to execute this operator are used to compute the reward for this action (Line 8). The reward is updated with the time and monetary cost to execute the operator and then the expected Q-value is updated by the Bellman Equation (3) with the updated reward (Line 9–11). The weights of the neural network are updated accordingly by the back-propagation method (Line 12). This process repeats for each operator in the QEP and terminates when all the operators in the QEP are executed. The query results are then sent to the user (Line 15).

Algorithm 1: Query Processing with Reinforcement Learning Based Re-Optimization (Re-OptRL)

Data: SQL query, Weight Profile wp, Reward Function R (), Learning rate α, Discount rate γ

Result: The query result set of the input query

1 t=0;
2 Result = $\emptyset$;
3 Q_t= 0;
4 **while** $QEP \neq \emptyset$ **do**
5   State S_t = convert QEP to a state vector;
6   Actiont=RunLearningModel (S_t, wp);
7   Result=Result $\cup$ execute (Op, $Action_t$);
8   QEP=QEP-Op;
9   Update R_t=R (wp, Actiont.time, Actiont.money));
10   Obtain Q-value of next state Q_{t+1} from the neural network;
11   Update Q-value of current state $Q_t = Bellman(Q_t, Q_{t+1}, R_t, \alpha, \gamma)$;
12   Update Weights in the neural network;
13   t=t+1;
14 **end**
15 **return** Result;

Fig. 4. Pseudo code of proposed algorithm (ReOptRL)

3.3 Reward Function

In ReOptRL, after an action is performed, the reward function is used to evaluate the action. This gives feedback on how the selected action performs to the learning model. The performed action with a high reward will be more likely to be selected again under the same state. The reward function plays a key role in the entire algorithm. In our algorithm, we would like the actions with low query execution time and monetary cost to be more likely chosen. To reflect this feature, here we define the reward function as follows:

$$Reward\ R = \frac{1}{1 + (W_t * T_{op}^q) + (W_m * M_{op}^q)} \tag{4}$$

where W_t and W_m are the time and monetary weights provided by the user, T_{op}^q and M_{op}^q is the time and monetary costs for executing the current operator op in query q. According to this reward function, the query is executed based on the user's preference which is either the user wants to spend more money for a better query execution time or vice versa. We call these preferences Weights. These weights defined by the user are called Weight Profile (wp), which is a two-dimensional vector and each dimension is a number between 0 to 1. Notice that, the user only needs to specify one dimension of the weight profile, the other dimension is computed with 1-Weight automatically. For example, if a user demands fast query response time and is willing to invest more money to achieve it, the weight profile for this user probably would be $<W_t = 0.9, W_m = 0.1>$. The detail can be found in our previous work [3]. This reward function is a monotonic decreasing function. With the increase of $(W_t * T_{op}^q) + (W_m * M_{op}^q)$, which is the total costs of executing a query operator, the reward decreases. Notice that, as $(W_t * T_{op}^q) + (W_m * M_{op}^q)$ approaches zero, the reward approaches positive infinity. When this situation happens, if an action A is performed with small total costs, then A will always be selected and performed, and all the other actions will be ignored. This is not desirable, and to keep the relationship of reward and total costs close to linear, we use $1 + (W_t * T_{op}^q) + (W_m * M_{op}^q)$ as the denominator in the reward function. In summary, if performing an action takes high costs, this action will be less likely to be chosen in the future.

4   Performance Evaluation

In this section, we first describe the hardware configuration, database benchmark, and parameters we used in our experimental model. We also introduce the algorithms we compared with our algorithm.

4.1   Hardware Configuration, Database Benchmark, and Parameter Values

There are two sets of machines that are used in our experiments. The first set consists of a single local machine used to train the machine learning model and to perform the query optimization. This local machine has an Intel i5 2500K Dual-Core processor running at 3 GHz with 16 GB DRAM. The second set consists of 10 dedicated Virtual Private Servers (VPSs) that are used for the deployment of the query execution engine. 5 of these VPSs, called small containers, have one Intel Xeon E5-2682 processor running at 2.5 GHz with 1 GB of DRAM. The other 5 VPSs, called large containers, each has two Intel Xeon E5-2682 processors running at 2.5 GHz with 2 GB of DRAM. The query optimizer and the query engine used in this experiment are modified from the open source database management system, PostgreSQL 8.4 [14]. The data are distributed among these VPSs. The queries and database tables are generated using the TPC-H benchmark [1]. There are eight database tables with a total size of 1,000 GB. We run 50,000 queries in total and these queries are generated by the query templates randomly selected from the 22 query templates from the TPC-H benchmark [1]. In the experiments, to update the Q-value using the Bellman equation as shown in Eq. (3) discussed in Sect. 3.2, we set the learning rate α as 0.1 and discount rate γ as 0.5.

4.2 Evaluation of ReOptRL

In this section, we compare the query processing performances obtained when the following query re-optimization algorithms are incorporated into query processing: 1) our proposed algorithm (denoted as **ReOptRL**); 2) the algorithm where a query re-optimization is conducted automatically after the execution of each stage in the query is completed (denoted as **ReOpt**), which we developed based on the state-of-art works [2,17]; 3) the algorithm where a query re-optimization is conducted by a supervised machine learning model decision (denoted as **ReOptML**); 4) the query re-optimization algorithm which uses sampling-based query estimation (denoted as **Sample**) proposed in [21]; and 5) the query processing algorithm that uses no re-optimization (denoted as **NoReOpt**).

We use NoReOpt as the baseline and the other algorithms are compared to the baseline. The two figures, Fig. 5 and Fig. 6, show the improvement of each algorithm over the baseline. From Fig. 5, we can see that, from the query execution time perspective, ReOptRL performs 39% better on average than NoReOpt, while ReOptML performs 27%, ReOpt 13% and Sample 1% better than NoReOpt. There are 15 out of 22 query types that have better performance if our algorithm is used. The best case is query 9 (Q9), for which our algorithms perform 50% better than NoReOpt. From Fig. 6, from the monetary cost perspective, query processing uses ReOptRL performs 52% better on average than NoReOpt, while ReOptML performs 27%, ReOpt 17%, and Sample 5% better than NoReOpt. There are 14 out of 22 query types that have better performance if our proposed algorithm is used. The best case is also Q9, for which our algorithms perform 56% better than NoReOpt.

The above results show that overall our proposed algorithm improves more time and monetary cost than the four algorithms, ReOpt, ReOptML, Sample, and NoReOpt. Especially, the monetary cost has a significant improvement (56% better than NoReOpt); however, for simple query types (1, 2, 3, 4, 6, 8, 10, 11) which are 8 query types out of 22 TPC-H query types, our algorithm does not improve their performance due to the overhead involved.

In these experiments, the Reinforcement Learning part of the query processing contributes to these improvements and it is beneficial in the following aspect. In all the three algorithms, ReOpt, ReOptML, and Sample, query re-optimization requires a lot of overhead data statistics that need to be accessed and updated frequently. In our proposed algorithms, no data statistics are needed, and re-optimization is based on the results of learning which can be decided quickly.

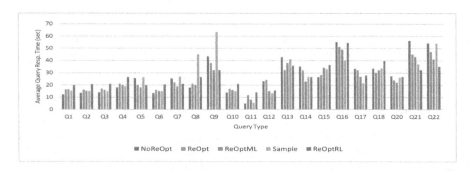

Fig. 5. Time cost performance of executing query using different algorithms

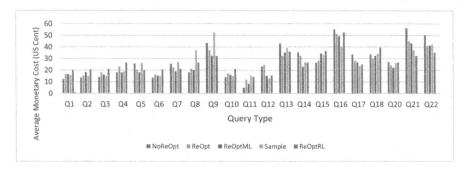

Fig. 6. Monetary cost performance of executing queries using different algorithms

5   Impacts of Weight Profiles

Our algorithm allows users to input their weight profiles and this feature is reflected in the reward function. We study how our proposed algorithm works under different weight profiles submitted by users. Figure 7 shows the performance of the monetary cost of each algorithm and the percentage of improvement of the monetary cost of each algorithm compared to the baseline on different weights. Notice that the Weight of Money in our application is (1-Weight of Time).

From Fig. 7, we can see that, when the weight of money increases, the monetary cost to execute the query decreases accordingly. Even when the weight of money is low, our algorithm still outperforms other algorithms on the monetary cost (by 10%). This happens because by using the learning model, our algorithm always chooses the right physical operator and container to execute the operator. Similarly, from the monetary cost perspective, our algorithm performs better on average than the other three algorithms.

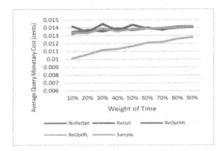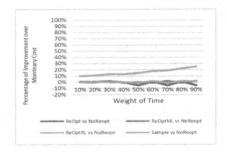

Fig. 7. Impacts of the weight of query execution time on the performance improvement of the algorithm over the baseline algorithm, NoReopt

The above results show that our algorithm can adapt to the user's weight profile very well. When the user has a high demand on either query execution time or money cost, our algorithm still outperforms the other algorithms.

6   Conclusion and Future Work

This paper presents an algorithm called ReOptRL for processing queries in a cloud database system taking both query execution time and money costs to be paid to the cloud provider into consideration. The algorithm uses a reinforcement learning-based model to decide the physical operator and machines to execute an operator from a query execution plan (QEP) of a query. The experimental performance evaluations using the TPC-H benchmark show that our proposed algorithm, ReOptRL, improves the query response time (from 12% to 39%) and monetary cost (from 17% to 56%) over the existing algorithms that use either no re-optimization, re-optimization after each stage in the query execution plan (QEP) is executed, supervised machine learning-based query re-optimization, or sample-based re-optimization.

While our studies have shown that reinforcement learning has positive impacts on the optimization and execution of each operator in a query, our algorithm currently does not consider the Service Agreement Level (SLA), which is an important feature of cloud database systems. For future work, we will investigate techniques to incorporate this feature into our algorithm by adjusting the reward function to consider SLA.

Acknowledgements. This work is partially supported by the National Science Foundation Award No. 1349285.

References

1. Barata, M., Bernardino, J., Furtado, P.: An overview of decision support benchmarks: TPC-DS, TPC-H and SSB. In: Rocha, A., Correia, A.M., Costanzo, S., Reis, L.P. (eds.) New Contributions in Information Systems and Technologies. AISC, vol. 353, pp. 619–628. Springer, Cham (2015). https://doi.org/10.1007/978-3-319-16486-1_61
2. Bruno, N., Jain, S., Zhou, J.: Continuous cloud-scale query optimization and processing. Proc. VLDB Endow. **6**(11), 961–972 (2013)
3. Helff, F., Gruenwald, L., d'Orazio, L.: Weighted sum model for multi-objective query optimization for mobile-cloud database environments. In: Proceedings of the Workshops of the EDBT/ICDT 2016 Joint Conference, vol. 1558 (2016)
4. Kabra, N., DeWitt, D.J.: Efficient mid-query re-optimization of sub-optimal query execution plans. In: SIGMOD 1998, pp.106–117 (1998)
5. Kipf, A., Kipf, T., Radke, B., Leis, V., Boncz, P.A., Kemper, A.: Learned cardinalities: estimating correlated joins with deep learning. In: 9th Biennial Conference on Innovative Data Systems Research, CIDR 2019 (2019)
6. Liu, H., Xu, M., Yu, Z., Corvinelli, V., Zuzarte, C.: Cardinality estimation using neural networks. In: CASCON 2015, pp. 53–59. IBM Corp. (2015)
7. Marcus, R., et al.: Neo: a learned query optimizer. Proc. VLDB Endow. **12**(11), 1705–1718 (2019)
8. Marcus, R., Papaemmanouil, O.: Deep reinforcement learning for join order enumeration. In: aiDM 2018 (2018)
9. Markl, V., Raman, V., Simmen, D., Lohman, G., Pirahesh, H., Cilimdzic, M.: Robust query processing through progressive optimization. In: Proceedings of the 2004 ACM SIGMOD International Conference on Management of Data, SIGMOD 2004, pp. 659–670 (2004)
10. Murphy, K.P.: Machine Learning: A Probabilistic Perspective. The MIT Press (2012)
11. Ohnishi, S., Uchibe, E., Yamaguchi, Y., Nakanishi, K., Yasui, Y., Ishii, S.: Constrained deep q-learning gradually approaching ordinary q-learning. Front. Neurorobot. **13**, 103 (2019)
12. Ortiz, J., Balazinska, M., Gehrke, J., Keerthi, S.S.: Learning state representations for query optimization with deep reinforcement learning. In: Proceedings of the 2nd Workshop on Data Management for End-To-End Machine Learning, DEEM 2018 (2018)
13. Park, Y., Tajik, A.S., Cafarella, M., Mozafari, B.: Database learning: toward a database that becomes smarter every time. In: Proceedings of the 2017 ACM International Conference on Management of Data, SIGMOD 2017, pp. 587–602 (2017)
14. PostgreSQL (2021). https://www.postgresql.org/
15. Stillger, M., Lohman, G.M., Markl, V., Kandil, M.: LEO, DB2's learning optimizer. In: Proceedings of the 27th International Conference on Very Large Data Bases, VLDB 2001, pp. 19–28 (2001)
16. Trummer, I., Moseley, S., Maram, D., Jo, S., Antonakakis, J.: SkinnerDB: regret-bounded query evaluation via reinforcement learning. Proc. VLDB Endow. **11**(12), 2074–2077 (2018)
17. Wang, C., Arani, Z., Gruenwald, L., d'Orazio, L.: Adaptive time, monetary cost aware query optimization on cloud database systems. In: IEEE International Conference on Big Data, Big Data 2018, pp. 3374–3382. IEEE (2018)

18. Wang, W., Zhang, M., Chen, G., Jagadish, H.V., Ooi, B.C., Tan, K.L.: Database meets deep learning: Challenges and opportunities. SIGMOD Rec. **45**(2), 17–22 (2016)

19. Wiering, M., van Otterlo, M.: Reinforcement Learning: State-of-the-Art. Springer, Heidelberg (2014). https://doi.org/10.1007/978-3-642-27645-3

20. Wu, C., et al.: Towards a learning optimizer for shared clouds. Proc. VLDB Endow. **12**(3), 210–222 (2018)

21. Wu, W., Naughton, J.F., Singh, H.: Sampling-based query re-optimization. In: SIGMOD 2016, pp. 1721–1736. Association for Computing Machinery (2016)

Top-K Formal Concepts for Identifying Positively and Negatively Correlated Biclusters

Amina Houari[1](✉)[iD] and Sadok Ben Yahia[2,3]

[1] University Mustapha STAMBOULI of Mascara, Mascara, Algeria
amina.houari@univ-mascara.dz
[2] Faculty of Siences of Tunis, LIPAH-LR11ES14, University of Tunis El Manar,
2092 Tunis, Tunisia
[3] Department of Software Science, Tallinn University of Technology, Tallinn, Estonia
sadok.ben@taltech.ee

Abstract. Formal Concept Analysis has been widely applied to identify differently expressed genes among microarray data. Top-K Formal
Concepts are identified as efficient in generating most important Formal
Concepts. To the best of our knowledge, no currently available algorithm
is able to perform this challenging task. Therefore, we introduce TOP-
BICMINER, a new method for mining biclusters from gene expression
data through Top-k Formal Concepts. It performs the extraction of the
sets of both positive and negative correlations biclusters. TOP-BICMINER
relies on Formal concept analysis as well as a specific discretization
method. Extensive experiments, carried out on real-life datasets, shed
light on TOP-BICMINER's ability to identify statistically and biologically
significant biclusters.

Keywords: Biclustering · Formal Concept Analysis · Top-K Formal
Concepts · Bioinformatics · DNA microarray data · Positive
correlation · Negative correlation

1 Introduction

Genes, proteins, metabolistes, etc. represent biological entities, a linked collection of these entities compose a biological network. Analysing biologically relevant knowledge and extracting efficient information from these entities is one
of important task in bioinformatics. In this context, DNA microarray technologies help to measure the expression levels of genes under experimental conditions
(simple). The discovery of transcriptional modules of genes that are co-regulated
in a set of experiments, is critical [6]. One way to do this is using clustering.
Biclustering, a particular clustering type, came to solve clustering's drawbacks
[6]. The main thrust of biclustering gene expression data stands in its ability to
identify groups of genes that behave in the same way under a subset of conditions. The majority of existing biclustering algorithms are based on greedy or

C. Attiogbé and S. Ben Yahia (Eds.): MEDI 2021, LNCS 12732, pp. 156–172, 2021.
https://doi.org/10.1007/978-3-030-78428-7_13

stochastic approaches, and provide, therefore, sub-higher-quality answers with restrictions of the structure, coherency as well as quality of biclusters [6,11]. Recent biclustering searches rely on pattern-mining [17].

In [15], we have presented a classification of biclustering algorithms where existing biclustering algorithms were grouped into three main streams : Systematic search-based biclustering, Stochastic search-based biclustering and Pattern-mining-based biclustering. The later includes:

1. Sequential-Pattern-Mining (SPM)-based approaches, used in order to extract order-preserving biclusters. Algorithms adopting this approach were given in [9] and [10].
2. Association Rules Mining (ARM)-based approaches: rely on finding all association rules that represent biclusters' samples/genes [12].
3. Formal Concept Analysis (FCA)-based approaches: a bicluster is considered as a formal concept that reflects the relation between objects and attributes [14,15,17].

Interestingly enough, most of the existing biclustering algorithms identify only positive correlation genes. Yet, recent biological studies have turned to a trend focusing on the notion of negative correlations. The authors in [25] studied in depth the negatively-correlated pattern. In a straightforward case, the expression values of some genes tend to be the complete opposite of the other ones, *i.e.* given two genes G1 and G2, under the same condition C, if both G1 and G2 are affected by C, while G1 goes up and G2 goes down, we can note that G1 and G2 highlight a negative correlation pattern. For example, the genes YLR367W and YKR057W of the Yeast microarray dataset have a similar disposition, but have a negative correlation pattern with the gene YML009C under 8 conditions [25]. Some suggest that these genes are a part of the protein translation and translocation processes and therefore should be grouped into the same cluster. Later on, several other algorithms were proposed [13,14,21].

In this work, we are particularly interested in the pattern-based biclustering algorithms especially those using Formal Concept Analysis (FCA)[8] to extract both positive and negative correlations biclusters.

FCA is a key method used for the analysis of object-attribute relationship and for knowledge representation [19]. FCA's mathematical settings have recently been shown to act as a powerful tool by providing a theoretical framework for the efficient resolution of many practical problems from data mining, software engineering and information retrieval [20].

In this respect, many approaches have been devoted to the extraction of biclusters using FCA. However, the latter's downside is their focus on one type of biclusters, extract overlapping ones, refrain from biological validation or only extract positive correlations biclusters.

In this paper, we introduce a new FCA-based algorithm for the biclustering of DNA microarray data, called TOP-BICMINER. This latter allows to observe the profile of each gene through all pairs of conditions by discretizing the original microarray data. Interestingly enough, TOP-BICMINER relies on a multi-criteria aggregation algorithm for getting out Top-K formal concepts in order to reduce the long list of formal concepts by selecting the most interesting Top-K ones. Undoubtedly, the entire set of formal concepts corpus is overwhelming and contains a large amount of information but only a small percentage of them contains useful information. Our proposed method heavily relies on the most well-known Technique for Order of Preference by Similarity to Ideal Solution (TOPSIS) [16].

By and large, the main contributions of this paper are as follows:

- We design an efficient FCA-based algorithm for extracting correlated genes called, TOP-BICMINER.
- We propose an algorithm, which is able to extract both biclusters of positive and negative correlations.
- We design the first algorithm, which relies on Top-K formal concepts.
- We prove the effectiveness of our method through extensive carried-out experiments on two real-life DNA microarray data. In fact, we extract statistically and biologically significant biclusters, showing competitive results in comparison with other popular biclustering algorithms.

The remainder of this paper is organized as follows. The next section sketches all definitions and notations that will be of use throughout this paper. Section 3 is dedicated to the description of the TOP-BICMINER algorithm. In Sect. 4, we provide the results of the application of our algorithm on real-life microarray datasets. Finally, Sect. 5 concludes this paper, and lists some issues for future work.

2   Key Notions

In this section, we present the main notions of need throughout the remainder of this paper;

Definition 1. (BICLUSTERING)
The biclustering problem focuses on the identification of the best biclusters of a given dataset. The best bicluster must fulfil a number of specific homogeneity and significance criteria (guaranteed through the use of a function to guide the search).

This leads us to the definition of a bicluster.

Definition 2. (BICLUSTER)
A bicluster is a subset of objects (genes) associated with a subset of attributes (conditions) in which these rows are co-expressed.
The bicluster associated with the matrix $M = (I, J)$ is a couple (A, B), such that $A \subseteq I$ and $B \subseteq J$, and (A, B) is maximal, i.e., if a bicluster (C, D) does not exist, with $A \subseteq C$ or $B \subseteq D$.

In the following, we recall some basic definitions borrowed from the FCA field. We start by presenting the notion of formal context.

Table 1. The quality assessment measures at a glance.

Name	Formula	Semantics
Coupling	$\left(|A_i \cap A_j| \times |B_i \cap B_j|\right) + \left(|A_i \cap A_j| + |B_i \cap B_j|\right)$	Assesses the reduction of the dependency between two formal concepts
Cohesion	$\left(|A_i| + |B_i|\right) + \left(|A_i| \times |B_i|\right)$	Assesses the increasing of the logical relationship between formal concepts
Stability	$\frac{|\{X \subseteq A \mid X' = B\}|}{2^{|A|}}$	This metric reflects the dependency of the intent on particular objects of the extent
Separation	$\frac{|A| \times |B|}{\left(\sum_{a \in A} |a'| + \sum_{b \in B} |b'|\right) - \left(|A| \times |B|\right)}$	Assesses the area covered by a formal concept
Distance	$\frac{1}{2}\left(\frac{|A \cup C| - |A \cap C|}{|\mathcal{O}|} + \frac{|B \cup D| - |B \cap D|}{|\mathcal{I}|}\right)$	Assesses the proximity of two formal concepts

Definition 3. (Formal context)
A formal context is a triplet $\mathcal{K} = (\mathcal{O}, \mathcal{I}, \mathcal{R})$, where $\mathcal{O}$ represents a finite set of objects, $\mathcal{I}$ is a finite set of items (or attributes) and $\mathcal{R}$ is a binary (incidence) relation, (i.e., $\mathcal{R} \subseteq \mathcal{O} \times \mathcal{I}$). Each couple $(o, i) \in \mathcal{R}$ expresses that the object $o \in \mathcal{O}$ contains the item $i \in \mathcal{I}$.

We use the prime symbol $'$ for the closure operators associated to a Galois connection like in A' and B'. The following definition introduces the formal concept.

Definition 4. (Formal concept)
A pair $\langle A, B \rangle \in \mathcal{O} \times \mathcal{I}$, of mutually corresponding subsets, i.e., $A = B'$ and $B = A'$, is called a formal concept, where A is called extent and B is called intent.

In the following, we present in Table 1, the most used Formal concept's assessment measures of the literature, namely coupling, cohesion, stability, separation and distance.

3 The Top-BicMiner Algorithm

The Top-BicMiner is a new algorithm for biclustering, aiming to extract positive and negative correlations biclusters from gene expression data. The Top-BicMiner algorithm proceeds by extracting the reduced set of formal concepts, which will be used to generate the Top-K ones using the multi-criteria decision analysis method (TOPSIS). The main originality of the Top-BicMiner algorithm is that it extracts positive and negative correlations biclusters. The quality of the unveiled formal concepts, is assessed through some metrics borrowed from the FCA community, to wit coupling, cohesion, stability, separation and distance in the sake of extracting the Top-K ones. The global architecture of the Top-BicMiner algorithm is sketched in Fig. 1. The different steps are thoroughly described in the following.

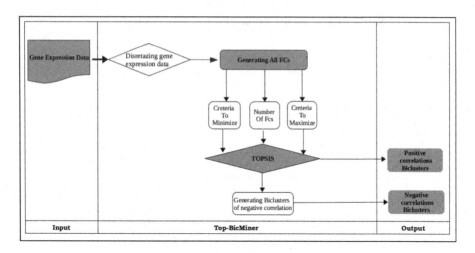

Fig. 1. TOP-BICMINER at a glance.

3.1   Step 1: Preprocessing of Gene Expression Data Matrix

Our method applies a preprocessing step to transform the original data matrix $\mathcal{M}1$ (gene expression data matrix) into two binary ones. First, we discretize the original data into a 3-state data matrix $\mathcal{M}2$ (behavior matrix). This latter is discretized into two binary data matrices: positive and negative. This step aims to highlight the trajectory patterns of genes. In microarray data analysis, we add genes into a bicluster whenever their trajectory patterns of expression levels are similar across a set of conditions.

Interestingly enough, our proposed discretization method enables monitoring the shape of profiles across genes and conditions. Furthermore, it maintains the same information of the expression levels' trajectory patterns.

In our case, the purpose of using the discretized matrix is to identify biclusters with positive and negative correlation genes. The whole data matrix $\mathcal{M}2$ provides useful information about the identification of biclusters.

Table 2. Example of gene expression matrix.

	$c1$	$c2$	$c3$	$c4$	$c5$
$g1$	4	5	3	6	1
$g2$	8	10	6	12	2
$g3$	3	3	3	3	3
$g4$	7	1	9	0	8
$g5$	14	2	18	0	16

This first step is split into two sub-steps:

1. First, we discretize the initial data matrix to generate the the 3-state data matrix . It consists in combining in pairs, for each gene, all condition's pairs. In the 3-state data matrix, each column represents the meaning of the variation of genes between a pair of conditions of $\mathcal{M}1$. It offers useful information for the identification of biclusters, *i.e.* up (1), down (−1) and no change (0). Indeed, $\mathcal{M}2$ represents the trajectory of genes between a pair of conditions in the data matrix $\mathcal{M}1$. $\mathcal{M}2$ is defined as follows:

$$\mathcal{M}2[i,j] = \begin{cases} 1 & \text{if} \quad \mathcal{M}1[i,l1] < \mathcal{M}1[i,l2] \\ -1 & \text{if} \quad \mathcal{M}1[i,l1] > \mathcal{M}1[i,l2] \\ 0 & \text{if} \quad \mathcal{M}1[i,l1] = \mathcal{M}1[i,l2] \end{cases} \tag{1}$$

with: $i \in [1\ldots n]$; $l1 \in [1\ldots m-1]$ and $l2 \in [2\ldots m]$; and $l2 > l1 + 1$;

2. For the second sub-step of the preprocessing phase, we build two binary data matrices in order to extract formal concepts from $\mathcal{M}3^+$ and $\mathcal{M}3^-$. This discretization is used in order to extract the genes that show an opposite change tendency over a subset of experimental conditions.

Formally, we define the binary matrices as follows:

$$\mathcal{M}3^+[i,j] = \begin{cases} 1 & \text{if} \quad \mathcal{M}2[i,j] = 1 \\ 0 & \text{otherwise} \end{cases} \tag{2}$$

$$\mathcal{M}3^-[i,j] = \begin{cases} 1 & \text{if} \quad \mathcal{M}2[i,j] = -1 \\ 0 & \text{otherwise} \end{cases} \tag{3}$$

Considering the data matrix given by Table 2, the preprocessing step is applied as follows. By applying Eq. 1, we represent the 3-state data matrix (as provided by Table 3).

Let $\mathcal{M}2$ be a 3-state data matrix (Table 3). By using Eq. 2 and Eq. 3, we obtain the binary matrices (Tables 4 and 5).

Table 3. 3-state data matrix ($\mathcal{M}2$).

	c1c2 (C1)	c1c3 (C2)	c1c4 (C3)	c1c5 (C4)	c2c3 (C5)	c2c4 (C6)	c2c5 (C7)	c3c4 (C8)	c3c5 (C9)	c4c5 (C10)
g1	1	−1	1	−1	−1	1	−1	1	−1	−1
g2	1	−1	1	−1	−1	1	−1	1	−1	−1
g3	0	0	0	0	0	0	0	0	0	0
g4	−1	1	−1	1	1	−1	1	−1	−1	1
g5	−1	1	−1	1	1	−1	1	−1	−1	1

Table 4. Positive binary data matrix ($\mathcal{M}3^{+}$).

	C1	C2	C3	C4	C5	C6	C7	C8	C9	C10
$g1$	1	0	1	0	0	1	0	1	0	0
$g2$	1	0	1	0	0	1	0	1	0	0
$g3$	0	0	0	0	0	0	0	0	0	0
$g4$	0	1	0	1	1	0	1	0	0	1
$g5$	0	1	0	1	1	0	1	0	0	1

Table 5. Negative binary data matrix ($\mathcal{M}3^{-}$).

	C1	C2	C3	C4	C5	C6	C7	C8	C9	C10
$g1$	0	1	0	1	1	0	1	0	1	1
$g2$	0	1	0	1	1	0	1	0	1	1
$g3$	0	0	0	0	0	0	0	0	0	0
$g4$	1	0	1	0	0	1	0	1	1	0
$g5$	1	0	1	0	0	1	0	1	1	0

3.2 Step 2: Extracting Formal Concepts

After the preprocessing step, we move to the step of formal-concept extraction[1] respectively from the matrices $\mathcal{M}3^{+}$ and $\mathcal{M}3^{-}$ (Tables 4 and 5).

3.3 Step 3: Extracting Top-K Formal Concepts

Due to the obtained number of extracted formal concepts (candidate biclusters), the main aim of our work is to tackle one of the most challenging issues within biclustering: reduce the number of obtained biclusters in order to validate them. In our previous work [14], we used only the stability measure [18] in order to remove the non-coherent formal concepts. In this work, we use a multi-criteria to be aggregated, namely, coupling, cohesion, stability, separation and distance. We have to maximize the following criteria: stability, cohesion and separation. In addition, the criteria to minimize are coupling and distance. In it's gene expression data application, (i) the stability is used to obtain biclusters with best dependency between genes and conditions, (ii) the cohesion, coupling, distance and separation are used to minimise the overlap between the obtained biclusters. Indeed, biclusters with high overlap have similar biological significance, i.e. the same biological function.

It is of paramount importance to design the Top-K extraction method, belonging to the relevance-oriented type, that usually extracts the first k formal concepts sequentially according to a ranking based on a aggregation of above

---

[1] The extraction of the formal concepts is carried out through the invocation of the efficient LCM algorithm [23].

mentioned assessment measures. These metrics are then aggregated through the use of a multi-criteria method to select the most pertinent formal concepts from candidates. In this step, we only consider the best obtained formal concepts ranked by the TOPSIS method [16]. The latter was chosen due to its simplicity, its ability to consider a non-limited number of alternatives and its ability to quickly rank the alternatives. In addition, the TOPSIS is an attractive method since it can incorporate relative weights for the criteria, which will allow, in our case, to assign different importance to the quality measures of formal concepts. In terms of performance, the TOPSIS method has been comparatively tested versus a number of other multi-attribute methods [24]. The main steps of TOPSIS are:

- **Step 1**: Form a normalized decision-matrix.
- **Step 2**: Construct the weighted normalized decision-matrix.
- **Step 3**: Determine the ideal solution and negative ideal solution.
- **Step 4**: Compute the distance of each alternative.
- **Step 5**: Compute the closeness coefficient.
- **Step 6**: Rank the alternatives according to their closeness coefficient.

3.4 Step 4: Extracting Correlated Genes

In this step, we present the problem of extracting biclusters of positive and negative correlations. We have the possibility to choose whether considering the biclusters of positive correlation that can be extracted or only extracting biclusters of negative correlations or both.

In the case of negatively correlated biclusters, these later represent the set of obtained formal concepts in the previous step, *i.e.* the Top-K Formal Concepts as a set of biclusters.

However, in the aim of extracting negatively correlated biclusters, we consider the Top-K formal concepts extracted during Step 3 to get negatively correlated genes for a given $\alpha 1^2$ only where we compute the proportion of similarity between coherent formal concepts from $\mathcal{M}3^+$ and $\mathcal{M}3^-$ in terms of conditions. Roughly speaking, ranked formal concepts having an intersection size greater than or equal to the threshold belong to the same bicluster. For genes, we consider the union of the two genes' sets.

3.5 Step 5: Extracting Maximal Negatively Correlated Genes

A bicluster is maximal whenever an object or an attribute cannot be added without violating the negative correlation criterion. We take into consideration biclusters having an intersection size in terms of genes greater than or equal to a given intersection threshold $\alpha 2$. Plainly speaking, we compute the intersection

---

[2] In fact, coherent formal concepts having an intersection size above or equal to the given threshold $\alpha 1$ belong to the same bicluster, while those with an intersection value below it, do not.

between two biclusters: If their similarity value is above or equal to $\alpha2$, we then generate a maximal bicluster.

4   Experimental Results

In this section, we present the results of applying our algorithm on two well-known real-life datasets. The evaluation of biclustering algorithms and the comparison between them are based upon two criteria: statistical and biological. We compare our algorithm versus the state-of-the-art biclustering algorithms, the MBA algorithm [3] which extracts biclusters with negative correlations, the Trimax algorithm[3] [17] and the NBF [14] which extract negatively correlated biclusters using FCA.

4.1   Description of Considered Datasets

During the carried out experiments, we used the following real-life gene expression datasets:

- *Yeast Cell-Cycle dataset:* The Yeast Cell-Cycle[4] dataset is very popular in the gene expression analysis community. It comprises 2884 genes and 17 conditions.
- *Human B-Cell Lymphoma dataset:* The Human B-Cell Lymphoma dataset contains 4026 genes and 96 conditions[5].

4.2   Study of Statistical Relevance

We use the coverage criterion as well as the *p-value* criterion, to evaluate the statistical relevance of our algorithm.

Coverage Criterion. We use the coverage criterion which is defined as the total number of cells in a microarray data matrix covered by the obtained biclusters. Validation using coverage is considered as worth of interest in the biclustering domain, since large coverage of a dataset is very important in several applications that rely on biclusters. In fact, the higher the number of highlighted correlations is, the greater the amount of extracted information. Consequently, the higher the coverage is, the lower the overlapping in the biclusters.

We compare the results of our algorithm versus those of Trimax [17], CC [7], BiMine [1], BicFinder [2] and NBF [14].

Table 6 (resp. Table 7) presents the coverage of the obtained biclusters. We can show that most of the algorithms have relatively close results. For the Human B-Cell Lymphoma (respectively Yeast Cell-Cycle) dataset, the biclusters extracted by our algorithm cover 100% (respectively 96.22%) of the genes,

---

[3] Available at https://github.com/mehdi-kaytoue/trimax.
[4] Available at http://arep.med.harvard.edu/biclustering/.
[5] Available at http://arep.med.harvard.edu/biclustering/.

100% of the conditions and 75.02% (respectively 79.08%) of the cells in the initial matrix. However, Trimax has a low performance as it covers only 8.50% of cells, 46.32% of genes and 11.46% of conditions. This implies that our algorithm can generate biclusters with high coverage.

It is also worth mentioning that for Yeast Cell-Cycle, the CC algorithm obtains the best results since it masks groups extracted with random values. Thus, it prohibits the genes/conditions that were previously discovered from being selected during the next search process. This type of mask leads to high coverage. Furthermore, The Yeast dataset only contains positive integer values. Thus, one can use the Mean Squared Residue (MSR) [7] to extract large biclusters. By contrast, the Human B-cell Lymphoma dataset contains integer values including negative ones. This means that the application of MSR on this dataset does not lead the extraction of large biclusters. This implies that our algorithm can generate biclusters with high coverage of a data matrix. This outstanding coverage is caused by the discretization phase as well as the extraction of biclusters without focusing on a specific type of biclusters.

Table 6. Human B-Cell Lymphoma coverage for different algorithms.

Algorithms	Total coverage	Gene coverage	Condition coverage
BiMine	8.93%	26.15%	100%
BicFinder	44.24%	55.89%	100%
CC	36.81%	91.58%	100%
Trimax	8.50%	46.32%	11.46%
NBF	73.75%	100%	100%
Top-BicMiner	**75.02%**	**100%**	**100%**

Table 7. Yeast Cell-Cycle coverage for different algorithms.

Algorithms	Total coverage	Gene coverage	Condition coverage
BiMine	13.36%	32.84%	100%
BicFinder	55.43%	76.93%	100%
CC	**81.47%**	**97.12%**	100%
Trimax	15.32%	22.09%	70.59%
NBF	77.17%	97.08%	100%
Top-BicMiner	79.08%	96.22%	100%

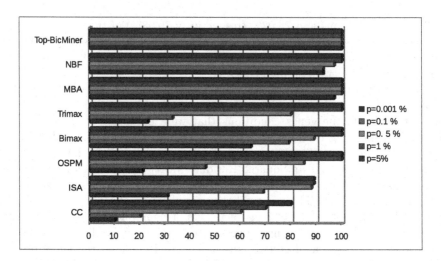

Fig. 2. Proportions of biclusters significantly enriched by GO annotations on the Yeast Cell-Cycle dataset.

P-Value Criterion. We rely on the P-value criterion to assess the quality of the extracted biclusters. To do so, we use the ***FuncAssociate***[6] web tool in order to compute the adjusted significance scores. The latter assess genes in each bicluster, indicating to which extent they match well with the different GO categories for each bicluster (adjusted *p-value*)[7]. The results of our algorithm are compared versus those of CC [7], ISA [5], OSPM [4] and Bimax [22]. We also compare our algorithm versus Trimax [17], MBA [3] and NBF[14].

The obtained results of the Yeast Cell Cycle dataset for the different adjusted *p-values* (p = 5%; 1%; 0.5%; 0.1%; 0.001%) for each algorithm over the percentage of total biclusters are depicted in Fig. 2. The TOP-BICMINER results show that 100% of the extracted biclusters are statistically significant with the adjusted *p-value* $p < 0.001\%$. While MBA achieve its best results when $p < 0.1$ and 97% with with the adjusted *p-value* $p < 0.001\%$. By contrast, Trimax achieves 23% of statistically significant biclusters when $p < 0.001\%$.

4.3   Biological Relevance

We use the biological criterion which allows measuring the quality of obtained biclusters, by checking whether the genes of a bicluster have common biological characteristics.

To evaluate the quality of the extracted biclusters and identify their biological annotations, we use ***GoTermFinder***[8] (Gene Ontology Term Finder). It searches for significant shared GO terms, used to describe the genes in a given list

---

[6] Available at http://llama.mshri.on.ca/funcassociate/.

[7] The best biclusters have an adjusted *p-value* less than 0.001%.

[8] Available at https://www.yeastgenome.org/goTermFinder.

Table 8. Significant GO terms (process, function, component) for two biclusters, extracted from Yeast Cell-Cycle dataset using TOP-BICMINER.

	Bicluster 1	Bicluster 2
Biological process	Cytoplasmic translation (53.1%, 7.80e–44)	Amide biosynthetic process (59.7%, 5.03e–19)
	Maturation of SSU-rRNA (32.1%, 6.30e–25)	Cleavage involved in rRNA processing (19.4%,1.14e–10)
	Gene expression (96.3%, 5.20e–35)	rRNA 5'-end processing (13.4%, 1.12e–08)
Molecular function	RNA binding (72.8%, 7.61e–30)	Structural constituent of ribosome (53.7%, 7.84e–35)
	Heterocyclic compound binding (74.1%, 1.16e–12)	Binding (77.6%, 5.81e–05)
	RNA-dependent ATPase activity(6.2%, 7.39e–06)	Organic cyclic compound binding (73.1%, 7.66e–10)
Cellular component	Intracellular ribonucleoprotein complex (97.5%, 3.11e–74)	Preribosome (47.8%, 2.22e–33)
	90S preribosome (29.6%, 7.94e–26)	Large ribosomal subunit (38.8%, 7.66e–20)
	Nucleolus (37.0%, 6.15e–17)	Cytosol (58.2%, 7.37e–12)

to help discovering what the genes may have in common. In fact, the biological criterion enables measuring the quality of the resulting biclusters, by checking whether the genes of a bicluster have common biological characteristics.

Figure 3 sketches the result obtained by *GOTermFinder* on a randomly selected bicluster. In fact, in GO nodes in the graph are color-coded according to their p-value, *i.e.* genes in the GO tree are associated with the GO term(s) to which they are directly annotated. Only significant hits with a p-value <= 0.01 and terms descended from them are included. As depicted in Fig. 3, the annotated genes **YBR297W, YBR299W, YBR298C** of our selected bicluster concerns the Gene Ontology term *"maltose metabolic process"* in terms of *biological process*. In fact, The GO is organized according to 3 axes: *biological process, molecular function* and *cellular component*[9].

We indicate in Table 8 the biological annotations of two randomly selected biclusters in terms of the above cited axis, where we report the most significant GO terms. For instance, with the first bicluster extracted from the Yeast Cell-Cycle dataset, the genes extracted by our algorithm concern the *Gene Ontology* term *"cytoplasmic translation"*, in terms of *Biological Process* with a *p-value* equal to $7.80e^{-44}$ (highly significant) and a background of 53.1%. The genes which represent the above mentioned background and annotated to the term *"cytoplasmic translation"* are: *YGL076C, YDR447C, YJL136C, YER117W, YDR012W, YDR064W, YJL190C, YLR388W, YDR450W, YGL147C, YDR382W, YDR025W, YKL156W, YLR185W, YLR048W, YHL001W, YGR162W, YKR059W, YLR325C, YIL052C, YHL033C, YDR418W, YGR214W, YHR064C,*

---

[9] http://geneontology.org/.

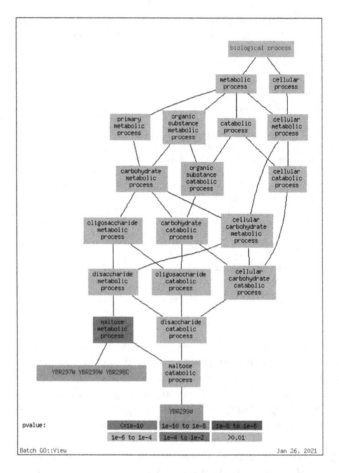

Fig. 3. Significant GO terms for a bicluster on Yeast Cell-Cycle dataset extracted by TOP-BICMINER.

YEL054C,   YFR031C-A,   YKL006W,   YER074W,   YJL138C, YDR471W,   YJR094W-A,   YLR340W,   YER025W, YHR021C, YLR367W, YDR500C, YLL045C, YLR061W, YJR145C, YGL030W, YLR344W, YLR264W, YHR203C.

Figure 4 (resp.5) illustrates the profile of a randomly selected bicluster obtained by our algorithm for the Yeast Cell-Cycle (resp. Human B-Cell Lymphoma) dataset. From these figures, negative patterns can be visually observed. In fact, different types of relationships between genes in one bicluster are plotted. Some genes are positively correlated (show similar patterns). However, these genes are negatively correlated (show opposite patterns) with other genes (Figs. 4 and 5).

The results on these real-life datasets indicate that our proposed algorithm can identify biclusters of positive and negative correlations with a high biological relevance.

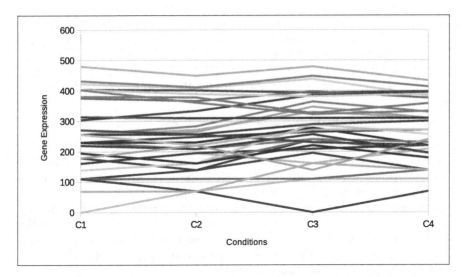

Fig. 4. The resulting bicluster profile obtained through the TOP-BICMINER algorithm on Yeast Cell-Cycle.

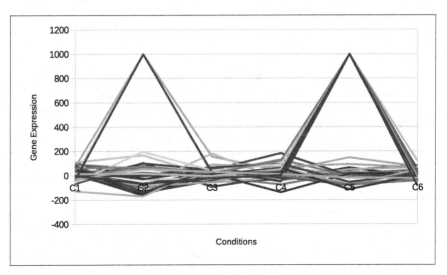

Fig. 5. The resulting bicluster profile obtained through the TOP-BICMINER algorithm on Human B-cell Lymphoma.

5   Conclusion

In this paper, we have introduced the TOP-BICMINER biclustering algorithm to discover both positively and negatively correlated genes from the gene expression data. TOP-BICMINER is an FCA-based biclustering algorithm for the extraction of biclusters through Top-k formal concepts and the use of the TOPSIS multi-criteria based method.

Our experiments showed that our algorithm outperforms its competitors in most of the cases for the particular datasets, , it permits extracting high-quality positively and negatively correlated biclusters. These biclusters have been evaluated with the GO annotations that check the biological significance of biclusters. The obtained results confirm the ability of TOP-BICMINER to extract significant biclusters.

As a future work, we plan to tackle the following issues :

1. **Scalability issues :** Future work will focus on the issue of extracting biclusters from big datasets, since big data mining is a new challenging task since computational requirements are difficult to provide. An interesting solution is to exploit distributed frameworks such as Spark.
2. **Pertinent coverage :** It is worth of interest that a reduced set of formal concepts can be extracted from a given formal context [20]. In fact, an interesting tackle of this issue is to find coverage of formal context by a minimal number of obtained formal concept with respect to the quality metrics (coupling, cohesion, stability, separation and distance).
3. **Biological knowledge :** Other avenues of future work also concern the discovery of biclusters by introducing biological knowledge during the extraction process.

References

1. Ayadi, W., Elloumi, M., Hao, J.K.: A biclustering algorithm based on a bicluster enumeration tree: application to DNA microarray data. BioData Mining **2**, 9 (2009)
2. Ayadi, W., Elloumi, M., Hao, J.K.: Bicfinder: a biclustering algorithm for microarray data analysis. Knowl. Inf. Syst. **30**(2), 341–358 (2012)
3. Ayadi, W., Hao, J.: A memetic algorithm for discovering negative correlation biclusters of DNA microarray data. Neurocomputing **145**, 14–22 (2014). https://doi.org/10.1016/j.neucom.2014.05.074
4. Ben-Dor, A., Chor, B., Karp, R.M., Yakhini, Z.: Discovering local structure in gene expression data: the order-preserving submatrix problem. J. Comput. Biol. **10**(3/4), 373–384 (2003)
5. Bergmann, S., Ihmels, J., Barkai, N.: Defining transcription modules using large-scale gene expression data. Bioinformatics **20**(13), 1993–2003 (2004)
6. Madeira, S.C., Oliveira, A.L.: Biclustering algorithms for biological data analysis: a survey. IEEE Trans. Comput. Biol. Bioinf. **1**, 24–45 (2004)
7. Cheng, Y., Church, G.M.: Biclustering of expression data. In: Proceedings of ISMB, UC San Diego, California, pp. 93–103 (2000)
8. Ganter, B., Wille, R.: Formal Concept Analysis - Mathematical Foundations. Springer, Heidelberg (1999). https://doi.org/10.1007/978-3-642-59830-2
9. Henriques, R., Antunes, C., Madeira, S.C.: Methods for the efficient discovery of large item-indexable sequential patterns. In: New Frontiers in Mining Complex Patterns - Second International Workshop, NFMCP 2013, Held in Conjunction with ECML-PKDD 2013, Prague, Czech Republic, 27 September 2013, Revised Selected Papers, pp. 100–116 (2013). https://doi.org/10.1007/978-3-319-08407-7_7
10. Henriques, R., Madeira, S.C.: Bicspam: flexible biclustering using sequential patterns. BMC Bioinf. **15**, 130 (2014). https://doi.org/10.1186/1471-2105-15-130

11. Henriques, R., Madeira, S.C.: Bic2pam: constraint-guided biclustering for biological data analysis with domain knowledge. Algorithms Molec. Biol. **11**, 23 (2016). https://doi.org/10.1186/s13015-016-0085-5

12. Houari, A., Ayadi, W., Yahia, S.B.: Discovering low overlapping biclusters in gene expression data through generic association rules. In: Bellatreche, L., Manolopoulos, Y. (eds.) MEDI 2015. LNCS, vol. 9344, pp. 139–153. Springer, Cham (2015). https://doi.org/10.1007/978-3-319-23781-7_12

13. Houari, A., Ayadi, W., Yahia, S.B.: Mining negative correlation biclusters from gene expression data using generic association rules. In: Zanni-Merk, C., Frydman, C.S., Toro, C., Hicks, Y., Howlett, R.J., Jain, L.C. (eds.) Knowledge-Based and Intelligent Information & Engineering Systems: Proceedings of the 21st International Conference KES-2017, Marseille, France, 6–8 September 2017, Procedia Computer Science, vol. 112, pp. 278–287. Elsevier (2017). https://doi.org/10.1016/j.procs.2017.08.262

14. Houari, A., Ayadi, W., Yahia, S.B.: NBF: an fca-based algorithm to identify negative correlation biclusters of DNA microarray data. In: Barolli, L., Takizawa, M., Enokido, T., Ogiela, M.R., Ogiela, L., Javaid, N. (eds.) 32nd IEEE International Conference on Advanced Information Networking and Applications, AINA 2018, Krakow, Poland, 16–18 May 2018, pp. 1003–1010. IEEE Computer Society (2018). https://doi.org/10.1109/AINA.2018.00146

15. Houari, A., Ayadi, W., Yahia, S.B.: A new fca-based method for identifying biclusters in gene expression data. Int. J. Mach. Learn. Cybern. **9**(11), 1879–1893 (2018). https://doi.org/10.1007/s13042-018-0794-9

16. Hwang, C.L., Yoon, K.: Methods for multiple attribute decision making. In: Multiple Attribute Decision Making, pp. 58–191. Springer, Heidelberg (1981). https://doi.org/10.1007/978-3-642-48318-9_3

17. Kaytoue, M., Kuznetsov, S.O., Macko, J., Napoli, A.: Biclustering meets triadic concept analysis. Ann. Math. Artif. Intell. **70**(1–2), 55–79 (2014). https://doi.org/10.1007/s10472-013-9379-1

18. Kuznetsov, S.O.: Stability as an estimate of degree of substantiation of hypotheses derived on the basis of operational similarity. Nauchno-Tekhnichekaya Informatisiya Seriya 2-Informatsionnye Protsessy I Sistemy (12), 21–29 (1990)

19. Li, X., Shao, M.-W., Zhao, X.-M.: Constructing lattice based on irreducible concepts. Int. J. Mach. Learn. Cybern. **8**(1), 109–122 (2016). https://doi.org/10.1007/s13042-016-0587-y

20. Mouakher, A., Ben Yahia, S.: Qualitycover: efficient binary relation coverage guided by induced knowledge quality. Inf. Sci. **355**, 58–73 (2016)

21. Nepomuceno, J.A., Troncoso, A., Aguilar-Ruiz, J.S.: Scatter search-based identification of local patterns with positive and negative correlations in gene expression data. Appl. Soft Comput. **35**, 637–651 (2015). https://doi.org/10.1016/j.asoc.2015.06.019

22. Prelic, A., et al.: A systematic comparison and evaluation of biclustering methods for gene expression data. Bioinformatics **22**(9), 1122–1129 (2006)

23. Uno, T., Asai, T., Uchida, Y., Arimura, H.: An efficient algorithm for enumerating closed patterns in transaction databases. In: Discovery Science, 7th International Conference, DS 2004, Padova, Italy, 2–5 October 2004, Proceedings, pp. 16–31 (2004). https://doi.org/10.1007/978-3-540-30214-8_2

24. Zanakis, S.H., Solomon, A., Wishart, N., Dublish, S.: Multi-attribute decision making: a simulation comparison of select methods. Eur. J. Oper. Res. **107**(3), 507–529 (1998)
25. Zhao, Y., Yu, J., Wang, G., Chen, L., Wang, B., Yu, G.: Maximal subspace coregulated gene clustering. IEEE Trans. Knowl. Data Eng. **20**(1), 83–98 (2008). https://doi.org/10.1109/TKDE.2007.190670

Data Management - Blockchains

Anonymization Methods of Structured Health Care Data: A Literature Review

Olga Vovk$^{(\boxtimes)}$ ⓘ, Gunnar Piho ⓘ, and Peeter Ross ⓘ

Tallinn University of Technology, Tallinn, Estonia
{olga.vovk,gunnar.piho,peeter.ross}@taltech.ee

Abstract. Anonymization of health data is one of the most important technics of sharing data for secondary purposes such as research and statistics while preserving people's privacy. This article is a systematic literature review of current methods that are used for the anonymization of health data. This research includes the most recent methods, from the articles published in 2017–2020. In total 1349 records were found and, according to the selection criteria, 21 articles were chosen for the final review. This full paper aims to provide a literature review on existing methods of anonymization structured health care data and identify existing issues with those methods.

Keywords: Anonymization · Health data · Structured data · Methods

1 Introduction

Modern technologies allow us to measure and collect a wide scope of data from the human body from blood pressure and to genomic data. Healthcare-related information can be presented in different forms, in Electronic Health Record (EHR) as diagnosis with code, medical images (e.g. X-ray and MRI images), or doctor's notes in free text. This data is stored not only within hospitals, labs, and other medical facilities but in personal devices that can monitor the physical or mental condition.

Digitalization brings us new opportunities to analyze data and make data-driven decisions. Alongside opportunities, new challenges come in place. One of the most important is to ensure privacy for personal data that is used for secondary purposes. We can achieve better outcomes by secondary use of health data. Nevertheless, it's important to follow legal requirements to ensure privacy protection for patients, and for this proper data anonymization tools and methods must be used. The application of anonymization methods allows using health data for secondary purposes without violation of a person's privacy. The additional challenge appears due to data heterogeneity and a combination of structured, semi-structured, and unstructured data.

This work has been conducted in the project "ICT programme" which was supported by the European Union through the European Social Fund.

2   Related Work

This research is focused on reviewing algorithms to anonymize structured health care data. The review of the algorithms has already been done by researchers before. The article [1] by Gkoulalas-Divanis A.et al. contains a survey of existing algorithms, however, this article was published in 2014 and it is expected that new algorithms were created since that time. The other article, [2] by Jayabalan M. et al. published in 2018, gives a more general overview of the data anonymization techniques. One more article that reviews privacy and security topic is [3] by Abouelmehdi, K., et al. This article includes data anonymization techniques, however, it mainly targeted other privacy and security questions as authentication, encryption, and legal regulation for data protection in different countries. Some other recent articles as [4] by Salas J. et al. and [5] by Ouafae et al. describe techniques of data anonymization, but, they mainly focus on the social network field, which may include certain information about people's health but significantly different from medical records. Moreover, the article [6] by Pawar A. et al. presents various anonymization techniques, with a focus on the comparison of existing algorithms. However, this article does not include the process of literature collection, which differs from the current systematic literature review on data anonymization.

Current research, performed in October-November 2020, aims to provide a systematic literature review of the most recent literature (2017–2020 years) about the existing anonymization techniques in health care.

3   Methodology

In this paper, we followed the guidelines for a systematic literature review provided by Kitchenham and Charters [7]. Research is performed in three stages: planning the review, conducting the review, and reporting the review.

A. Planning the review. The current literature review aims to answer the following research questions:

> *RQ1. What are the methods of data anonymization?*
> *RQ2. What are the challenges and issues in using those methods?*

B. Conducting the review. To find relevant literature following databases were used: PubMed, IEEE Explore Digital Library, Science Direct, and Springer-Link. The search criteria were that all of the keywords "anonymization" or "de-identification", "health", and "data" had to appear in the title, abstract, or keywords of the articles. The search was performed within articles published between 2017 and 2020 years. Only papers written in English were included in the research. The search process is shown in Fig. 1. Further article selection we performed using a four-step process. At the first step, we selected articles based on the title. We excluded duplicated articles and articles written in languages other than English. In the second step, we excluded articles that do not address current research interest based on a general overview, e.g. articles from another field. Too short and low-quality papers were excluded. In the third step, the article's title, abstract, keywords, and conclusion were examined. Articles that are not relevant for current research, based on title, abstract, keywords were not included further. In the final step, the full text of the articles were examined. Papers that did not address the research question were excluded. Also, articles written by the same research group on the same dataset were excluded.

C. Reporting the review. The findings of this literature review are presented in the Result section.

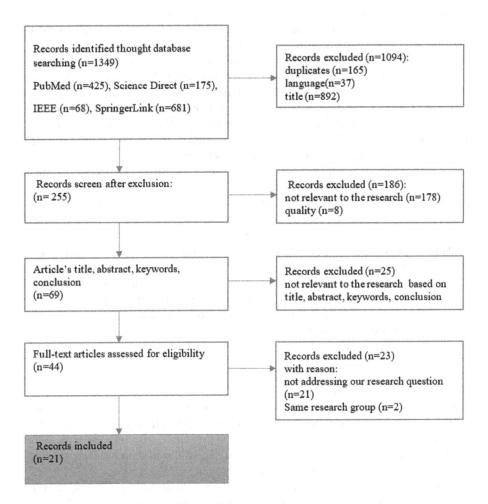

Fig. 1. Selection process.

4 Results

In this section, we summarize the research findings related to the questions stated in the Methodology section. A total of 21 articles were selected in the review. All of those 21 articles answer to the RQ1 and 19 articles answer to the RQ2. Selected papers are presented in Table 1.

4.1 RQ1

All of the 21 articles, selected in the review, address RQ1. Table 2 shows the appearance of the anonymization methods within the selected articles. This table presents all methods mentioned in the article including the main methods to which the article is dedicated and other methods that are used for comparison. All of the selected articles mention k-anonymity, which is the most well-known and commonly used method in data anonymization.

k-anonymity is used as a base to develop other methods such as (p, α)-sensitive k-anonymity, θ-sensitive k-anonymity, (k, k^m) -anonymity or to compare with different methods such as cryptographic algorithms. l-diversity and t-closeness, extensions of k-anonymity, are also well known and widely used methods, mentioned in 15 and 13 articles respectively. Also, cryptographic algorithms, a novel solution in the data anonymization field appears in 4 articles. Methods in differential privacy, ϵ-differential privacy, in particular, appear 3 times. Pseudonymization (tokenization/hashing) also appear in articles 3 times. Other methods are mentioned in 1–2 articles.

Table 2 shows the list of anonymization methods mentioned in publications selected for the review as well as articles that mention those methods. Overall, 33 methods were mentioned in selected articles and 12 out of those methods are presented as the main methods in articles. The last column shows the list of articles where corresponding methods are introduced as main. Articles in Table 2 are referenced according to the list of articles presented in Table 1.

4.2 RQ2

The challenges and issues in the anonymization of health care data mentioned in the analyzed articles as well as methods that have those problems are summarized in Table 3. As we can conclude from this table, all methods have some common issues. Those issues are privacy risk, the utility of data, and anonymization being not only technical problem.

The main challenge in data anonymization, mentioned in the majority of articles is the right balance between privacy preservation and data utility. Minimizing privacy risks while maintaining data quality, which includes decreasing data loss and preserving data trustfulness are the most important issues to overcome when applying certain methods. The risk of re-identification should be taking into account when choosing an algorithm. Privacy preservation includes several concerns as described below in the Discussion section [9,12,23].

Table 1. Selected articles.

Num.	Article	Year
1	Prasser, Fabian, Johanna Eicher, Raffael Bild, Helmut Spengler, and Klaus A. Kuhn. "A tool for optimizing de-identified health data for use in statistical classification" [8]	2017
2	Khan, Razaullah, Xiaofeng Tao, Adeel Anjum, Tehsin Kanwal, Abid Khan, and Carsten Maple. "θ-Sensitive k-Anonymity: An Anonymization Model for IoT based Electronic Health Records" [9]	2020
3	Li, Xiao-Bai, and Jialun Qin. "Anonymizing and sharing medical text records" [10]	2017
4	Rajendran, Keerthana, Manoj Jayabalan, and Muhammad Ehsan Rana. "A Study on k-anonymity, l-diversity, and t-closeness Techniques" [11]	2017
5	Aminifar, Amin, Yngve Lamo, Ka I. Pun, and Fazle Rabbi. "A Practical Methodology for Anonymization of Structured Health Data" [12]	2019
6	Pawar, Ambika, Swati Ahirrao, and Prathamesh P. Churi. "Anonymization techniques for protecting privacy: A survey" [6]	2018
7	Jayabalan, Manoj, and Muhammad Ehsan Rana. "Anonymizing healthcare records: A study of privacy preserving data publishing techniques" [2]	2018
8	Anjum, Adeel, Kim-Kwang Raymond Choo, Abid Khan, Asma Haroon, Sangeen Khan, Samee U. Khan, Naveed Ahmad, and Basit Raza. "An efficient privacy mechanism for electronic health records" [13]	2018
9	Majeed, Abdul. "Attribute-centric anonymization scheme for improving user privacy and utility of publishing e-health data" [14]	2019
10	Poulis, Giorgos, Grigorios Loukides, Spiros Skiadopoulos, and Aris Gkoulalas-Divanis. "Anonymizing datasets with demographics and diagnosis codes in the presence of utility constraints" [15]	2017
11	Ribeiro, Sérgio, and Emilio Nakamura. "Privacy Protection with Pseudonymization and Anonymization In a Health IoT System: Results from OCARIoT" [16]	2019
12	Prasser, Fabian, Florian Kohlmayer, Helmut Spengler, and Klaus A. Kuhn. "A scalable and pragmatic method for the safe sharing of high-quality health data" [17]	2017
13	Hsiao, Mei-Hui, Wen-Yang Lin, Kuang-Yung Hsu, and Zih-Xun Shen. "On Anonymizing Medical Microdata with Large-Scale Missing Values-A Case Study with the FAERS Dataset" [18]	2019
14	Lu, Yang, Richard O. Sinnott, Kain Verspoor, and Udaya Parampalli. "Privacy- preserving access control in electronic health record linkage" [19]	2018
15	Saeed, Rashad, and Azhar Rauf. "Anatomization through generalization (AG): A hybrid privacy-preserving approach to prevent membership, identity and semantic similarity disclosure attacks" [20]	2018
16	Lee, Hyukki, Soohyung Kim, Jong Wook Kim, and Yon Dohn Chung. "Utility- preserving anonymization for health data publishing" [21]	2017
17	Lee, Hyukki, and Yon Dohn Chung. "Differentially private release of medical microdata: an efficient and practical approach for preserving informative attribute values" [22]	2020
18	Kanwal, Tehsin, Adeel Anjum, and Abid Khan. "Privacy preservation in e-health cloud: taxonomy, privacy requirements, feasibility analysis, and opportunities" [23]	2020
19	Abouelmehdi, Karim, Abderrahim Beni-Hessane, and Hayat Khaloufi. "Big healthcare data: preserving security and privacy" [24]	2018
20	Neumann, Geoffrey K., Paul Grace, Daniel Burns, and Mike Surridge. "Pseudonymization risk analysis in distributed systems" [25]	2019
21	Kumar, Anil, and Ravinder Kumar. "Privacy Preservation of Electronic Health Record: Current Status and Future Direction" [26]	2020

Table 2. Anonymization methods.

Num	Anonymization method	Article	Article where method is main
1	k-anonymity	1, 2, 3, 4, 5, 6, 7, 8, 9, 10, 11, 12, 13, 14, 15, 16, 17, 18, 19, 20, 21	1, 3, 4, 6, 7, 8, 9, 12, 15, 16, 18, 19, 21
2	l-diversity	1, 3, 4, 5, 6, 7, 8, 9, 13, 14, 15, 17, 18, 19, 21	1, 4, 6, 7, 19
3	t-closeness	1, 3, 4, 5, 6, 7, 8, 9, 14, 17, 18, 19, 21	1, 4, 6, 7, 19
4	p-sensitive k-anonymity	2, 7, 8, 17, 21	–
5	$p+$-sensitive k-anonymity	2, 8, 18, 21	–
6	Cryptographic algorithms (RSA, ElGamal, DES, AES)	5, 9, 18, 21	5, 18, 21
7	Balanced $p+$-sensitive k-anonymity	2, 8, 9	8
8	(p, α)-sensitive k-anonymity	2, 8, 18	–
9	Pseudonymization, tokenization, hashing	6, 11, 19	6, 11, 18
10	ϵ-differential privacy	6, 17, 18	6, 17, 18
11	(ϵ, m)-Anonymity	8, 15, 18	–
12	k-join-anonymity model	8, 21	–
13	m-invariance	7, 18	–
14	(α, k)-Anonymity	8, 18	–
15	Closed l-diversity/Closed l-diversification	13	13
16	Semantic linkage k-anonymity (SLKA)	14	14
17	θ-sensitive k-anonymity	2	2
18	(k, e)-anonymity	8	–
19	S-diversity	6	–
20	(X, Y)-Privacy	8	–
21	(k, k^m)-anonymity	10	10
22	h, k, p-Coherence	10	–
23	ρ-Uncertainty	10	–
24	Privacy-constrained anonymity	10	–
25	PS-rule based anonymity	10	–
26	(l, e) diversity	15	15
27	h-ceiling	16	–
28	$(1, k)$-anonymity	18	–
29	(k, k)-anonymity	18	–
30	$(k, 1)$-anonymity	18	–
31	k-Map	18	–
32	σ-Presence	18	–
33	c-Confident σ-presence	18	–

Table 3. Challenges in health care data anonymization.

Issue	Article	Methods
Privacy risk	1, 5, 6, 7, 8, 9, 10, 11, 12, 13, 14, 15, 16, 17	Applicable to all methods
Utility of data	1, 4, 5, 6, 7, 8, 9, 10, 13, 14, 15, 16, 17	Applicable to all methods
Vulnerability to different attacks	2, 4, 6, 8, 9, 15, 18	k-anonymity k-Map (k, k)-anonymity $(1, k)$-anonymity $(k, 1)$-anonymity k^m-anonymity Privacy-constrained anonymity k-join-anonymity Cryptographic algorithms k-diversity (a, k)-anonymity t-Closeness p-Sensitive k-anonymity θ-sensitive k-anonymity h, k, p-Coherence PS-rule based anonymity ρ-Uncertainty p-sensitive k-anonymity (p, α)-sensitive k-anonymity $p+$-sensitive k-anonymity Balanced $p+$-sensitive k-anonymity (l, e)-Diversity σ-Presence c-Confident σ-presence (X, Y)-Privacy model
Different methods for different types of data (micro-data, big data, transaction data)	9, 10, 11, 12, 13	h,k,p-Coherence ρ-uncertainty PS-rule based anonymity m-invariance k-join-anonymity S-diversity (α, k)-Anonymity (k, e)-anonymity (X, Y)-Privacy model (l, e) diversity h-ceiling
Linkage attack	7, 9, 10, 14	k-anonymity k-Map l-diversity t-Closeness (α, k)-Anonymity p-Sensitive k-anonymity (k, k)-anonymity
Trustfulness of data	10, 16, 18, 20	ϵ-differential privacy
Not only technical problem	11, 17, 20	Applicable to all methods
Computational resources requirements	5, 18, 21	Cryptographic algorithms
Difficult to implement in real-life data	4	t-Closeness

5   Discussion

In this section we will discuss findings described in the Result section.

5.1   Main Challenges in Data Anonymization

The very big issue is the potential vulnerabilities of anonymization methods to attacks. Several of the reviews articles mention this problem. The different types of attacks can be divided into 3 main categories as shown in Fig. 2.

Identity disclosure means that person can be re-identified in the dataset. Attribute disclosure is revealing sensitive information about the person, for example, disclosing the diagnosis or treatment. Membership disclosure is information about a person's data being included in a certain dataset on not. Figure 2 shows anonymization methods that are used to protect data from the most common types of attacks. In reviewed papers were mentioned 21 methods, that are vulnerable to at least one type of attack. In reality, this number may be even

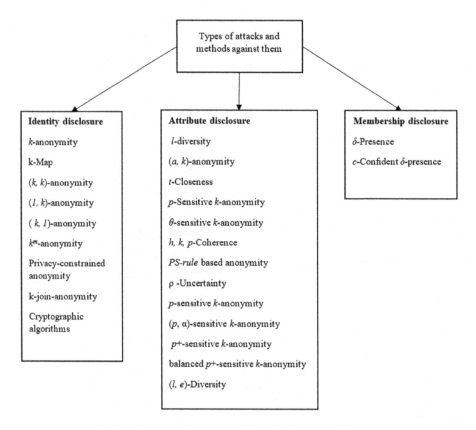

Fig. 2. Types of attacks and methods against them.

bigger, since some new methods introduced in the last 3 years are not fully investigated for possible vulnerabilities.

One of the main problems in data anonymization is the absence of a universal method that can protect from potential attacks while preserving data utility. Even k-anonymity, the most widely used method, can be vulnerable to some attacks, such as homogeneity attack or background knowledge attack [27]. In the k-anonymous dataset, each person cannot be distinguished from at least k-1 other individuals [28]. In other words, k individuals have the same characteristic and the greater is k, the lower is the risk of identification and vice versa. For example, if k=2, there is a 50% chance of identification of the person, because only 2 people have the same attributes [29].

p-sensitive k-anonymity is based on using a nearest neighbour search algorithm and offers protection from the attribute disclosure later were improved to p+-sensitive k-anonymity and (p, α)-sensitive k-anonymity [2,13]. Those models use top-down specialization to generate an anonymous dataset, that must contain at least p distinct categories for each sensitive attribute against each quasi-identifier group [13]. Some more advanced methods p+-sensitive k-anonymity and balanced p+-sensitive k-anonymity based on k-anonymity significantly contribute to privacy protection, however, they contain certain vulnerabilities as well and do not protect against the sensitive variance attack and categorical similarity attack. To overcome this problem, researchers [9] offer the θ-sensitive k-anonymity privacy model. The θ-Sensitivity is the product of variance that represents the diversity (σ^2) and observation 1 (μ).

Although k-anonymity works against identity disclosure, it does not protect from attribute disclosure. For this reason, methods such as l-diversity and t-closeness were introduced. l-diversity and t-closeness along with other methods (a, k)-anonymity and p-sensitive-k-anonymity are used against attribute disclosure. l-diversity intent to solve sensitive attributes disclosure by adding intra-group diversity for sensitive attributes. This method requires to have several attributes with different values within the group [6,27].

t-closeness is used to overcome the drawback of l-diversity by reducing the granularity in the represented data. According to this method, the equivalence group has the same distribution of sensitive values as in original data [6,30]. However, it is highly difficult to achieve in real data. Closed l-diversity and Closed l-diversification methods were introduced in [18] to be applied in micro-datasets where k-anonymity and l-diversity do not provide sufficient privacy protection.

Certain methods are intended to be used in specific cases and with particular types of datasets. For example, m-invariance is a privacy protection technique for the re-publication of dynamic datasets with sensitive personal information [2]. Also, k-join-anonymity permits more effective generalization to reduce the information loss on microdata. It was designed to provide the same level of accuracy as k-anonymity, however, it may be vulnerable to attribute disclosure [13]. S-diversity overcomes similarity attacks in the worst-case scenario by using a clustering algorithm [6]. (α, k)-anonymity created to protect identifications and relationships to sensitive information in data, but the risk of attribute disclosure exists due to the possibility of linkage attacks [13]. (k, e)-anonymity is based on the separation of published sensitive values and quasi-identifiers into different tables. The accuracy is improved because quasi-identifiers don't need to be generalized. However, it is implemented on one-time publication and the risk of probabilistic attack is in place [13]. (l, e) diversity method also provides anonymization through the separation of quasi-identifiers and sensitive attributes into different tables. This method uses bucketization, with ID for buckets and each bucket with sensitive attribute must contain at least l distinct values and semantic similarity between two values should not be greater than e [23]. The main difference of (ϵ, m)-Anonymity model is that it deals with numeric sensitive data, but cannot be applicable for categorical sensitive attributes. (X, Y)-Privacy model focuses on anonymizing sequential releases. This method intends to be implemented for the republication of static microdata table to ensure that the current data cannot be linked to the previous release [13,31]. However, this model was reported to be vulnerable to minimality attacks, forward attacks, and backward attacks [32].

The other big challenge in anonymization in health care is that various dataset types such as microdata and large scale data require different approaches. Problems of large datasets, those containing both demographic information (age, gender, zip code) and diagnosis code, named RT-datasets, are discussed in the [17].

5.2 Advantages and Disadvantages of Anonymization Methods

The main problems identified with existing methods are reduction of accuracy and excessive information loss. Also, methods that use adding noise may influence data truthfulness, which has a negative impact on the data utility. The proposed method (k, k^m)-anonymity that is based on k-anonymization principles, intent to solve those issues. (k, k^m) -anonymity provides privacy protection in the situation when the attacker has previous knowledge about demographics. m is a quasi-identifier shared with up to k individuals in the dataset. This means if an attacker knows m, no more than k individuals can be identified. It is important to notice that an RT-dataset may be k and k^m but not (k, k^m), which means a different level of data protection [15].

Another model, k-map is similar to k-anonymity, however, performed on larger datasets. k-map contains less restriction allowing publishing more detailed information and preserving data utility, which on the other side provides less privacy protection. $(1, k)$-anonymity, $(k, 1)$-anonymity and (k, k)-anonymity follows similar principles as k-map and k-anonymity. Models differ in the assumption of the attacker's capability to re-identify data. Those models allow to save better data utility, however, provide weaker privacy compared to k-anonymity [1].

Balanced $p+$-sensitive k-anonymity was proposed by [13] as improvement, because p-sensitive p-anonymity, (p, α)-sensitive k-anonymity and $p+$-sensitive k-anonymity do not provide adequate privacy protection to the end-users. One more anonymization method, called the utility-preserving model uses h-ceiling to prevent overgeneralization. It implies that the level of generalization is limited to h. Another method based on k-anonymity is Semantic linkage k-anonymity (SLKA) [19]. This method is intended to decrease the probability of matching data and to prevent linkage attack.

Other methods, briefly mentioned as used for data anonymization are h, k, p-Coherence, ρ-uncertainty and PS-rule (Privacy Sensitive rule) based anonymity, used for publishing transaction data for research purposes. Privacy-constrained, anonymity proposed a method to anonymize patient-specific clinical profiles, used to avoid linkage between genomic sequence and other clinical information that may lead to the identification of the patient. The author of ϵ-differential privacy offered instead of comparing prior probability and the posterior probability, compare the risk with and without published record. ϵ-differential privacy is based on the statement that the removal or addition of a single record in the database should not significantly affect the outcome of the query [22,33]. The other two models, σ-Presence and c-Confident σ-Presence, are used against membership disclosure. This means it is not disclosed whether information about an individual is presented in the dataset or not [34]. For example, the attacker has statistical information on how many male-female, single-married, young-old people are presented in the database, but don't know if there is a married young woman in the database.

Different approach to data anonymization is presented in [12,14,23,26]. In those studies, cryptographic algorithms are used for data protection. Paper [26] mention different encryption techniques (hashing, AES, as well as RSA, DES, and Diffie–Hellman key exchange algorithms), however, they are considered in context secure data storage than anonymization for data exchange. A similar approach is described in [23]. The work shows how cryptographic algorithms can be used for cloud security. Research [14] also states that cryptographic algorithms are more secure compare to non-cryptographic using in a cloud environment. As in previous research, this is used for data storage and retrieval however not anonymized sharing.

Paper [12] offers an innovative approach for using cryptographic algorithms not just for secure data storage and transition but specifically for anonymization. To achieve that, it is proposed mapping unique record of raw data, using the function to other unique data, different from initial raw data. The one-way function must be implemented to ensure that raw data must not be reversible, to ensure that the adversary cannot mark data back. The study shows that the proposed method is efficient in regards to data utility. In total 4 cryptographic (2 symmetric and 2 asymmetric) algorithms were used showing better results compared to the k-anonymity based technique providing privacy with minimized data loss. Anonymized data must be different enough from initial data to protect against attacks. Although cryptographic techniques provide security, their implementation requires additional computational power, storage space, and infrastructure [23,26].

6   Conclusion

This paper gives an overview of existing methods, used for structured health data anonymization. We summarized anonymization methods identified in relevant literature in 2017–2020. We identified widely used methods along with new and prospective methods. We also discuss the advantages and disadvantages of selected methods. This research was conducted following the guidelines for systematic literature reviews presented by Kitchenham and Charters [7].

We had the following questions: RQ1. What are the methods of data anonymization? RQ2. What are the challenges and issues in using those methods? RQ1 shows that a variety of methods exist to anonymize structured health care data. Methods that are discussed the most are k-anonymity, l-diversity and t-closeness. Those methods, especially the most well-known k-anonymity, became a background to build other anonymization methods. All of the methods identified in RQ1 have their benefits and limitation. It was discovered, that most of the methods mentioned in this literature review are build based on k-anonymity principles. However, there is also another approach - anonymization using cryptographic algorithms.

This review shows that at least 7 new anonymization methods for health data were presented in 2017–2020. Most of them are the improvement of existing methods, such as balanced p+-sensitive k-anonymity, Closed l-diversity, SLKA, θ-sensitive k-anonymity and (k, k^m) -anonymity or, they are using existing methods to solve anonymization problems like cryptographic algorithms. This tendency shows that the field is developing and more improved algorithms can be expected in upcoming years.

RQ2 shows that various methods are applied in the different datasets (micro- and big data, transaction data, etc.) and despite the variety of methods there is no universal solution that can be applied in all situations. The main issues identified in the RQ2 are finding the right balance between data privacy and utility. It has been shown, that many methods that provide privacy protection lead to sufficient data loss, which consequently has a negative impact on research quality that uses this data.

Moreover, anonymization methods may protect from one type of attack but remain vulnerable to others. For example, k-anonymity and some other methods based on it, protect from identity disclosure, however, may be vulnerable to attribute and membership disclosure. The other identified problem is that certain methods, for instance, t-closeness are very difficult to achieve in a real dataset. To apply this method each group must be well-represented in the dataset which usually is not the case in the real dataset. The other issue is the possibility that an attacker can perform a linkage attack, where the person can be identified by linking anonymized dataset with background information, for example, gender, salary, or zip code. The issues with trustfulness of data may appear in methods in case of applying substitution, permutation, or adding noise. Those techniques are often used for micro-data anonymization, however, may have a negative impact on data quality. Cryptographic algorithms show promising result in anonymization field, however, require certain computational power and resources to be implemented.

Besides, it was identified that anonymization is not only a technical problem, it also requires guidelines. Some articles mention different legal requirements in data protection and data anonymization in different countries. Anonymized data is out of the scope of personal data protection legislation. Nevertheless, the law requires implementing sufficient measures, to ensure that people cannot be identified. However, it is still decided case-by-case what measures are optimal solution [35]. We believe that more comprehensive approach must be applied towards health data anonymization and it should be solved by a combination of proper methods and regulations. Additional work must be done not only on improving methods, but on the framework on using them to harmonize the data anonymization field.

The article has certain limitation. Certain databases such as Scopus and Web of Science were not included in the search. However, it is planned for further research. Also, in future work, we plan to focus the research on comparing practical tools in which those methods are implemented, compare their efficiency and usability in real health care data.

References

1. Gkoulalas-Divanis, A., Loukides, G., Sun, J.: Publishing data from electronic health records while preserving privacy: a survey of algorithms. J. Biomed. Inf. **50**, 4–19 (2014)
2. Jayabalan, M., Rana, M.E.: Anonymizing healthcare records: a study of privacy preserving data publishing techniques. Adv. Sci. Lett. **24**(3), 1694–1697 (2018)
3. Sosu, R.N.A., Quist-Aphetsi, K., Nana, L.: A decentralized cryptographic blockchain approach for health information system. In: International Conference on Computing, Computational Modelling and Applications (ICCMA), Cape Coast, Ghana, pp. 120–1204 (2019)
4. Salas, J., Domingo-Ferrer, J.: Some basics on privacy techniques, anonymization and their big data challenges. Math. Comput. Sci. **12**(3), 263–274 (2018)
5. Ouafae, B., Mariam, R., Oumaima, L., Abdelouahid, L.: Data anonymization in social networks (2020)
6. Pawar, A., Ahirrao, S., Churi, P.P.: Anonymization techniques for protecting privacy: a survey. In: 2018 IEEE Punecon, pp. 1–6. IEEE (2018)
7. Kitchenham, B., Charters, S.: Guidelines for performing systematic literature reviews in software engineering (2007)
8. Prasser, F., Eicher, J., Bild, R., Spengler, H., Kuhn, K.A.: A tool for optimizing de-identified health data for use in statistical classification. In: 2017 IEEE 30th International Symposium on Computer-Based Medical Systems (CBMS), pp. 169–174. IEEE (2017)
9. Khan, R., Tao, X., Anjum, A., Kanwal, T., Khan, A., Maple, C., et al.: θ-sensitive k-anonymity: an anonymization model for iot based electronic health records. Electronics **9**(5), 716 (2020)
10. Li, X.-B., Qin, J.: Anonymizing and sharing medical text records. Inf. Syst. Res. **28**(2), 332–352 (2017)
11. Rajendran, K., Jayabalan, M., Rana, M.E.: A study on k-anonymity, l-diversity, and t-closeness techniques. IJCSNS **17**(12), 172 (2017)

12. Aminifar, A., Lamo, Y., Pun, K.I., Rabbi, F.: A practical methodology for anonymization of structured health data (2019)
13. Anjum, A., et al.: An efficient privacy mechanism for electronic health records. Comput. Secur. **72**, 196–211 (2018)
14. Majeed, A.: Attribute-centric anonymization scheme for improving user privacy and utility of publishing e-health data. J. King Saud Univ.-Comput. Inf. Sci. **31**(4), 426–435 (2019)
15. Poulis, G., Loukides, G., Skiadopoulos, S., Gkoulalas-Divanis, A.: Anonymizing datasets with demographics and diagnosis codes in the presence of utility constraints. J. Biomed. Inf. **65**, 76–96 (2017)
16. Ribeiro, S.L., Nakamura, E.T.: Privacy protection with pseudonymization and anonymization in a health iot system: results from ocariot. In: 2019 IEEE 19th International Conference on Bioinformatics and Bioengineering (BIBE), pp. 904–908. IEEE (2019)
17. Prasser, F., Kohlmayer, F., Spengler, H., Kuhn, K.A.: A scalable and pragmatic method for the safe sharing of high-quality health data. IEEE J. Biomed. Health Inf. **22**(2), 611–622 (2017)
18. Hsiao, M.H., Lin, W.Y., Hsu, K.Y., Shen, Z.X.: On anonymizing medical microdata with large-scale missing values-a case study with the faers dataset. In: 2019 41st Annual International Conference of the IEEE Engineering in Medicine and Biology Society (EMBC), pp. 6505–6508. IEEE (2019)
19. Lu, Y., Sinnott, R.O., Verspoor, K., Parampalli, U.: Privacy-preserving access control in electronic health record linkage. In: 2018 17th IEEE International Conference On Trust, Security And Privacy In Computing And Communications/12th IEEE International Conference On Big Data Science And Engineering (TrustCom/BigDataSE), pp. 1079–1090. IEEE (2018)
20. Saeed, R., Rauf, A.: Anatomization through generalization (ag): a hybrid privacy-preserving approach to prevent membership, identity and semantic similarity disclosure attacks. In: 2018 International Conference on Computing, Mathematics and Engineering Technologies (iCoMET), pp. 1–7. IEEE (2018)
21. Lee, H., Kim, S., Kim, J.W., Chung, Y.D.: Utility-preserving anonymization for health data publishing. BMC Med. Inf. Decis. Mak. **17**(1), 104 (2017)
22. Lee, H., Chung, Y.D.: Differentially private release of medical microdata: an efficient and practical approach for preserving informative attribute values. BMC Med. Inf. Decis. Mak. **20**(1), 1–15 (2020)
23. Kanwal, T., Anjum, A., Khan, A.: Privacy preservation in e-health cloud: taxonomy, privacy requirements, feasibility analysis, and opportunities. Cluster Comput. **24**(1), 293–317 (2020). https://doi.org/10.1007/s10586-020-03106-1
24. Abouelmehdi, K., Beni-Hessane, A., Khaloufi, H.: Big healthcare data: preserving security and privacy. J. Big Data **5**(1), 1–18 (2018). https://doi.org/10.1186/s40537-017-0110-7
25. Neumann, G.K., Grace, P., Burns, D., Surridge, M.: Pseudonymization risk analysis in distributed systems. J. Internet Serv. Appl. **10**(1), 1–16 (2019). https://doi.org/10.1186/s13174-018-0098-z
26. Kumar, A., Kumar, R.: Privacy preservation of electronic health record: current status and future direction. In: Gupta, B.B., Perez, G.M., Agrawal, D.P., Gupta, D. (eds.) Handbook of Computer Networks and Cyber Security, pp. 715–739. Springer, Cham (2020). https://doi.org/10.1007/978-3-030-22277-2_28
27. Machanavajjhala, A., Kifer, D., Gehrke, J., Venkitasubramaniam, M.: l-diversity: privacy beyond k-anonymity. ACM Trans. Knowl. Disc. Data (TKDD) **1**(1), 3-es (2007)

28. Sweeney, L.: k-anonymity: a model for protecting privacy. Int. J. Uncertainty Fuzziness Knowl.-Based Syst. **10**(05), 557–570 (2002)
29. El Emam, K., Arbuckle, L.: Anonymizing health data: case studies and methods to get you started. O'Reilly Media, Inc. (2013)
30. Li, N., Li, T., Venkatasubramanian, S.: t-closeness: Privacy beyond k-anonymity and l-diversity. In: 2007 IEEE 23rd International Conference on Data Engineering, pp. 106–115. IEEE (2007)
31. Wang, X., McCallum, A.: Topics over time: a non-markov continuous-time model of topical trends. In: Proceedings of the 12th ACM SIGKDD International Conference on Knowledge Discovery and Data Mining, pp. 424–433 (2006)
32. Wong, R.C.W., Fu, A.W.C., Wang, K., Pei, J.: Minimality attack in privacy preserving data publishing. In: Proceedings of the 33rd International Conference on Very Large Data Bases, pp. 543–554 (2007)
33. Dwork, C.: Differential privacy: a survey of results. In: Agrawal, M., Du, D., Duan, Z., Li, A. (eds.) TAMC 2008. LNCS, vol. 4978, pp. 1–19. Springer, Heidelberg (2008). https://doi.org/10.1007/978-3-540-79228-4_1
34. Rohilla, S., Bhardwaj, M.: Efficient anonymization algorithms to prevent generalized losses and membership disclosure in microdata. Am. J. Data Mining Knowl. Disc. **2**(2), 54–61 (2017)
35. European Commission. Article 29 working party opinion 05/2014 on anonymisation techniques

Categorical Modeling of Multi-model Data: One Model to Rule Them All

Martin Svoboda[✉], Pavel Čontoš, and Irena Holubová

Faculty of Mathematics and Physics, Charles University, Prague, Czech Republic
{svoboda,contos,holubova}@ksi.mff.cuni.cz

Abstract. As most of the DBMSs have become multi-model, there have occurred plenty of related issues. One of them is a design of a multi-model application, where the step from the conceptual layer to a set of distinct interlinked logical models is not straightforward. We propose an approach based on category theory, which provides a unified view of the data and a solid mathematical basis for their management. We propose a schema and instance categories covering popular models and we show how an ER model can be transformed to such a categorical layer. We also outline the whole framework based on the categorical model.

Keywords: Multi-model data · Category theory · Conceptual modeling

1 Introduction

Most of the database management systems (DBMSs) of various types have recently followed the Gartner predictions [4] of support of multiple data models.

Example 1. Consider an example of a multi-model scenario in Fig. 1. Social network of customers is captured using a graph. Additional information, such as their credit limit, is recorded in a relational table. Orders submitted by customers are stored as JSON documents. A column family maintains the history of all orders, and a key/value mapping maintains current shopping carts. A cross-model query might return *friends of customers who ordered any item with a price higher than 180*. □

Adding another data model to an existing DBMS, though simple and natural in the idea, is quite complex, as the combined models (and respective systems) often have contradictory features. Currently, there exist more than 20 representatives of *multi-model databases* [8], involving well-known traditional relational and also novel NoSQL tools. Such a situation is difficult for users who want to develop a multi-model database application. According to the traditional recommendations, they would first create a conceptual schema (e.g., using ER or UML). Then, there exist verified approaches how to transform such a schema into, e.g., the relational model. However, the step from an ER/UML conceptual model to virtually any possible (not standardized) combination of multiple models is not that straightforward.

Supported by the GAČR project no. 20-22276S.

C. Attiogbé and S. Ben Yahia (Eds.): MEDI 2021, LNCS 12732, pp. 190–198, 2021.
https://doi.org/10.1007/978-3-030-78428-7_15

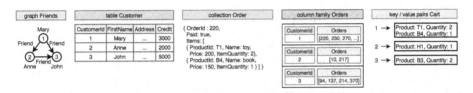

Fig. 1. A sample multi-model scenario

For this purpose, we need a representation that would allow us to (1) capture all the existing models, preferably in the same and definitely in a standard way; (2) query across multiple interconnected models; (3) perform correct and complete evolution management, i.e., propagation of changes; (4) enable data migration without complex reorganizations; and (5) permit integration of new data models. In this paper, we propose a solution based on *category theory* [2], a theory sufficiently general and providing a strong mathematical background for further management of the represented data. The main contributions are:

- We define schema and instance categories covering all known data models.
- We provide the ER-to-category transformation algorithm.
- We outline the whole database framework based on the categorical model.

Paper Outline: In Sect. 2, we overview the related work and in Sect. 3 basics of category theory. We then describe the ER-to-category transformation in Sect. 4 and the whole framework in Sect. 5. In Sect. 6, we conclude.

2   Related Work

While the multi-model DBMSs [8] (more or less painfully) provide an extension of the original data structures of a single core model, there also exist more general approaches. E.g., the *NoSQL Abstract Model* (*NoAM*) [1] represents the data as named collections of blocks consisting of a set of entries. *Associative arrays* [5] are defined as mappings from pairs of unique (column and row) keys to values.

For the higher levels of abstraction, there also exist a few proposals. *TyphonML* [3] enables us to specify conceptual entities, their attributes, relations, and datatypes, and map them to different single-model DBMSs of a polystore. Similarly in [6], an ER schema is partitioned and then mapped to different data models. However, none of them provides a detailed specification of management of inter-model references, support for overlapping, cross-model querying, etc.

The idea of exploiting category theory to represent data models is not new. Most of the approaches, denoted as *bottom-up*, start from a single logical model (namely relational [9], or object-relational, i.e., hierarchies of classes [10]) and define a respective schema category and operations using standard categorical approaches (such as functors). A *top-down* approach from [7] defines a schema category covering various conceptual modeling approaches, yet only with respect to the most common model of that time, i.e., the relational one.

3   Preliminary Concepts

A category $\mathbf{C} = (\mathcal{O}, \mathcal{M}, \circ)$ consists of a set of objects $\mathcal{O}$ (graph vertices), a set of morphisms $\mathcal{M}$ (directed edges), and a composition operation $\circ$ for the morphisms. Each morphism is modeled/depicted as an *arrow* $f : A \rightarrow B$, where $A, B \in \mathcal{O}$, A being referenced to as a *domain* and B as a *codomain*. Whenever $f, g \in \mathcal{M}$ are two morphisms $f : A \rightarrow B$ and $g : B \rightarrow C$, it must hold that $g \circ f \in \mathcal{M}$, i.e., morphisms can be composed using the $\circ$ operation and the composite $g \circ f$ must also be a morphism of the category (i.e., transitivity is required). Moreover, $\circ$ must be associative, i.e., $h \circ (g \circ f) = (h \circ g) \circ f$ for any morphisms $f, g, h \in \mathcal{M}$, $f : A \rightarrow B, g : B \rightarrow C$, and $h : C \rightarrow D$. Finally, for every object A, there must exist an *identity* morphism 1_A such that $f \circ 1_A = f = 1_B \circ f$ for any $f : A \rightarrow B$ (serving as a unit with respect to the composition).

Example 2. **Rel** is a category where objects represent sets and morphisms are binary relations over these sets. For morphisms $f : A \rightarrow B$ and $g : B \rightarrow C$, $(a, c) \in g \circ f$ whenever there exists at least one value $b \in B$ such that $(a, b) \in f$ and $(b, c) \in g$.   □

4   Schema Representation

We understand our solution as a means for schema and data representation. So, following a top-down approach, we describe how to translate a given ER schema into an equivalent categorical representation. The aim of a *schema category* $\mathbf{S}$ is to grasp the structure of the data and basic integrity constraints. An *instance category* $\mathbf{I}$ represents a particular database state, i.e., a database instance filled with data conforming to schema $\mathbf{S}$. Both $\mathbf{S}$ and $\mathbf{I}$ are identical as for their structure, they differ in the internal contents of their objects and morphisms.

4.1   Schema Category

Schema category is defined as $\mathbf{S} = (\mathcal{O}_\mathbf{S}, \mathcal{M}_\mathbf{S}, \circ_\mathbf{S})$. Objects in $\mathcal{O}_\mathbf{S}$ represent individual entity types, attributes, and relationship types in the input ER schema. To simplify explanation, we distinguish *entity*, *attribute*, and *relationship* objects, as well as *attribute*, *identifier*, *relationship*, and *hierarchy* morphisms. Next, we assume that $\mathbb{A}$ is the active domain of attribute names, $\mathbb{E}$ the domain of entity type names, and $\mathbb{R}$ the domain of relationship type names.

Each object $o \in \mathcal{O}_\mathbf{S}$ will internally be modeled as a pair $(superid, ids)$. $superid \subseteq \mathbb{A}$ will be a *super-identifier* of a given object, i.e., a set of attributes using which the respective instances can be uniquely identified. $ids \subseteq \mathcal{P}(superid)$ will be a set of individual *identifiers* a given object is associated with. It must hold that $superid \supseteq \bigcup_{id \in ids} id$. In case of entity and attribute objects, equality will hold. In case of relationship objects, there can be additional attributes.

Example 3. Having an entity type `Person` with one simple identifier `PersonId`, one composed identifier consisting of two attributes `FirstName` and `LastName`, and additional attributes `Age` and `Address`, the corresponding entity object `Person`$_\mathbf{S}$ will be modeled as a pair $(superid, ids)$, where $superid = \{$`PersonId`, `FirstName`, `LastName`$\}$ and $ids = \{\{$`PersonId`$\}, \{$`FirstName`, `LastName`$\}\}$.   □

Next, each morphism $m \in \mathcal{M}_\mathbf{S}$ will be a pair (min, max) modeling the traditional concept of relationship cardinalities and attribute multiplicities: $min \in \{0, 1\}$ for the lower bound, $max \in \{1, *\}$ for the upper bound. For each object $o \in \mathcal{O}_\mathbf{S}$, its identity morphism is defined as $1_o = (1, 1)$. Having two morphisms $m_1 = (min_1, max_1)$ and $m_2 = (min_2, max_2)$ s.t. $m_1, m_2 \in \mathcal{M}_\mathbf{S}$, their mutual composite is evaluated as $m_2 \circ_\mathbf{S} m_1 = (\min(min_1, min_2), \max(max_1, max_2))$.

4.2   Instance Categories

Each instance category $\mathbf{I} = (\mathcal{O}_\mathbf{I}, \mathcal{M}_\mathbf{I}, \circ_\mathbf{I})$ represents a particular data instance of a schema $\mathbf{S}$. Assuming that $\mathbb{V}$ is the domain of all possible values of attributes, object $o_\mathbf{I} = \{t_1, t_2, \ldots, t_n\} \in \mathcal{O}_\mathbf{I}$ for some $n \in \mathbb{N}_0$ will be modeled as a set of tuples, each represented as a function $t_i : superid \rightarrow \mathbb{V}$ for any $i \in \mathbb{N}, 0 < i \leq n$. The tuples are unordered, unique, and with all attributes specified.

Example 4. An instance object $\mathtt{Person}_\mathbf{I}$ for $\mathtt{Person}_\mathbf{S}$ from Example 3 can, e.g., be:

$$\mathtt{Person}_\mathbf{I} = \{\{(\mathtt{PersonId}, 1), (\mathtt{FirstName}, \mathtt{Mary}), (\mathtt{LastName}, \ldots)\},$$
$$\{(\mathtt{PersonId}, 2), (\mathtt{FirstName}, \mathtt{Anne}), (\mathtt{LastName}, \ldots)\}\}.$$

$\square$

As for the morphisms, they act as binary relations (i.e., they abide by the principles of the **Rel** category). For schema morphism $m_\mathbf{S} = (min, max) \in \mathcal{M}_\mathbf{S}$ the cardinality condition must be satisfied. Identity morphism 1_o for each object $o \in \mathcal{O}_\mathbf{I}$ is defined as a function $1_o = \{(t, t) \mid t \in o\}$. Finally, the composition operation $\circ_\mathbf{I}$ corresponds to the composition in **Rel**.

4.3   Schema Translation

To simplify the description, we assume that $\mathbb{A}$, $\mathbb{E}$, and $\mathbb{R}$ are mutually distinct. Hence, names of category objects can be taken from the original ER items.

In visualizations, we will label all objects with their names, and entity objects with identifiers. We will not include loops for identity morphisms, and we will only include morphisms we explicitly need to construct (implicitly assuming the transitive closure over the composition). Names of morphisms and cardinalitites $(1, 1)$ will be omitted. We also suppose that a will stand for the actual value of variable a and that names cannot contain the·symbol to be used as a separator. E.g., if $a = \mathtt{Customer}$ and $i = 1$, expression a·i is interpreted as $\mathtt{Customer·1}$.

We will now describe how the ER constructs are transformed into objects and morphisms of $\mathbf{S}$. There may exist dependencies between ER entity types, but they need to be acyclic. So, hierarchies are transformed by starting at their root nodes and proceeding toward their leaves, weak entity identifiers can be transformed only when all the entity types they depend on are already resolved. Once all the entity types are processed, relationship types can be processed.

Entity Types. Let $e \in \mathbb{E}$ be an entity type and $a_1, \ldots, a_m$ its attributes for some $m \in \mathbb{N}_0$. Let each of these attributes $a_i \in \mathbb{A}$ be associated with a particular multiplicity (min_i, max_i). In case a given attribute a_i is a structured attribute, let then $a_i^1, \ldots, a_i^{k_i}$ be its individual sub-attributes for some $k_i \in \mathbb{N}$. Finally, let $id_1, \ldots, id_n$ be all the (strong) identifiers of e for some $n \in \mathbb{N}_0$, each modeled as a set of at least one attribute it consists of. It must hold that $id_j \subseteq \{a_1, \ldots, a_m\}$ and $|id_j| \geq 1$ for any $j \in \{1, \ldots, n\}$, as well as that only attributes with the trivial multiplicity can be involved in the identifiers, i.e., that $\forall a_i \in \bigcup_{j=1}^{n} id_j$ the multiplicity of a_i must only be $(1,1)$. Note that there may not be even a single identifier now in case e is a descendant in a hierarchy and/or contains a weak identifier. Transformation of both these constructs will be discussed later.

First, an *entity* object e is created for e s.t. $\mathsf{e} = (superid, ids)$. Assuming that $u(a_i) = \{a_i\}$ for any unstructured attribute a_i and $u(a_i) = \{a_i^1, \ldots, a_i^{k_i}\}$ otherwise, we can write $ids = \{id_1', \ldots, id_n'\}$ with $id_j' = \bigcup_{a_i \in id_j} u(a_i)$ for each $j \in \{1, \ldots, n\}$, and $superid = \bigcup_{id' \in ids} id'$. In other words, identifiers are preserved, just structured ones are unfolded to the corresponding sub-attributes.

Next, each attribute a_i with $i \in \{1, \ldots, m\}$ is processed. In case a_i is unstructured, an *attribute* object $\mathsf{a_i} = (\{a_i\}, \{\{a_i\}\})$ is created. When a_i is structured, the resulting *attribute* object equals to $\mathsf{a_i} = (\{a_i^1, \ldots, a_i^{k_i}\}, \{\{a_i^1, \ldots, a_i^{k_i}\}\})$. For each of these sub-attributes a_i^j, $j \in \{1, \ldots, k_i\}$, another *attribute* object $\mathsf{a_i^j} = (\{a_i^j\}, \{\{a_i^j\}\})$ is produced, together with an attribute morphism $\mathsf{a_i} \cdot \mathsf{a_i^j} = (1, 1) : \mathsf{a_i} \to \mathsf{a_i^j}$ binding it with its base attribute a_i. Finally, an attribute morphism $\mathsf{e} \cdot \mathsf{a_i} = (min_i, max_i) : \mathsf{e} \to \mathsf{a_i}$ is yielded, binding a_i with e.

Example 5. Entity type `Person` depicted in Fig. 2 is transformed to an entity object `Person` having $superid = \{\texttt{PersonId}, \texttt{FirstName}, \texttt{LastName}\}$ and $ids = \{\{\texttt{PersonId}\}, \{\texttt{Firstname}, \texttt{LastName}\}\}$. Its attributes are transformed to the respective attribute objects `Email`, `FirstName`, etc., together with the given cardinalities. In case of the structured attribute `Address`, there is an attribute object `Address`, as well as attribute objects `Street`, `City`, and `PostalCode` for its individual sub-attributes. ☐

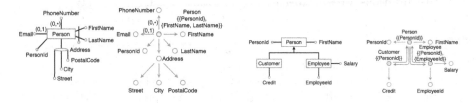

Fig. 2. Translation of entity types   **Fig. 3.** Translation of ISA hierarchies

ISA Hierarchies. Let us have a descendant entity type c and its parent entity type p (multiple inheritance is not permitted). Let $\mathsf{c} = (superid_c, ids_c)$ be the corresponding partially resolved entity object for c, and $\mathsf{p} = (superid_p, ids_p)$ the fully resolved entity object for p. We need to take all the individual identifiers of p and add them to the existing identifiers of c. In particular, $ids_c = ids_c \cup ids_p$, and $superid_c = superid_c \cup (\bigcup_{id \in ids_p} id)$. Note that only identifiers are inherited, not ordinary attributes. They remain associated only with entity objects they are locally a part of. Finally, we also need to mutually bind both the entity types with *hierarchy* morphisms – for technical reasons, in both directions, i.e., morphisms $\mathsf{c \cdot p \cdot up} = (1,1) : \mathsf{c} \to \mathsf{p}$ and $\mathsf{c \cdot p \cdot down} = (1,1) : \mathsf{p} \to \mathsf{c}$ are constructed.

Example 6. The translation of an ISA hierarchy in Fig. 3 produces entity objects `Person`, `Customer`, and `Employee`, each with respective attribute objects and morphisms. Notice the inherited identifier {`PersonId`} in objects for the descendants.   □

Weak Identifiers. Let us have an entity type e with at least one weak identifier such that all the individual entity types involved in the relationship types forming these weak identifiers are already entirely resolved.

In this sense, let $\mathsf{e} = (superid, ids)$ be the current entity object for our weak entity type $e \in \mathbb{E}$ and v be a particular weak identifier we are about to transform. Let $a_1, \ldots, a_n$ for some $n \in \mathbb{N}_0$ be attributes of e involved in this weak identifier, and $r_1, \ldots, r_m$ for some $m \in \mathbb{N}$, $m \geq 1$ be all the relationship types involved in e s.t. $R_v \subseteq \{1, \ldots, m\}$, $R_v \neq \emptyset$ are indicies of relationship types actually involved in this weak identifier, each necessarily with cardinality $(1,1)$ toward e. For each such relationship type r_i, $i \in R_v$ we assume that $e_i^1, \ldots, e_i^{k_i}$ for some $k_i \in \mathbb{N}$, $k_i \geq 2$ are all the individual entity types participating in r_i (once for each occurrence, including e itself) and $\forall j \in \{1, \ldots, k_i\}$, it then holds that (min_i^j, max_i^j) is the cardinality of e_i^j in r_i and $\mathsf{e}_i^j = (superid_i^j, ids_i^j)$ is the already fully resolved entity object for e_i^j (except for e as such).

Within the context of a particular relationship type r_i, $i \in R_v$, it may happen that some of the involved entity types may have identifiers composed of attributes with the same names (as a consequence of, e.g., the inheritance of identifiers in hierarchies). For this purpose, let us introduce $m(ids_i^j) = \{\{\mathsf{a \cdot i \cdot j} \mid a \in id\} \mid id \in ids_i^j\}$ as an auxiliary set of marked identifiers for entity type e_i^j within r_i for each suitable i and j, i.e., for $i \in R_v$ and $j \in \{1, \ldots, k_i\}$ (except for e).

For each participating relationship type r_i, $i \in R_v$, let us now find all sets of attributes by which r_i may contribute to the weak identifier we are resolving. In case there exists at least one involved entity type e_i^j (except for e) with cardinality 1 in the upper boundary max_i^j, we define $ids_i = \{id \mid id \in m(ids_i^j) \wedge j \in \{1, \ldots, k_i\} \wedge e_i^j \neq e \wedge max_i^j = 1\}$. Otherwise, i.e., when $\forall j \in \{1, \ldots, k_i\}$ it holds that $max_i^j = *$ (except for e), we put $ids_i = \{\bigcup_{j=1}^{k_i} id^j \mid id^j \in m(ids_i^j) \wedge e_i^j \neq e\}$. For the purpose of the translation of the relationship type r_i itself, as described later, let us also denote all the additional (here omitted) attributes that will later on form instances of a given relationship type as $a(r_i) = \{\mathsf{a \cdot i \cdot j} \mid a \in id \wedge id \in m(ids_i^j) \wedge j \in \{1, \ldots, k_i\} \wedge e_i^j \neq e \wedge max_i^j = *\}$ in the former case, and $a(r_i) = \emptyset$ in the latter.

Now, assuming that $ids' = \{(\bigcup_{x=1}^n u(a_x)) \cup (\bigcup_{i \in R_v} id_i) \mid id_i \in ids_i\}$, we can put $ids = ids \cup ids'$ and $superid = superid \cup (\bigcup_{id \in ids'} id)$.

Example 7. Figure 4 depicts the translation of a weak entity type `Product`, resulting into an entity object `Product` with respective attribute objects and morphisms. Note how the identifier {`ProductNo,ManufactNo·1·1`} was generated.   □

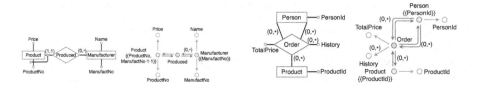

Fig. 4. Translation of weak entity types   **Fig. 5.** Translation of relationship types

Relationship Types. Let $r \in \mathbb{R}$ be a relationship type and $e_1, \ldots, e_m$ be the participating entity types for some $m \in \mathbb{N}$, $m \geq 2$. Let each of these entity types $e_i \in \mathbb{E}$ be associated with a cardinality (min_i, max_i), and $\mathbf{e_i} = (superid_i, ids_i)$ be the corresponding fully resolved entity object for e_i. Finally, let $a_1, \ldots, a_n$ be attributes associated with r for some $n \in \mathbb{N}_0$. For analogous reasons as in the previous section, let $n(ids_i) = \{\{\mathbf{a} \cdot \mathbf{i} \mid a \in id\} \mid id \in ids_i\}$ for each i. *Relationship* object $\mathbf{r} = (superid, ids)$ for r is constructed as follows: If r participates in a weak identifier for some e_i, we put $ids = ids_i$ and $superid = superid_i \cup a(r)$, as defined in the previous section. Else, if there exists at least one entity type e_i s.t. $max_i = 1$, we define $ids = \{id \mid id \in n(ids_i) \wedge i \in \{1, \ldots, m\} \wedge max_i = 1\}$. Otherwise, i.e., when $\forall i \in \{1, \ldots, m\}$ it holds that $max_i = *$, we put $ids = \{\bigcup_{i=1}^{m} id_i \mid id_i \in n(ids_i)\}$, i.e., identifiers of all the involved relationship types need to be incorporated. Under all circumstances, $superid = \bigcup_{i=1}^{m} \bigcup_{id_i \in n(ids_i)} id_i$ so that instances of $\mathbf{r}$ in the instance category can be correctly represented and fully materialized.

Next, for $\forall i \in \{1, \ldots, m\}$, two relationship morphisms are created, one for each direction. In particular, $\mathbf{r} \cdot \mathbf{e_i} \cdot \mathbf{in} = (min_i, max_i) : \mathbf{e_i} \to \mathbf{r}$, and $\mathbf{r} \cdot \mathbf{e_i} \cdot \mathbf{out} = (1, 1) : \mathbf{r} \to \mathbf{e_i}$. Finally, attributes $a_1, \ldots, a_n$ associated with r will be transformed in the same way as attributes associated with entity types.

Example 8. Relationship type `Order` between two entity types `Person` and entity type `Product` depicted in Fig. 5 is transformed to a relationship object `Order` and a set of relationship morphisms, together with respective cardinalities, connecting the relationship object `Order` with the respective entity objects `Person` and `Product`.   □

5   Categorical Framework

As depicted in Fig. 6, the whole categorical framework involves three related parts. On the left, there is the input ER schema of the reality from Fig. 1. It is automatically translated (see Sect. 4.3) to the schema category (see Sect. 4.1), and, having the particular data instances, also the instance category (see Sect. 4.2). The schema category is the core part of the framework, being an abstraction for the unified representation of the (combination of) logical model(s). The instance category is an auxiliary data structure, not just for the purpose of, e.g., representation of (intermediate) results of queries.

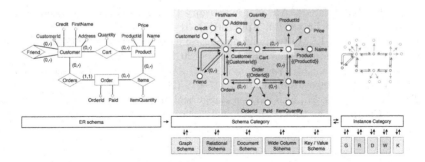

Fig. 6. Architecture of the categorical framework

As also depicted in the figure by the colors, the user is expected to denote (either in the original ER schema or in the schema category graph) the (possibly overlapping) parts corresponding to particular data models. The category is then mapped to a particular logical multi-model schema.

6 Conclusion

The main added value of the described categorical approach is that we enable the user to connect the conceptual schema with the logical layer of a particular multi-model DBMS. We also gain the following main outcomes: 1) universal applicability to any multi-model DBMS, 2) support for overlapping models, 3) natural combination with a conceptual query language, and 4) straightforward evolution management. On the other hand, there remain several open questions, such as the compact representation of the schema category, the definition of the conceptual query language itself, or the problem of reverse engineering.

References

1. Atzeni, P., Bugiotti, F., Cabibbo, L., Torlone, R.: Data modeling in the NoSQL world. Comput. Stand. Interfaces **67**, 103–149 (2020)
2. Barr, M., Wells, C.: Category Theory for Computing Science, vol. 49. Prentice Hall New York (1990)
3. Basciani, F., Di Rocco, J., Di Ruscio, D., Pierantonio, A., Iovino, L.: TyphonML: a modeling environment to develop hybrid polystores. In: MODELS 2020. ACM, New York (2020)
4. Feinberg, D., Adrian, M., Heudecker, N., Ronthal, A.M., Palanca, T.: Gartner magic quadrant for operational database management systems, 12 Oct 2015

5. Kepner, J., Chaidez, J., Gadepally, V., Jansen, H.: Associative Arrays: Unified Mathematics for Spreadsheets, Databases, Matrices, and Graphs. CoRR abs/1501.05709 (2015)
6. Kolonko, M., Müllenbach, S.: Polyglot persistence in conceptual modeling for information analysis. In: ACIT 2020, pp. 590–594 (2020)
7. Lippe, E., Ter Hofstede, A.H.M.: A category theory approach to conceptual data modeling. RAIRO-Theor. Inf. Appl. **30**(1), 31–79 (1996)
8. Lu, J., Holubová, I.: Multi-model databases: a new journey to handle the variety of data. ACM Comput. Surv. **52**(3), 1–38 (2019)
9. Schultz, P., Spivak, D.I., Vasilakopoulou, C., Wisnesky, R.: Algebraic databases. Theory Appl. Categories **32**(16–19), 547–619 (2017)
10. Tuijn, C., Gyssens, M.: CGOOD - a categorical graph-oriented object data model. Theor. Comput. Sci. **160**(1–2), 217–239 (1996)

Guaranteeing Information Integrity Through Blockchains for Smart Cities

Walid Miloud Dahmane[1], Samir Ouchani[2]([⊠]) [iD], and Hafida Bouarfa[1]

[1] Computer Science Department, Saad Dahlab University, Blida, Algeria
[2] LINEACT CESI, Aix-en-Provence, France
souchani@cesi.fr

Abstract. Given the threats that the smart city faces especially tampering the integrity of information, it has become necessary to integrate more robust and decentralized technologies that ensure transparency and sustainability of the system. Blockchain is a technology initially directed to limit the manipulation in financial transactions. One of the most popular currencies adopting this technology is Bitcoin, the latter has been very successful due to the high protection it provides. In this work, we integrate blockchain technology to become the mainstay for protecting all types of information that is collected by smart devices within a smart city. We give the blockchain structure and its internal components, and accordingly we propose an architecture for the IoT system that is compatible with the constrained devices. We explain the steps of communications between nodes, highlight the limitations, and propose solutions to them. Then, we conclude the work with experiments to show the effectiveness and the validity of the proposed architecture. The obtained results were more satisfactory, which encourage us to apply it in reality.

Keywords: Blockchain · Smart city · IoT · Cryptography · Digital signature · Hashing

1 Introduction

The world suffers from constant threats that affect the integrity of the information distributed in the city and particularly in the smart city. In the 2016 US presidential elections, there are many proofs that outsiders has carried out cyberattacks [1], this kind of attack may harm the stability of the country and spread distrust among citizens. The appropriation of sensitive information such as election results and false promotions before the elections by foreign countries may push the target countries to become a colony. There are many false stories on Internet sites that aim to increase advertising sales [2], the politicians spread misinformation during the 2009 healthcare debate [3], so, the smart city information is the result of unknown people, which forces us to collect them within a safe and reliable system. In February 2005, it was discovered that hackers were able to penetrate the system of iOS, after which 130 million devices were sold in

© Springer Nature Switzerland AG 2021
C. Attiogbé and S. Ben Yahia (Eds.): MEDI 2021, LNCS 12732, pp. 199–212, 2021.
https://doi.org/10.1007/978-3-030-78428-7_16

2014, this process allows the hackers to access the personal data of users [4]. We conclude that, the databases should be encrypted with a hard-to-crack technology that guarantees users' privacy, it is also required that the communication channels must be protected by protocols in order to avoid the sniffing attacks [5] which aim to steal the information and encryption keys.

The blockchain [6] is a digital technology, which was exploited in 2008 by an unknown person, its pseudonym is Satoshi Nakamoto to create the Bitcoin currency. This technology depends on decentralization, meaning that all the parties included in the network have the right to see the content of the information stored in the blockchain, and they also have the right to validate or reject the operations. The blockchain is difficult to falsify due to its copies are distributed to all members of the network, the mining operations need a period of time for each new block, and all blocks are related with each other using hash technology, so, the modification in one block requires a modification of the owned blockchains and all distributed blockchains on the network, but this operation is difficult to achieve. These features guarantee to the users the transparency and the integrity of data.

The researchers trends to introduce the blockchain in many fields. The Office of the National Coordinator for Health Information Technology lunched challenges to provide ideas for integrating blockchain technology in the healthcare field and the applications already created in this field. They focused on the challenges related to privacy, and technical barriers related to data storage and distribution. For this, they interested on validation, auditing, and authorization [7]. In the field of banking and financial services, this technology has been integrated in order to create a safe and continuous system [8]. In addition, online education faces obstacles such as the absence of the privacy, lack of certified results certificate, and robot's participation. Thus, blockchain technology can be combined to solve these problems [9].

The structure of the blockchain and its framework depends on the purpose of the application's domain. Many questions about the possibility of adopting this successful technology in the smart city, in order to mitigate the cyber attacks which threat the integrity of the information, especially after the terrible number of the cyber criminals. The proposed framework should be without or with few weaknesses, scalable and compatible with the characteristics of IoT devices such as the limitation of the capacities, latency, IoT protocols, WSNs, etc. The framework develops two concepts, the *independence* of the users and their lack of an intermediary party, also the possibility *detect* the criminals and prevent their goals, because the dominance of the attackers in the network (more than 50% attackers) makes the network loss its reliability. So the technology is double-edged, despite the great protection it grants, it is difficult to recover it in case of damage. Hence, our proposed solution includes the advantages of the blockchain and provides mechanisms to increase its robustness [19,20].

This contribution is organized as follows. Section 2 gives a set of contributions that have common points with our issue. Section 3 explains our proposed framework for the smart city information. Before conclusions, Sect. 4 checks the ability to realize the framework in the reality.

2   Related Work

In this section, we review the literature related to the blockchain technology in various fields including their weaknesses and strengths.

The blockchain technology is used by many domains, for example Raikwar et al. [10] adopted it to achieve the security of the insurance platform, where the transactions processes as smart contract. They implemented the framework on Hyperledger fabric and the results showed the need to chose the blockachain parameters in order to optimize the network latency. In addition, the database does not respect the privacy because the recorded data are without encryption. Also, Liu et al. [11] proposed to create a reliable system that checks the data integrity without a third party. Also, Li et al. [12] gave a crowdsourcing system, which receive the tasks from the requester and share them between the workers to solving them, the framework doesn't consist on third part. The tests show that the system is scalable and applicable. Cebe et al. [13] used blockchain technology for the forensics of the accident vehicles. It is composed of a forensic daemon inside the vehicle which receives the information from the Event Data Recorders (EDR) and broadcasts Basic Safety Messages (BSM). The forensic daemon publishes the EDR and BSM to the insurance company and the car manufacturers. These latter collects those data to analysis its. Unfortunately, the framework does not focus on the types of wireless communication technologies that require high data transmission speed and protection.

Sergii et al. [14] applied the Rolling Blockchain concept to the WSNs deployed in the smart city. Its proposed network is considered as distributed server managing the blockchain of their sub-cluster and the total blockchain. The measurement monitored from the WSNs have stored in their own blockchain, checked from all the node of the cluster and shared with the nodes remains. Since WSNs have a low capacity memory, the size of the blockchain depends on the parameters of the "worst" memory node. They gave the mathematical model for the complete chain and its segment that is removed from the original chain. They constructed a linear distribution of sensors in order to conclude if the network find a new path between two WSNs after the randomly removing links. Effectively, when they augmented the level of attack (proportion of edges removed), the network always creates an alternative paths until its break down. However, the experiments did not test the integrity of the recorded data in their proposed blockchain structure. In addition, the recording of the blockchain in the WSN makes the network constrained by the worst sensor. Hence, the proposed network structure imposes the sensors to apply the blockchain operations (verification, confirmation and storing) that affect energy storing, processing and the memorizing capacities which are limited in the sensors.

Jia et al. [15] concerned with increasing the level of protection on a crowd sensing network consisting of three parts: intelligent crowd sensing networks, confusion mechanisms, and blockchain. The crowd sensing network contains sensors which collect the users information that will be sent to the confusion mechanism, this latter regroups the sensors into 10 nodes, one of them is miner which creates new block of information. The confusion mechanism integrates the received data in the blockchain, and it gives the users virtual coins and puts the encrypted data in the server. After that, the server stores the users information and motivates the sensor to collect the information. They encode the user information using Confusion Mechanism Encode Algorithm (CMA-E) and hashes the blockchain data by Merkle tree algorithm. They created an information storage system through android application that records the data. The encoding algorithm is not strong (it can be broken) because it is based on symmetric cryptography technique.

The goal of novo et al. [16] is to propose a decentralized access control system for IoT devices by using the blockchain technology. The system is composed from WSNs, managers are responsible for the access control permissions, agent nodes to deploy the smart contracts, a smart contract contains all the operations allowed in the access management system, and a blockchain network which can be readable from all but only written by the private nodes and management hubs through an interface that translates the CoAP message received from IoT devices. The solution had an issue with the delay of the blockchain network to release access control information.

Nagothu et al. [17] suggested a secure smart service enclosing a microservices model and a blockchain mechanism. The goal is to make a reliable decentralized system and give a tamper proof of data in the insecure system. Each microservice records its collected information in its dedicated database, then the master database combines the records. After that, the miner node extracts its hash which will be added in the new block of the blockchain. They use the smart contract to give the authorized access to the videos captured by the surveillance cameras. The fogs are near to the edge process in the real time, while the cloud computing performs high protection tasks as the reorganization and discovering the malicious intents. The contribution lacks the application and analysis of the obtained results to confirm the effectiveness of the proposed hypothesis. They do not give examples to support the hypothesis, such as object-recognition algorithms, security protocols and hashing mechanisms.

3   A Blockchain-Based Integrity Framework

A smart city has different types of information/data that can change the structure of blockchain, e.g., measurement of sensor buildings, buying and selling operations, police reports, etc. First, we show the deployed blockchain structure.

3.1   Blockchain Structure

Figure 1 represents our structure that fits with a smart city need, and it is is composed of six fields.

Previous Hash. It is a copy of the previous block hash, as for the genesis block is distinguished by zero.

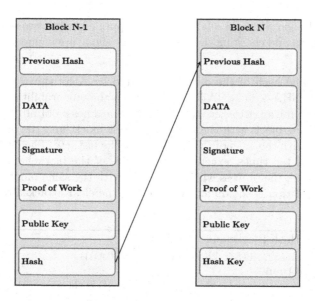

Fig. 1. Blockchain structure.

Data. Its form depends on its domain of use (e.g., WSN information, worker's personal information, etc.). Data can be readable from all network users, otherwise, the user encrypts its clear data by its RSA public key. Consequently, he should be the only one who can decrypt the message.

Signature. The data privacy guarantees that its owner is the only one who created this block, this condition is achieved through the digital signature, it is a digital fingerprint that confirms the identity of the user. The digital signature $\mathbb{S}$ is produced through encrypting the hash $\mathbb{H}$ of the block by the RSA private key (d,n), where; $\mathbb{S} = \mathbb{H}^d \bmod n$. To validate the authentication, the decryption of the signature by the RSA public key (e,n) must equal the hash of the block, i.e.; $\mathbb{H} = \mathbb{S}^e \bmod n$.

Proof of Work. It is a value used to create a hash that satisfies a pre-existing condition. It is difficult to find this value in some conditions due to the many generated possibilities. In this case, miner tools characterized with great generating power are needed. The goal is to reduce the number of attackers. Of course, the degree of difficulty of the condition is related to the sensitivity of the information, where, the more sensitive the information, the harder the condition. In some information of less importance, this part can be excluded. For example, what is the value (proof of work) for a resulting hash starts with 72 zeros? The hashing process must be calculated 2^{71} times, and a normal computer takes thousands of years to find this hash.

Public Key. The approach consists of RSA public key, it is better if its size is 2048 bits or 4096 bits (high-strength key, very high-strength key respectively) according to the sensitivity of information. The public key has two roles, encrypting personal data and checking the validity of the signature included in the block by other users.

Data Hash. The hash of the data is generated by Algorithm 1 where it uses SHA-1, SHA-2, SHA-3 and SHA-5 due to prevent the vulnerabilities of each one, so, it is difficult for an attacker to damage the weaknesses of all [18]. It hashes the hash of a new block with the hash of the last block, and the obtained hash will be hashed with the hash of the block before the last, etc. The hash operation will be repeated ten times, except when the size of the blockchain is less than ten blocks, where, the process will be repeated with the same number of blocks, the goal of repetition is the difficulty of finding the clear attacked data.

Algorithm 1. Hash Blockchain Algorithm.

1: $Hash \leftarrow SHA5(SHA3(SHA2(SHA1(New\_DATA))))$
2: **if** $NB >= 10$ **then**
3:     $n \leftarrow 9$ ▷ NB is the size of the Blockchain. .
4: **else**
5:     $n \leftarrow NB - 1$
6: **end if**
7: **for** $i \leftarrow NB, NB - n, i -- $ **do**
8:     $Hash \leftarrow SHA5(SHA3(SHA2(SHA1(Hash + Hash\_Block[i]))))$
9: **end for**
10: **return** $Hash$

3.2   Blockchain Network

The blockchain network included in the smart city is a distributed network where each part can connect to the other as illustrated in Fig. 2 where each part contains the following components.

IoT Device. It is an electric device that can *receive, send, process, store, encrypt* and *decrypt* data. Mostly, it has low storage, processing and energy consumption capacities. As the case of smartphones, tablets and WSNs, those characteristics do not allow them to store the blockchain in its limited memory.

Manager. It is an unconstrained device that can be under many types of hardware like server, computer or raspberry that featured by high capacities, which provide a blockchain to its assigned IoT device. Moreover, the validation and confirmation processes are provided by this component. Since, the MAC address is a unique ID in all the network devices and can identify 2^{48} objects, it can be considered as the best identification, especially with the increasing number of IoT devices. Thus, the manager records in its memory the MAC of the registered devices with their private and public keys.

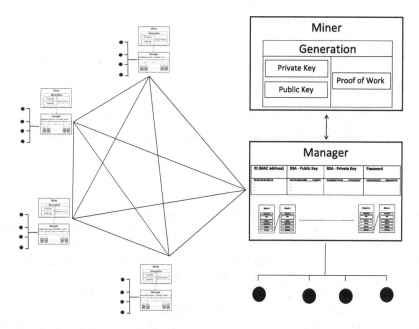

Fig. 2. Blockchain network.

Miner. In our proposed network, the mining component has two roles, the first is to create a valid hash through hashing the data and the proof of work together, the generated hash have to respect a condition predefined. The time length of the operation depends on the condition set. The difficulty of the mining process depends on the sensitivity of the information, it prevents ordinary users from generating random blocks. The second role is to create the RSA keys for each user.

Advisor. It is the first installed component in the blockchain network, and it has two functions:

- **Record New Manager:** this operation is summarized in Fig. 3(a) and it has:

• Record me and give me the dominant blockchain.
• Request the blockchain.
• Get the blockchain.
• Grant the dominant blockchain.

- **Grant the Mangers Addresses:** all managers that need to validate their new blocks require the addresses of the other network managers. This task is applied by the advisor as illustrated in Fig. 3(b).

• Store the managers addresses.
• Request the managers addresses.
• Provided the addresses.

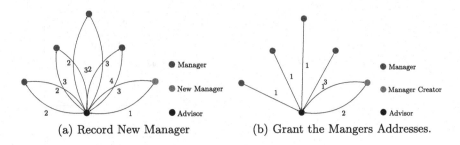

(a) Record New Manager   (b) Grant the Mangers Addresses.

Fig. 3. Advisor tasks.

Figure 4(a) illustrates the phases applied when a new manager wants to join a network of the blockchain. First, the new manager contacts the advisor, this letter records its address, then, it extracts the dominant blockchain from the other managers, we did not make the advisor as provider of blockchain for two main reasons, the first is to maintain the decentralization of the network and the second reason is to avoid an attack on advisor, which will disrupt the smooth flow of the process. Finally, the advisor provides the blockchain to the new manager.

As shown in the sequence diagram of Fig. 4(b), the IoT device sends a request of register to the manager, that checks the existence of the device in its memory. If this operation is a first inscription, the manager requests the miner about the public and the private key. Then, the miner generates a random pair of keys for the new device. The manager records this pair of keys with the ID (MAC address) of the subscribed device, and gives a password through hashing the private key. Then, it sends its password with a success message.

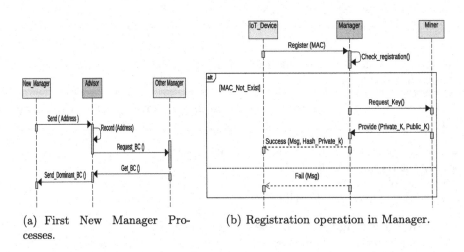

(a) First New Manager Processes.   (b) Registration operation in Manager.

Fig. 4. IoT components behaviors.

3.3   The Dominance of Fraud

Our proposed blockchain network depends on the dominant blockchain in the network. This concept may impose a weakness when the malicious users dominate the majority of the network (or 51% malicious). The attackers will participate a single malicious blockchain that serves their interests, causing a loss of network reliability. This risk occurs in two cases, when spreading malicious managers in the network, or when hacking of managers.

We propose Validation through Confidence - Algorithm (VCA) Algorithm 2 against this type of threat. It should be installed before the attack in all the managers of the network. The algorithm consists of the "confidences criteria variable" given to the managers. This variable is increasing if the verification processes are correct and do not violate the majority of the chosen managers. The variable will be converted into rank in order to classify it with the others manager.

VCA defines the interesting ranks and their ratios of managers which will test the new block, the chosen random managers will be recorded in a table. It passes the new block to the managers in order to count the managers which accept this block, if the ratio of the acceptation decision equals the ratio of rejecting decision, the algorithm will extract a new random population. If the ratio of the acceptation decision is greater, the new block will be shared on the network to add it with all managers. In addition, it increases the confidence criteria of the managers which accept the new block. Then, it deletes the addresses of the managers which reject it from the advisor and it initializes their confidence criteria. Also, if the ratio of the rejecting decision is the major, the new block will be ignored. In addition, it increases the confidence criteria of the managers which reject the new block and it deletes the addresses of the managers which accept it from the advisor and it initializes their confidence criteria.

4   Implementation and Experiments

This experiment is developed in a JAVA environment, we avoided other frameworks like Ethereum, Hyperledger, etc., because they did not support our new architecture and operations. Our decentralized network is composed of four managers related by the advisor (Fig. 5(b)), where each has a miner and two IoT devices (Fig. 5(a)).

The generated information by the IoT devices is encrypted through its public key which is provided by the miners, the blockchains of the network are installed in the managers. The P2P communication depends on the sockets. The hash Algorithm 1 searches of the proof of work that produces with the data a hash which respects this condition: every hash generated should begin with at least zero. To enhance the integrity of the data and the privacy of the users, the size of the RSA private keys consists of 2048 bits. We applied a test scenario (Fig. 6), where the managers joined by the advisor to get the dominant blockchain, the IoT devices registered in theirs mangers, the data was encrypted, validated with the managers, and added in the network blockchains.

Algorithm 2. Validation through the Confidence Algorithm (VCA)

1: **procedure** VCA(NewBlock)
2:    Define the rank and its ratio   ▷ e.g: Rank_A ← 60% ; Rank_B ← 20%.
3:    **do**
4:        **for** $i \leftarrow 1, N$ **do**   ▷ N: The number of the nodes chosen.
5:            Confidence [i] ← Random(Rank) ▷ Fill the confidence table by random nodes.
6:        **end for**
7:        **for** $i \leftarrow 1, N$ **do**
8:            Res ← Block_accept(Confidence [i], NewBlock)
9:            **if** $Res == True$ **then**
10:                decision_Accept ++ ▷ Count the nodes which accept the new block.
11:            **end if**
12:        **end for**
13:    **while** $decision\_Accept\_ratio == 50$
14:    **if** $decision\_Accept\_ratio > 50$ **then**
15:        Add_Block(NewBlock)
16:        Increase_Node()   ▷ Increase the confidence criteria of the nodes which accept the new block.
17:        Delete_Node()   ▷ Delete nodes which reject the new block.
18:    **else**
19:        Increase_Node()   ▷ Increase the confidence criteria of the nodes which reject the new block.
20:        Delete_Node()   ▷ Delete nodes which accept the new block.
21:    **end if**
22: **end procedure**

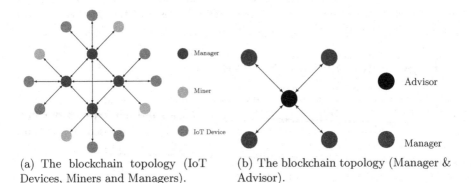

(a) The blockchain topology (IoT Devices, Miners and Managers).

(b) The blockchain topology (Manager & Advisor).

Fig. 5. Topology of the blockchain network.

4.1 Integrity Attack

In this experiment, we have a malicious person who changes the data blockchain of one manager in order to create a blockchain which serves his personal interests (Fig. 7(a)). As a next step, the damaged manager receives a request from an

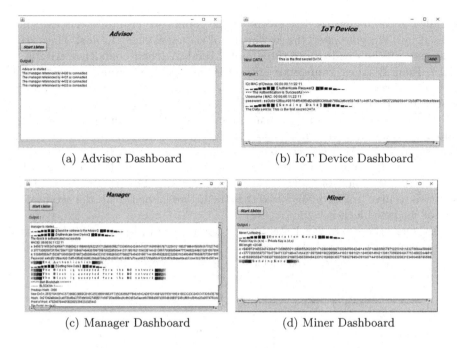

(a) Advisor Dashboard   (b) IoT Device Dashboard

(c) Manager Dashboard   (d) Miner Dashboard

Fig. 6. Registration, authentication and creation block scenarios.

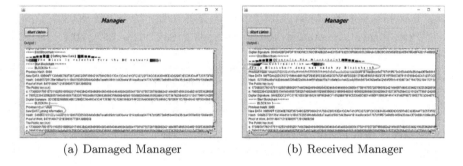

(a) Damaged Manager   (b) Received Manager

Fig. 7. Integrity attack.

IoT device that wants to add its data in the blockchains network. Then, the damaged manager shares the new block with the other managers in order to get the validation from them. Thus, the network managers find the discrepancy in its blockchain with their blockchains. As a result, the new block will be rejected (Fig. 7(b)).

4.2   The Dominance of Attackers

The dominance of the fraud (more than of 50% attackers) in the network makes the original blockchain lose its value. To prevent this type of threat, VCA was installed in the managers. It consists on the confidence criteria attributed, initialised by 1, to guarantee the legitimacy of the managers, through which, the managers will be divided into three ranks, Rank_A (their confidence criteria is high), Rank_B (their confidence criteria is average), and Rank_C (their confidence criteria is low). For each rank, VCA considers the following values: R_A = 80%, R_B = 60%, and R_C = 10%.

We fixed the number of legitimates managers to ten for all the tests as illustrated in Table 1. In the first phase, the network creates a blockchain of ten blocks and all blocks were accepted from 100% of managers. In the other tests, we increased the number of attackers which tampered our blockchain in order to convert it to a new one and make it the dominant. The results obtained are shown as a curve in Fig. 8. We found that the higher is the threat, the acceptance rate of the new block is low.

Although the threat level was high in the the third, fourth, fifth and sixth tests (because the ratio of the threat more or equal to 50%), the proposed VCA was able to accept the created block by a legitimate user. It is only rejected in

Table 1. The ratios of the acceptation of each test.

Test	T1	T2	T3	T4	T5	T6	T7
Number of legitimates	10	10	10	10	10	10	10
Number of all managers	10	15	20	25	40	50	60
Percentage of attackers	0%	30%	50%	60%	75%	40%	83.3%
Percentage of acceptation	100%	100%	83%	83%	62%	55%	50%

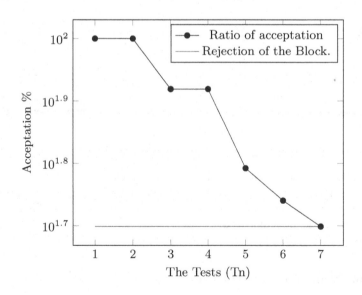

Fig. 8. Ratios of acceptation of the new block for each test.

the seventh test, and this does not mean that VCA has failed in this last test. In this case, a new configuration is created with the values (R_A = 80%, R_B = 60% and R_C = 5%), and consequently the new block was accepted with 71%. Thus, most of attackers are in the rank C, which means, we should emphasize this rank, and focus on a safer rank such as rank A.

5   Conclusion

Smart city sustainability depends on developing a reliable framework that ensures the information integrity. This work developed a framework for protecting smart city information through blockchain technology, where we suggested the effective components and their functionalities. Also, we proposed a validation algorithm through confidence to ensure data integrity against attacks. In addition, we implemented and experimented the framework by showing that the proposed architecture is secure and operates without conflicting.

In the near future, we seek to expand the work by adding to our proposed blockchain architecture a smart contract in order to manage more protection principles such as authentication and the access controls. Further, we target to use formal methods tools to prove the correctness of the proposed framework and the soundness of the developed algorithms. In addition, we aim to deploy the solution for public services such healthcare and education.

References

1. Background to "Assessing Russian Activities and Intentions in Recent US Elections": The Analytic Process and Cyber Incident Attribution, 6 January 2017. https://digital.library.unt.edu/ark:/67531/metadc94926
2. Allcott, H.: Gentzkow, Matthew: Social media and fake news in the 2016 election. J. Econ. Perspect. **31**(2), 211–236 (2017)
3. Lewandowsky, S., Ecker, U.K.H., Seifert, C.M., Schwarz, N., Cook, J.: Misinformation and its correction: continued influence and successful debiasing. Psychol. Sci. Public Interest **13**(3), 106–131 (2012)
4. Fuchs, M.H., Kenney, C., Perina, A., VanDoorn, F.: Why Americans should care about Russian hacking; center for American progress: Washington. DC, USA (2017)
5. Prabadevi, B., Jeyanthi, N.: A review on various sniffing attacks and its mitigation techniques. Indonesian J. Electr. Eng. Comput. Sci. **12**, 1117–1125 (2018)
6. Lesavre, L., Varin, P., Mell, P., Davidson, M., Shook, J.: A Taxonomic Approach to Understanding Emerging Blockchain Identity Management Systems (2019)
7. Angraal, S., Krumholz, H., Schulz, W.: Blockchain Technology: Applications in Health Care. Circulation: Cardiovascular Quality and Outcomes 10, e003800 (2017). doi: 10.1161/CIRCOUTCOMES.117.003800
8. Treleaven, P., Gendal Brown, R., Yang, D.: Blockchain Technology in Financ. Computer **50**(9), 14–17 (2017). https://doi.org/10.1109/MC.2017.3571047
9. Sun, H., Wang, X., Wang, X.: Application of blockchain technology in online education. Int. J. Emerging Technol. Learn. (iJET) **13**(10), 252–259 (2018)

10. Raikwar, M., Mazumdar, S., Ruj, S., Sen Gupta, S., Chattopadhyay, A., Lam, K.: A blockchain framework for insurance processes. In: 2018 9th IFIP International Conference on New Technologies, Mobility and Security (NTMS), pp. 1–4 (2018)

11. Liu, B., Yu, X.L., Chen, S., Xu, X., Zhu, L.: Blockchain based data integrity service framework for IoT data. In: 2017 IEEE International Conference on Web Services (ICWS), Honolulu, HI, pp. 468–475 (2017)

12. Li, M., et al.: CrowdBC: a blockchain-based decentralized framework for crowdsourcing. IEEE Trans. Parallel Distrib. Syst. 30(6), 1251–1266 (2019)

13. Cebe, M., Erdin, E., Akkaya, K., Aksu, H., Uluagac, S.: Block4Forensic: an integrated lightweight blockchain framework for forensics applications of connected vehicles. IEEE Comm. Mag. 56(10), 50–57 (2018)

14. Kushch, S., Prieto-Castrillo, F.: Blockchain for dynamic nodesin a smart city. In: 2019 IEEE 5th World Forum on Internet of Things (WF-IoT), April 2019

15. Jia, B., Zhou, T., Li, W., Liu, Z., Zhang, J.: A blockchain-based location privacy protection incentive mechanism in crowd sensing networks. Sensors 18, 3894 (2018)

16. Novo, O.: Blockchain Meets IoT: An Architecture for Scalable Access Management in IoT. 5, 1184–1195 (2018). https://doi.org/10.1109/JIOT.2018.2812239

17. Nagothu, D., Xu, R., Nikouei, S.Y., Chen, Y.: A microservice-enabled architecture for smart surveillance using blockchain technology. In: IEEE International Smart Cities Conference (ISC2), Kansas City, MO, USA 2018, pp. 1–4 (2018)

18. Maetouq, A.: Comparison of hash function algorithms against attacks: a review. Inte. J. Adv. Comput. Sci. Appl. (IJACSA) 9(8) (2018)

19. Ouchani, S.: Ensuring the functional correctness of IoT through formal modeling and verification. In: MEDI, pp. 401–417 (2018)

20. Ouchani, S., Aït Mohamed, O.: A formal verification framework for Bluespec System Verilog. FDL, Mourad Debbabi, pp. 1–7 (2013)

A Blockchain-Based Platform
for the e-Procurement Management
in the Public Sector

Hasna Elalaoui Elabdallaoui[(✉)], Abdelaziz Elfazziki, and Mohamed Sadgal

Computer Science Department, Faculty of Sciences Semlalia, Cadi Ayyad University,
Marrakech, Morocco
h.elalaoui@edu.uca.ac.ma, {elfazziki,sadgal}@uca.ma

Abstract. Public procurement represents an important part of the countries' budgets. It is impossible to effectively solve the problem of public expenditure without establishing a rational and transparent public procurement management system. These markets are an essential government function for the delivery of goods and services to citizens. The overall success of an automated public procurement function leads to progress and economic growth for the country. In this paper, we analyze the potential of blockchain technology to improve the efficiency, ease, and transparency of public procurement in the case of Morocco and identify the current challenges facing public procurement, namely the lack of trust and transparency among critical stakeholders in the procurement process. To solve these problems, we propose a blockchain-based infrastructure to improve the mechanism of public procurement. The use of this technology will reduce the time spent processing documentation, reduce the degree of corruption in the public procurement process, by creating reputation lists accessible to all participants. Also, the use of smart contracts makes it possible to minimize the number of intermediaries in the conclusion of public contracts.

Keywords: Blockchain · Public procurement · Smart contracts ·
Interoperability · Project-monitoring

1 Introduction

The scale and scope of corruption in public procurement are not new. Countries and localities have tried to counteract this behavior for years by codifying norms, prohibitions, and formal processes, but this has often only increased the costs of corruption rather than fully deterring it. On the other hand, weak rule of law or limited political will often hamper the effectiveness of such measures. Automatic management of public procurement based on blockchain technology will offer governments the potential to clean up the corruption process thanks to the contributions of digitization and the particular characteristics of blockchain. This technology enables permanent and tamper-proof record-keeping, transparency, real-time audibility, and automated "smart contracts". This increases uniformity, objectivity, and transparency. For example, the

C. Attiogbé and S. Ben Yahia (Eds.): MEDI 2021, LNCS 12732, pp. 213–223, 2021.
https://doi.org/10.1007/978-3-030-78428-7_17

blockchain [1] makes it more difficult to delete records of offers and public comments or to modify offers once submitted. This decentralizes decision-making, oversight, and record-keeping improves transparency, and delegates power to authorities who may be prone to corruption. Thus, blockchain technology creates a decentralized platform for validating transactions, data, and information that are independent of any third-party control in a verifiable, secured transparent and permanent set-up. Therefore, blockchain technology has the potential to be adopted to offer solutions for public e-procurement.

In this context, we are trying to develop a platform that allows authorities and bidders to conduct bidding and evaluation processes automatically in an Ethereum environment while allowing third parties such as decision-makers and citizens to monitor and report risky activities in real-time.

In this paper, we propose a blockchain-based e-procurement platform that takes into account key success parameters in public procurement using morocco as a case study. We commence with a thorough analysis of the procurement system in morocco. Based on the findings of this analysis, the platform elaboration crops a set of concrete recommendations as a concept study.

The remainder of the paper is structured as follows. In Sect. 2, we explain the public procurement principle and we present the current procedure of managing public contracts in Morocco. Then, a brief literature review about the use of the blockchain technology in the public procurement is presented in Sect. 3. We elaborate the proposed infrastructure in Sect. 4. The implementation of this infrastructure as a DApp is shown in Sect. 5 and a we finish with a conclusion in Sect. 6.

2   Brief Literature Review: Blockchain in Tendering Process

Public procurement is an essential element in government activities that provide added value for citizens. Blockchain applications are an emerging field of scientific research allowing several questions to be answered. Technology research is very diverse and previous work includes applications in cryptocurrency and centralized banking systems to combat corruption in the financial sector [2].

According to [3], blockchain technology has great potential to dramatically improve procurement systems, especially concerning data integration between business functions. Thus, the infusion of blockchain technology is driving a positive change in the way organizations operate, creating more opportunities for growth, innovation, and expansion on a global scale.

The authors in [4] see that Russia suffers from poor quality public procurement management problems that require in-depth study and assessment of the level of efficiency. They envision that the use of blockchain in public procurement will reduce the time spent processing documentation, reduce the degree of corruption in the public procurement process, by creating reputation lists accessible to all participants in this process. Also, the use of a smart contract makes it possible to minimize the number of intermediaries in the conclusion of public contracts.

Public procurement in Nigeria is nonetheless affected by current challenges such as lack of trust and transparency among stakeholders in the procurement process. To address these issues, a blockchain-based [5] framework is developed to enable interoperability of information systems involved in the procurement process, increase citizen participation in determining project requirements, and enable monitoring and reporting of more transparent projects.

In the same perspective, the bid selection process in Indonesia is full of competition. The authors in [6] must be supported by efficient and transparent technology so that it does not cause fraud and suspicion among bidders. They estimate that by adopting a blockchain, electronic tendering can reduce the source of fraud, namely the manipulation of databases. Their research explored the process of developing a blockchain-based system.

3   Process of Awarding Open Tenders in Morocco

Public procurement is a major economic pillar with considerable weight, which explains the great interest of governments, international institutions, companies, and other organizations in the governance of public procurement. We cannot begin to discuss public procurement management without looking at public procurement procedures and the general process of open tendering.

The preparation of a public contract goes through several phases. It begins with the operation of determining the need to be satisfied and ends with the publication of the call for tenders as shown in Fig. 1.

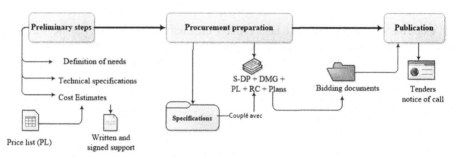

Fig. 1. Process of preparation and publication of a call for tenders

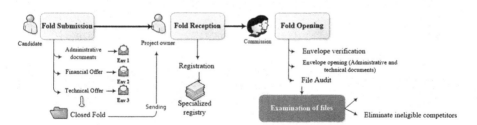

Fig. 2. Submissions' reception and opening

At the launch of the call for tenders, each competitor is required to submit a complete file in a closed envelope as regulated by Article 29 of the Public Procurement Code[1].

---

[1] https://www.finances.gov.ma/Publication/depp/2014/decret_2_12_349.pdf.

These envelopes are then received and opened by a commission one by one in a public session as shown in the Fig. 2. The call for tenders' commission shall conduct the examination and evaluation of the technical offers and any other technical documents, if requested, behind closed doors. Thereafter, the same commission examines the financial offers in an open session. These evaluations concern only the candidates admitted at the end of the examination of all the provided documents. They are carried out according to the process shown in Fig. 3.

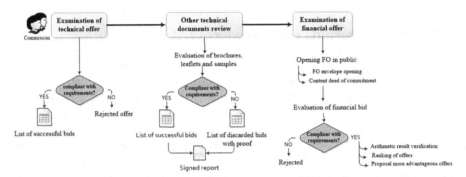

Fig. 3. Examination and offers evaluation

Once the successful bids have been ranked, the final step is to select the most advantageous bid (the lowest bid), make the results of the bid review and make them available to the public, and finally award the contract to the successful bidder according to the process shown in the Fig. 4.

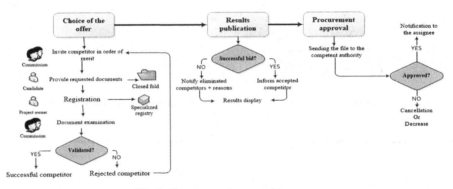

Fig. 4. Procurement approval process.

4   The Proposed Blockchain-Based Platform for Public Procurement

It is unfair to see how multitudes of people start up different businesses to provide services to governments, but cannot get a chance to put their skills to use because of the bribes, favoritism, and corruption that exist. Deploying the blockchain would mean time is revoked for patronage, bribes, and family ties. The blockchain is always linked to transparency, an aspect that is lacking in the bidding process.

Therefore, we believe that blockchain technology, coupled with smart contracts, is the solution that can mitigate the problems related to public procurement. As defined, a blockchain can have main practical interests [7–10], such as guaranteeing:

- Automation and immutability with tamper-proof files and submissions.
- Impartiality with decentralized ownership and authority of information.
- Real-time transparency and auditability with information that can be accessed by authorized or unauthorized access. This transparency, coupled with ease of access, allows a holistic view of award results.
- The deployment of smart contracts drastically reduces manually managed processes, in particular the review and evaluation of bidders' files.

The objective of this contribution is to replicate the public procurement process by dematerializing some of the common steps involved in managing calls for tenders. This dematerialization, based on the blockchain, is mainly implemented by the execution of the functions of smart contracts as explained in Table 1.

Table 1. Alignment of the traditional procedure vs. blockchain procedure

Traditional phase	Dematerialized phase	Implementation
Registration	Registration of suppliers and administrators	Account creation with private/public key pair and database storage
Folds sending	Submission of the offer	Preparation of the transaction
Fold reception	Deadline Verification	Executing the 'isOnDelay' function of the 'Register_Submission' contract
	Receipt of bids	Insertion in the blockchain: Execution of the 'Register' function of the 'Register-Submission' contract
Opening of bids in public session	Verification of compliance	Execution of the 'compliance_review' function of the 'Assess_offer' contract

(continued)

Table 1. (*continued*)

Traditional phase	Dematerialized phase	Implementation
Review and evaluation of bids submitted	Review and evaluation of bids submitted	Execution of the function 'OF_OT_assessment' of the contract 'Assess_offer'
Choice of the offer	Awarding of the contract	Execution of the function 'Attribution' of the contract 'Assess_offer'
Publication of results	Insertion in the database of successful bids	Launch of the verification_confirmed event of the second contract

4.1 Presentation of the Blockchain-Based Infrastructure

The proposed infrastructure is composed of four main components as shown in the Fig. 5. We explain the role of each component as follows:

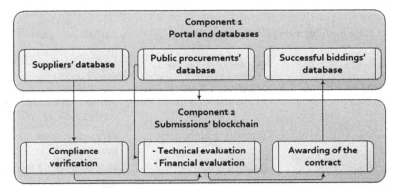

Fig. 5. The infrastructure components

The Submissions' Blockchain

The bids are governed by a blockchain to avoid any falsification or alteration in the proposed offers. This chain stores all the applications and offers submitted by competing suppliers. This includes all supporting documents that can attest that the bid meets the criteria requested by the buyer and its suitability for a published offer.

We chose Ethereum as a secure and reliable decentralized system for the storage and transaction of information [11]. In Ethereum, there are two types of accounts: user accounts allowing transactions; and smart contract accounts that can run a program when they receive a transaction containing a message.

The bidding block is a private Ethereum blockchain [11] with permission to which only authorized persons can have access (administrator); who is responsible for the examination of all bids. This blockchain contains smart contracts to open and/or close the bids within the given time frame and codes for the verification and evaluation of the bids.

The Proposed Tender Management Process

The company/supplier submits its offer and once the submission deadline has been reached, it cannot be changed. In essence, the tender process should be completely confidential. In this regard, adopting smart contracts is essential to eliminate questionable transactions and favoritism. The smart contract also includes a code indicating how the evaluation of the offers is carried out. This code is immutable and irreversible. Figure 6 illustrates the process for managing offers.

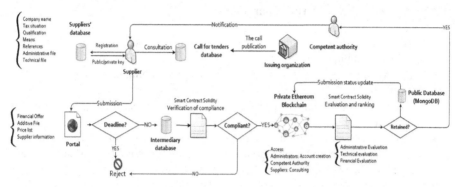

Fig. 6. Proposed tender management process

This process has the potential to ensure secure and transparent management. Although it is difficult to implement, it is uncertain whether the process is supported without human intermediation. In essence, governments can guarantee access to information to citizens and provide a safe and fair process that allows for the proper allocation of bids to deserving bidders.

4.2 Smart Contracts

In this work, smart contracts are widely used to autonomously and securely manage the procurement process. This process includes the following treatments: Deadline for submission, Bid Registration, Compliance Verification, Technical and financial evaluation, and the Award of the contract. Once all the scores have been added up, the contract is awarded to the bidder with the highest and lowest score among the successful bids. All these processes will constitute the functions of the two smart contracts integrated with the bidding block, whose attributes, functions, and events are:

1. The 'Register_Submission' contract's parameters:

- address public bidder: Bidder's address
- Uint startTime: Necessary attribute to store the submission date.

- isOnDelay(uint time): A modifier that verifies whether the submission was made on time or past the 'time' deadline.
- Launch_verification(address from, address to):Event launched when the transaction is recorded. It allows to pass the transaction to the second contract for validation.
- Register_Submission() public: A constructor generating a quote by assigning its bidder's address and initiating the submission date.
- Register(address receiver) isOnDelay(date): This function allows, if the deadline is respected, to save the submission in the blockchain and to launch the 'launch_verification' event

2. **The 'Assess_Offer' contract's parameters:**

- Struct submission { Address bidder; Uint score;}: A structure gathering all the details of a submission
- mapping (address = > Submission): associates a submission's address with its information.
- Assess_Offer() public: Constructor initiating the verification process (with 0 as a score for all bids).
- OF_OT_assessment () public isReviewed(): The function that assigns a score to the technical offer and one to the financial offer of each bid that has successfully passed the first phase.
- Compliance_review() public returns (bool): This function allows to check the conformity of the bids by returning true if the parts are well supplied and increments the evaluation score of each bid.
- Attribution () public returns (address[]) isAssessed(): this function allows to recognize the contract awarder according to the ranking of scores. It returns the first 3 bids that have been previously evaluated. At the end, it launches the event 'Verification_Confirmed'.
- isReviewed(): verifies whether the submission has passed the compliance verification phase.
- isAssessed(): verifies whether the bid has successfully passed the technical and financial evaluation.
- Verification_confirmed (address from, address to, uint amount): Event launched at the end of the evaluation process to publish the results in the database and display them publicly on the portal.

5 A DApp for the Public Procurement

DApps can run on a P2P network as well as on a blockchain network. In this case, data is stored in an encrypted and transparent manner. In the context of public procurement, a decentralized application will be placed on a blockchain to manage bidders' bids. Once a bid is posted, no one - including the creators of the application - will be able to delete the submitted bid. The modification will be possible by the bidder, but the original bid can never be altered. To correctly illustrate the sequence of actions in the decentralized application, we present, step by step, in Fig. 7, the complete life cycle of a bidding transaction and its validation.

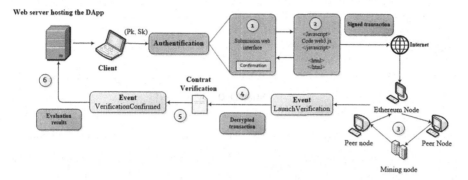

Fig. 7. Lifecycle of a bidding transaction with the decentralized application

5.1 The DApp Implementation

For the implementation of the application and its evaluation, we sequence a set of steps. A first step is to install and configure the private blockchain to develop smart contracts and deploy them locally. The following tools have been chosen and installed for very specific reasons:

- Truffle: a very powerful framework facilitating the development of the client-side application, the interaction with the contracts, and their deployment in any network.
- MetaMask: a plugin allowing to transform any browser into a blockchain browser. Besides, it also allows the management of accounts and Ether funds for the payment of transactions.

We chose Ropsten Test Net Ethereum network to deploy the contracts because it is the most decentralized and most similar to MainNet. It provides 10 Ethereum accounts with their addresses and private keys.

The decentralized application was deployed on an Ethereum test network as explained in the previous section. In this network, we sent about 100 transactions to simulate candidate bids participating. This Ethereum network was configured in such a way that each block had a maximum capacity of approximately 3,000,000 gas units. The storage in a block is divided with 22% of the capacity allocated to the 'Register_Submission' contract, 54% to the second smart contract, and the rest for the storage of hashed data.

Figure 8 shows the total gas consumption of each validation phase of a transaction concerning the number of participants, from the registration of a participant to its offer evaluation, if selected.

We also conducted a brief cost analysis of the two contracts execution on the virtual machine and the Ethereum network, examining the gas requirements for a full invocation

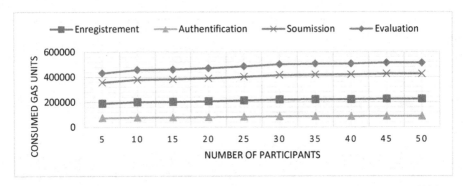

Fig. 8. Units of gas consumed for each phase of the process according to the number of bidders

of each smart contract. The results are presented in Table 2. These costs can be converted into U.S. dollars (or any other currency) to have a total estimate of the operations integrated into the Ethereum blockchain.

Table 2. Gas consumption costs of the operations of the two smart contracts

Contrat	Name of the operation	Cost of the transaction	Cost of execution	Total gas consumed
Register_Submission	Contract creation	219 535	124 371	343 906
	Constructor	62 279	41 007	103 286
	Deadline checking	22 299	1027	85 605
	Submission storage	71 745	10 650	82 395
Assess_Offer	Contract creation	222 546	139 401	361 947
	Constructor	74 120	45 125	119245
	Compliance verification	18 523	2001	20 524
	Offers' evaluation	21 744	3250	24 994
	Awarding of the contract	65 478	11 450	76 928
Total				1 218 839

6 Conclusion

This work presents a practical implementation of a prototype of a public procurement management solution and more specifically the open tendering process using blockchain technology and smart contracts as a means of recording and verifying submitted bids. Based on the observation that this sector is rife with fraud and incidents of corruption, and that a considerable number of bidders are excluded or treated unequally, we have designed a new mechanism that allows for greater transparency and flexibility in the processing of applications at a very low cost.

Ethereum and smart contracts are progressive achievements from the blockchain, transformed into a versatile response to some of the related problems in today's world, and which could allow wide use of the blockchain very soon.

References

1. Zheng, Z., Xie, S., Dai, H., Chen, X., Wang, H.: An overview of blockchain technology: architecture, consensus, and future trends. In: Proceedings - 2017 IEEE 6th International Congress Big Data, BigData Congress, pp. 557–564 (2017). https://doi.org/10.1109/BigDataCongress.2017.85.
2. Nicholson, J.: The library as a facilitator: how bitcoin and block chain technology can aid developing nations. Ser. Libr. (2017). https://doi.org/10.1080/0361526X.2017.1374229
3. Weissbarth, R.: Procurement 4.0: Are you ready for the digital revolution ? Strategy& (2016)
4. Kosyan, N.G., Mil'kina, I.V.: Blockchain in the public procurement system, E-Management (2019). https://doi.org/10.26425/2658-3445-2019-1-33-41.
5. Akaba, T.I., Norta, A., Udokwu, C., Draheim, D.: A framework for the adoption of blockchain-based e-procurement systems in the public sector. In: Hattingh, M., Matthee, M., Smuts, H., Pappas, I., Dwivedi, Y.K., Mäntymäki, M. (eds.) I3E 2020. LNCS, vol. 12066, pp. 3–14. Springer, Cham (2020). https://doi.org/10.1007/978-3-030-44999-5_1
6. Yutia, S.N., Rahardjo, B.: Design of a blockchain-based e-tendering system: a case study in LPSE. In: Proceeding - 2019 Int. Conf. ICT Smart Soc. Innov. Transform. Towar. Smart Reg. ICISS 2019 (2019). https://doi.org/10.1109/ICISS48059.2019.8969824.
7. Ølnes, S., Ubacht, J., Janssen, M.: Blockchain in government : Benefits and implications of distributed ledger technology for information sharing Blockchain in government: benefits and implications of distributed ledger technology for information sharing. Gov. Inf. Q. **34**, 355–364 (2017). https://doi.org/10.1016/j.giq.2017.09.007
8. Janssen, M.: Blockchain applications in government Challenges of Blockchain Technology Adoption for e-Government : A Systematic Literature Review (2018). https://doi.org/10.1145/3209281.3209329
9. Aristidou, C., Marcou, E.: Blockchain standards and government applications. J. ICT Stand. **7**, 287–312 (2019). https://doi.org/10.13052/jicts2245-800X.736.
10. Ojo, A., Adebayo, S.: Blockchain as a next generation government information infrastructure: a review of initiatives in D5 countries. In: Ojo, A., Millard, J. (eds.) Government 3.0 – Next Generation Government Technology Infrastructure and Services. PAIT, vol. 32, pp. 283–298. Springer, Cham (2017). https://doi.org/10.1007/978-3-319-63743-3_11
11. Founder, G.W., Gavin, E.: Ethereum: a secure decentralised generalised transaction ledger (2017)

Databases and Ontologies

COVIDonto: An Ontology Model for Acquisition and Sharing of COVID-19 Data

Jean Vincent Fonou-Dombeu$^{(\boxtimes)}$, Thimershen Achary, Emma Genders, Shiv Mahabeer, and Shivani Mahashakti Pillay

School of Mathematics, Statistics and Computer Science,
University of KwaZulu-Natal King Edward Avenue, Scottsville, Pietermaritzburg
3209, South Africa
`fonoudombeuj@ukzn.ac.za,`
`{216005583,217057683,217022172,217039130}@stu.ukzn.ac.za`

Abstract. The collection and sharing of accurate data is paramount to the fight against COVID-19. However, the health system in many countries is fragmented. Furthermore, because no one was prepared for COVID-19, manual information systems have been put in place in many health facilities to collect and record COVID-19 data. This reality brings many challenges such as delay, inaccuracy and inconsistency in the COVID-19 data collected for the control and monitoring of the pandemic. Recent studies have developed ontologies for COVID-19 data modeling and acquisition. However, the scopes of these ontologies have been the modeling of patients, available medical infrastructures, and biology and biomedical aspects of COVID-19. This study extends these existing ontologies to develop the COVID-19 ontology (COVIDonto) to model the origin, symptoms, spread and treatment of COVID-19. The NeOn methodology was followed to gather data from secondary sources to formalize the COVIDonto ontology in Description Logics (DLs). The COVIDonto ontology was implemented in a machine-executable form with the Web Ontology Language (OWL) in Protégé ontology editor. The COVIDonto ontology is a formal executable model of COVID-19 that can be leveraged in web-based applications to integrate health facilities in a country for the automatic acquisition and sharing of COVID-19 data. Moreover, the COVIDonto could serve as a medium for cross-border interoperability of government systems of various countries and facilitate data sharing in the fight against the COVID-19 pandemic.

Keywords: COVID-19 · Ontology · Pandemic · Protégé · Web Ontology Language

1 Introduction

The collection and sharing of accurate data is paramount to the fight against COVID-19 [1]. Reliable data is important for the day to day tracking and tracing

C. Attiogbé and S. Ben Yahia (Eds.): MEDI 2021, LNCS 12732, pp. 227–240, 2021.
https://doi.org/10.1007/978-3-030-78428-7_18

of cases of infections [2]. The daily availability of accurate data on testing, infections, and fatalities would be of great benefit for the monitoring and control of the pandemic at local/regional, country/central and global levels [1]. However, the health systems of many countries around the world are fragmented and do not have integrated systems between private and public hospitals, clinics, health centers, etc. for the electronic capturing and sharing of COVID-19 data in real-time.

Horgan et al. [2] reported the disconnection between the states, clinics and public health systems in the US. The authors explained that there are inconsistencies in the COVID-19 data being collected in the US, due to decentralized health systems and the lack of common definitions of the data as well as common governance and reporting rules. Makulec [3] added that the US health system is highly decentralized and managed at country and state levels; this poses serious challenges in the collection and merging of health data across all US states. Similar criticism was made for the Swiss government; in fact, due to a decentralized health system, weak infrastructure and inappropriate methodology, the Swiss government has failed to accurately record cases of COVID-19 infection [2]. Although progress has been made in the UK in the digitalization of medical records which has so far provided some efficiency in the collection and sharing of COVID-19 data amongst health professionals, the full integration of public hospitals and clinics is still to be achieved [2]. Horgan et al. [2] also reported disparities in the health care systems in Canada; they explained how the Ontario province has a fully digitalized health system, whereas, the Quebec province health system is still partially digitalized. Makulec [3] further explained that each province in Canada has its own health system; making a total of 13 heterogeneous health systems in Canada. The author further reported the extended digitalization of health records in the Canada health systems that facilitate the sharing of medical data across provinces, but indicated that the collection of COVID-19 data for contact tracing remained paper-based. Furthermore, there is inconsistencies in the COVID-19 data being collected due to the absence of a common data model for representing patients in digital health records; this undermines the interoperability of proprietary systems of health facilities and prevents the historical tracing of signs, symptoms and diagnostic tests of a patient [2].

Attempts have been made in many countries to use apps to collect COVID-19 data from citizens; however, only a low percentage of citizens download apps; it was reported how only 13% of people has downloaded an app launched by the Singapore government and that in the UK, the government is hoping to reach 50% of the population with the COVID-19 app [4]. In many countries, COVID-19 data are currently being recorded manually by health workers at local/regional levels and faxed/scanned to central government where they are aggregated. This has been witnessed in the US and Switzerland [2]. Bazzoli [5] reported that manual paper-based processes are still being used in public health agencies in the US to gather and submit daily COVID-19 data. In Switzerland, cantons are capturing new COVID-19 cases on forms; these forms are then faxed to the central government where they are manually aggregated [2]. Furthermore, the collected COVID-19 data are recorded in different formats at various Health entities [1]. The abovementioned shortcomings in the COVID-19 data collection systems in developed countries including US, UK and Switzerland can certainly be observed in other industrialized nations as well as many developing countries where there is little or no digitalization of medical records. This reality brings many challenges such as delay, inaccuracy and inconsistency in the COVID-19 data collected for the control and monitoring of the pandemic.

Recent studies [6–8] have attempted to use Semantic Web technologies for COVID-19 data modeling and acquisition. The authors in [6] proposed the use of semantic web services and Big Data technologies to extract COVID-19 data from multiple sources in real-time for statistics and reporting. The COVID-19 disease pattern is modelled with DLs in [7] with the aim of providing unambiguous information about the spread of COVID-19. Ontologies are used in [8] to develop an online repository for real-time analysis of COVID-19 trials. Ontologies have been developed to model various aspects of COVID-19 pandemic in [9–12]; however, the authors interests were on the modeling of patients [9,11] and medical infrastructure [9], the biology [10] and biomedical [10,12] aspects of COVID-19. This study extends these existing ontologies to develop the COVID-19 ontology (COVIDonto) to model the origin, symptoms, spread and treatment of COVID-19.

The NeOn methodology [13] was applied to gather data from secondary sources to formalize the COVIDonto ontology in Description Logics (DLs). DLs is a family of logic-based knowledge representation languages that provide features for representing ontology's elements including concepts/classes, properties/roles, instances/individuals and their combinations with operators such as union, intersection, negation, etc., as well as restrictions/constrains on these elements [14]. The modelling of an ontology in DLs facilitates its implementation in OWL because the syntax of the OWL language is based on DLs. The proposed COVIDonto ontology was implemented in a machine-executable form with the OWL language in Protégé ontology editor. Protégé is an open source ontology editing platform that can be downloaded free of charge from the internet to develop ontologies. The proposed COVIDonto ontology is a formal executable model of COVID-19 that can be leveraged in web-based applications to integrate health facilities in a country for the automatic acquisition and sharing of COVID-19 data. Furthermore, in this era of Big data and machine learning, the COVIDonto ontology can be used to store relevant COVID-19 data that can serve to draw various statistical analyses for research and decision making. Moreover, the COVIDonto could serve as a medium for cross-border interoperability of government systems of various countries and facilitate data sharing in the fight against the COVID-19 pandemic.

The rest of the paper is structured as follows. Section 2 discusses related studies. The methodology used to develop the COVIDonto ontology is presented in Sect. 3. The experiments and discussions are given in Sects. 4 and 5 and a conclusion ends the paper in the last section.

2   Related Work

Authors have developed ontologies to model diseases and pandemics. In [15], the Disease Ontology (DO) is presented; it models the diseases that affect human beings in the form of taxonomy or inheritance relationships between disease types. The DO is a large ontology that integrates concepts from the Unified Medical Language System (UMLS), Systematized Nomenclature of Medicine Clinical Terms (SNOMED CT) and International Classification of Diseases (ICD-9) [15].

Another relevant ontology is the Infectious Disease Ontology (IDO) [16]. IDO is constituted of various modules that model the domain of infectious diseases in all aspects. One prominent module of IDO is the IDO Core which defines a number of diseases and pathogen neutral terms related to infectious diseases. IDO Core is based on Open Biomedical and Biological Ontologies (OBO) Foundry principles [17]. The terms used in IDO Core are aligned under the Basic Formal Ontology (BFO), which is an upper level framework used by the ontologies that form the OBO Foundry [18]. By using BFO as its upper level framework, IDO core is automatically integrated and interoperable with more than 300 ontologies that use BFO. Furthermore, any ontology that extends the IDO Core is also interoperable with these ontologies.

The IDO Core has been extended in a number of ontologies such as the Coronavirus Infectious Disease Ontology (CIDO) [19] and Brucellosis Ontology (IDOBRU) [20]. The CIDO ontology covers information related to coronavirus diseases such as the transmission, prevention, diagnosis, and so forth [19]; but it does not address the COVID-19 disease specifically. The CIDO ontology is relevant as it is a good source of terms that can be reused in the COVIDonto ontology proposed in this study. The IDOBRU ontology models information related to the brucellosis domain [20]. Brucellosis is an infectious disease caused by bacteria. The IDOBRU ontology contains terms related to symptoms and treatment aligned under IDO Core; this makes it relevant to this study. In fact, the terms symptom and treatment and their positions in the IDO Core hierarchy can be reuse in the COVIDonto ontology proposed in this work. The COVIDonto developed in this work reuses terms from both CIDO and IDOBRU and can therefore be seen as another extension of IDO Core.

Recent studies [9–12] have developed ontologies to model various aspects of COVID-19 pandemic. In [9], the authors developed an ontology, namely, CODO to model COVID-19 cases with emphasis on the patient and medical infrastructures; Sargsyan et al. [10] modeled the biological aspects of COVID-19 through the molecular and cellular entities as well as medical and epidemiological concepts; an ontology for clinical decision support to COVID-19 patients was developed in [11]; authors in [12] extended the Infectious Disease Ontology (IDO) to represent the biomedical aspects of COVID-19. The scopes of the above studies were on the modeling of patients [9,11] and medical facilities [9], the biology [10] and biomedical [10,12] aspects of COVID-19. This study extends the existing ontologies to model the origin, symptoms, spread and treatments of COVID-19. Furthermore, the proposed COVIDonto can be merged with these existing ontologies to build a more comprehensive ontology for the COVID-19 pandemic.

The materials and methods of the study are presented and applied in the next section.

3   Materials and Methods

The NeOn ontology development methodology [13] was used to develop the COVIDonto ontology. The NeOn methodology provides a glossary of processes and activities, as well as guidelines for developing ontologies. Different scenarios are proposed in the NeOn methodology [13], each scenario represents a different approach to developing ontologies. This work follows a mixture of scenario 4: reusing and re-engineering ontological resources and scenario 2: reusing and re-engineering non-ontological resources. Non-ontological resources include related published articles, research reports and websites, whereas, ontological resources are the related ontologies and taxonomies discussed in the related work section above.

3.1   Data Collection

Following the guidelines of the NeOn methodology [13], data was collected from ontological and non-ontological resources on four key information areas including the origin, symptoms, spread and treatment of COVID-19. These four areas were guided by the WHO [21] and broadly provide the scope for the COVIDonto ontology. Data about the origin and classification of COVID-19 was gathered from [11,15,21,22]. Data about the symptoms of COVID-19 was gathered from [21,23]. Data about the spread of COVID-19 was gathered from [21,24]. Data about the treatment of COVID-19 was gathered from [21,25,26]. The collected data was used to formalized the COVIDonto in the following subsection.

3.2   Construction of Vocabulary of the COVIDonto Ontology

Initially, a set of competency questions taken from the WHO website on COVID-19 [21] were considered. In ontology engineering, competency questions are questions in natural languages that are used to specify the requirements and delimit the scope of an ontology [27,28]. The competency questions selected from the WHO website for the COVIDonto ontology, include "What is COVID-19?", "What are the symptoms of COVID-19?", "How does COVID-19 spread?" and "Is there a vaccine, drug or treatment for COVID-19?" [21].

Thereafter, the knowledge fulfilling the aforementioned competency questions was extracted from the data sources. This knowledge was thoroughly analyzed to model concepts that represent various aspects of COVID-19 as well as the relations between these concepts. The origin of COVID-19, that is, the date and place where the COVID-19 pandemic appeared for the first time was modeled with concepts such as date, location, province, country as well as the biological origin of COVID-19, represented with concepts including virus, SARS-COV-2,

Sarbecovirus, Riboveria, Nidovirales, Coronavidae, Orthocoronavidae, and Beta-coronavirus; these concepts were adopted from related ontologies discussed in the related work section. The concepts that model the symptoms of COVID-19 include the people who may be infected by COVID-19 such as the adolescents, adults and elderlies; the risk classification such as no, low or high risk; various symptom classifications used in the COVID-19 literature including less, most, serious, mild, moderate and severe symptoms. The spread of COVID-19 was modeled with concepts that represent (1) the modes of infection such as medium and material, various types of media including droplet and radio waves, and materials including surface, plastic, stainless steel, cardboard and copper, (2) the parts of a person that can be infected by COVID-19 such as body part, hand and face, and (3) the preventive measures such as disposable mask, self-isolation and quarantine, and so forth. The treatment of COVID-19 was modelled with concepts such as medicine, treatment, medical equipment, various types of treatment including monitoring, symptomatic, pneumonia and severe pneumonia treatments, various classification of patients such as mild, moderate and severe case patients, the types of medicines including oral and injectable medicines, other relevant concepts such as emergency sign, symptom progression, and so forth. For each concept obtained, an exhaustive list of relations or statements that link the concepts to other information or knowledge in COVID-19 domain was formulated; for instance, for the less common symptom concept, it was listed the less common symptoms of COVID-19 including aches and pains, nasal congestion, headache, conjunctivitis, sore throat, diarrhea, loss of taste, loss of smell, rash on skin, Discoloration of fingers, discoloration of toes and so forth. In total, 163 statements in natural language were formulated to comprehensively describe the COVID-19 pandemic in terms of the concepts and relations.

After the modeling of concepts and relations, the next steps in the formalization of the COVIDonto ontology consisted to further analyze the relations to identify the properties and instances of the ontology. This process resulted in 38 properties or relationships in the COVIDonto ontology. The properties are the semantic relations between pair of concepts or between concepts and instances of the COVIDonto ontology. These properties are mainly unary, binary and taxonomy relations between the concepts and instances of the COVIDonto ontology. For instance, the property *has lung problems* is used in the COVIDonto ontology to represent the relationship between the concepts *Person* and *High Risk*, to mean that a *Person* who has lung problems is at *High Risk* of contracting COVID-19. Another example is the property *spread through* which models the relationship between the instances *SARS-COV-2* and *"oral discharge of infected person"*; meaning that *SARS-COV-2* spreads through oral discharge of infected person.

The taxonomy of relations in the COVIDonto ontology is modeled with the *"is a"* property. The *"is a"* property represents the inheritance relations between the concepts of the COVIDonto ontology. An example is the *"is a"* relation between the concepts *Elderly* and *Person*, to represent that an *Elderly* is a

Person. Another example is the "is a" relation between the concepts *Riboveria* and *Virus*, to say that *Riboveria* is a *Virus*. For instance, the *"is a"* relations are used to represent the COVID-19 symptoms and their categorization within the COVIDonto ontology. The physical symptoms of COVID-19 are classified into less common, most common and serious symptoms. The list of symptoms in each category is also modeled. For instance, the most common symptoms of COVID-19 are tiredness, dry cough and fever and the severe symptoms include loss of speech, chest pain, loss of movement and difficulty breathing.

Table 1. Part of FOL representation of axioms of the COVIDonto ontology

	FOL formulas
1	$\forall x(Person(x) \land InfectedWith(x, COVID19) \rightarrow \exists y(symptom(y) \land (experiences(x, y) \lor \forall z(\neg symptom(z) \land \neg experiences(x, z))))$
2	$\exists x \exists y(Person(x) \land MedicalProblem(y) \land hasUnderlying(x, y))$
3	$\forall x(Person(x) \land hasUnderlyingMedicalProblem(x) \rightarrow \exists y(HighRisk(y) \land hasARisk(x, y)))$
4	$\forall x(Person(x) \land hasUnderlyingMedicalProblem(x) \rightarrow (hasHighBloodPressure(x) \lor hasHeartProblems(x) \lor hasLungProblems(x) \lor hasDiabetes(x) \lor hasCancer(x)))$
5	$\forall x(Elderly(x) \rightarrow \exists y(HighRisk(y) \land hasARisk(x, y)))$
6	$\forall x(Adolescent(x) \rightarrow \exists y(LowRisk(y)hasARisk(x, y)))$
7	$\forall x(Symptoms(x) \rightarrow \exists y(Day(y) \land startBetween(x, y) \land (\geq 10y \leq 14))$
8	$\forall x(Person(x) \rightarrow \exists y(DisposableMask(y) \land ShouldNotReuse(x, y)))$
9	$\forall x \forall z \forall m(Person(x) \land Hand(z) \land Face(m) \land touch(x, m) \land touch(x, ContaminatedSurface) \land \neg washing(x, z) \rightarrow InfectedPerson(x))$
10	$\forall x \forall y(Person(x) \land COVID19Symptom(y) \land Experience(x, y) \rightarrow \exists z(Home(z) \land RemainsAt(x, z) \land isInSelfIsolation(x)))$
11	$\forall x \forall y(IsolatedPerson(x) \land COVID19Symptom(y) \land ExposedTo(x, COVID19) \land \neg Experience(x, y) \rightarrow isInQuatantine(x))$
12	$\forall x \exists y(Patient(x) \land presentEmergencySign(x, y) \land EmergencySigns(y)) \rightarrow isTreatedWith(x, AirwayManagement) \land isTreatedWith(x, OxygenTherapy))$
13	$\forall x \exists y(EmergencySign(y) \land presentsEmergencySign(x, y) \rightarrow Patient(x))$

Instances or individuals of the COVIDonto ontology were also gathered. Instances or individuals in an ontology are the occurrences of concepts or properties in the ontology. For example, the instances *China* and *Wuhan* are the occurrences of the concepts *Country* and *Province* in the COVIDonto ontology, respectively. Other examples are *"oral discharge of infected person"* and *"nasal discharge of infected person"* which are instances or occurrences of the concept *Droplet.* The list of concepts/classes, properties/roles and instances/individuals gathered, constituted the vocabulary of the COVIDonto ontology.

Table 2. Part of DLs representation of axioms of the COVIDonto ontology

	DLs Axioms
1	$Person \sqcap \exists InfectedWith.COVID19 \sqsubseteq \exists experiences.symptom \sqcup forall does NotExperience.Symptom$
2	$Person \sqcap \exists hasUnderlying.MedicalProblem$
3	$Person \sqcap \exists hasUnderlying.MedicalProblem \sqsubseteq \exists hasARisk.HighRisk$
4	$Person \sqcap \exists hasUnderlying.MedicalProblem \sqsubseteq \exists has.(HighBloodPressure \vee HeartProblem \vee LungProblem \vee Diabete \vee Cancer)$
5	$Elderly \sqsubseteq \exists hasARisk.HighRisk$
6	$Adolescent \sqsubseteq \exists hasARisk.LowRisk$
7	$Symptoms \sqsubseteq \geq 1\exists startBetween.Day \wedge \leq 14\exists startBetween.Day$
8	$Person \sqsubseteq \forall shouldNotReuse.DisposableMask$
9	$Person \wedge touches.\{ContaminatedSurface\} \wedge \exists touches.Face \wedge \neg washing.Hand \sqsubseteq InfectedPerson$
10	$Person(x) \wedge \exists experience.COVID19Symptom \wedge \exists remainsAt.Home \sqsubseteq \exists isIn.SelfIsolation$
11	$IsolatedPerson \wedge \exists experience.COVID19Symptom \wedge \exists exposedTo.COVID19 \sqsubseteq \exists isIn.Quatantine$
12	$Patient \wedge \exists present.EmergencySign \sqsubseteq \exists isTreatedWith.(AirwayManagement \vee OxygenTherapy)$
13	$\exists presents.EmergencySign \sqsubseteq Patient$

3.3 Logic-Based Representation of the COVIDonto Ontology

The abovementioned 163 statements describing the COVID-19 pandemic were
further represented in logic-based syntaxes including First-order-logic (FOL) and
DLs, to form the axioms of the COVIDonto ontology. FOL is an extension of
the propositional logic; it enables to represent things in the world with objects,
relations/predicates and functions. The FOL syntax expresses knowledge the
same as in natural language; therefore, it is an intermediate step from natural
language statements to DLs concepts. Table 1 presents the FOL formulas for
some statements that describe the COVID-19 pandemic and are modeled in
the COVIDonto ontology. The FOL formulas in Table 1 represent the following
knowledge of COVID-19 in the COVIDonto ontology:

- Any person infected with COVID-19 experience some symptoms or can expe-
  rience no symptoms (index 1 in Table 1)
- A person can have underlying medical problems (index 2 in Table 1)
- Elderly people are at a high risk (index 5 in Table 1)
- A person who has underlying medical problems is at a high risk (index 3 in
  Table 1)
- A person who has underlying medical problems has high blood pressure, heart
  problems, lung problems, diabetes, or cancer (index 4 in Table 1)
- Symptoms start between 1 to 14 d (index 7 in Table 1)
- Person should not reuse disposable mask (index 8 in Table 1)

- A person who touches contaminated surface and touches face and not washing hand, is an infected person (index 9 in Table 1)
- A person who experiences COVID-19 symptom and remains at home, is in self-isolation (index 10 in Table 1)
- A person who does not experience COVID-19 symptoms and is exposed to COVID-19 and is isolated, is in-Quarantine (index 11 in Table 1)
- Patients who present emergency signs are treated with airway management or oxygen therapy (index 12 in Table 1)
- Individuals who present emergency signs or at least one emergency sign are patients (index 13 in Table 1)

The statements of the COVID-19 pandemic were further represented in DLs in preparation for their implementation in Protégé. Table 2 presents the corresponding DLs concepts of FOL formulas in Table 1. The experimental results are presented and discussed next.

Fig. 1. Concepts/Classes of COVIDonto ontology in Protégé

4   Experimental Results

This section presents the implementation of the COVIDonto ontology in terms of the computer and software environments that were used in the experiments, and the presentation and discussion of results.

4.1   Computer and Software Environments

The COVIDonto was developed in Protégé version 5.5 on a computer with the following characteristics: Lenovo V310-151KB, Windows 10 Pro operating System, x64-based processor with 2.50 GHz CPU and 4,00 GB RAM.

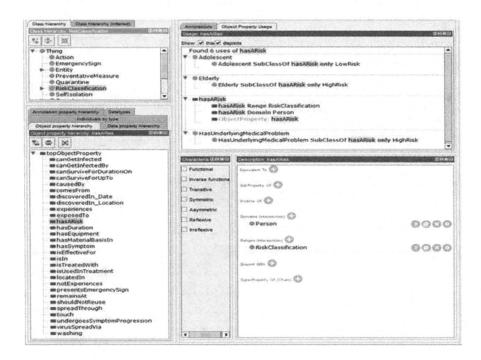

Fig. 2. Properties of COVIDONTO ontology in Protégé

4.2   Results and Discussions

Figure 1 shows the concept/class hierarchy of the COVIDonto ontology on the left panel in a Protégé screenshot. In particular, the concept/class SeriousSymptom is highlighted in the left panel and its subclasses are chest pain, difficult breathing, loss of movement and loss of speech. Both the left and right panels of Fig. 1 further show the various categorizations of COVID-19 symptoms, namely, less common, most common, serious and severe symptoms. The left panel depicts that the less common symptoms of COVID-19 are aches and pains, conjunctivitis, diarrhea, discoloration of fingers, discoloration of toes, headache, loss smell, loss of taste, rash on skin and sore throat. The most common symptoms of COVID-19 as indicated on the left panel of Fig. 1 include dry cough, fever and tiredness. Moreover, the bottom right panel of Fig. 1 shows that serious symptoms of COVID-19 are physical symptoms and can be treated with medicine.

The properties of the COVIDonto ontology are displayed in the bottom left panel of Fig. 2. In particular, the property hasARisk is highlighted and its usage in the ontology is given in the top left panel of Fig. 2. It is represented in the COVIDonto ontology that both adolescents and elderlies are at risk of contracting COVID-19 and that adolescents are at low risk, whereas, elderlies are at high risk.

The resulting COVIDonto ontology constitutes a knowledge base system that may be utilized to record useful data on the COVID-19 pandemic such as the data on testing, patients and treatments. In fact, the COVIDonto ontology includes the concept person which in turn is categorized into person with underlying medical condition, infected person, isolated person and age group of person, as shown in Fig. 3. Figure 3, further shows that a person with underlying medical problems is a person who has lung problems, and/or cancer, diabetes, high blood pressure and heart problem. Similarly, Fig. 3 tells that an infected person is either a mild, moderate or severe case patient. Lastly, Fig. 3 indicates that a person is either an adolescent, adult or elderly. All these concepts in Fig. 3 that represent a person in the COVIDonto ontology can record useful testing and patients' data.

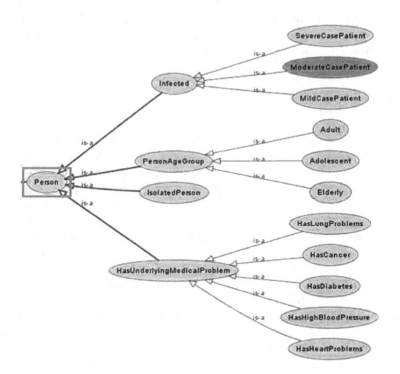

Fig. 3. Concept person and categorization in the COVIDonto ontology

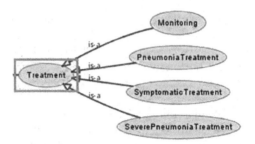

Fig. 4. Concept treatment and categorization in COVIDonto ontology

The data on the various treatments of patients infected with COVID-19 can also be recorded in the COVIDonto ontology through the concept treatment as shown in Fig. 4. In fact, Fig. 4 depicts that a treatment of a COVID-19 patient is either a monitoring, pneumonia, symptomatic or severe pneumonia treatment.

In light of the above, the COVIDonto ontology proposed in this study is a formal model that represent structured knowledge on COVID-19; it can be used to record relevant data on COVID-19. This data may be useful to governments, scientists, medical personnel, etc. in the management of and fight against the pandemic. However, the COVIDonto ontology need to be deployed in a real world application to provide the abovementioned benefits in the management of the COVID-19 pandemic.

5   Conclusion

A formal and executable ontology for COVID-19 named COVIDonto was developed in this study. The NeOn ontology development methodology was applied to develop the COVIDonto ontology. Data was collected from various sources on different aspects of COVID-19 such as the origin COVID-19, its symptoms, spread and treatment. The COVIDonto ontology is a knowledge base system for COVID-19 data that can be used to record patients' information such their medical conditions, symptoms (mild, moderate and severe), type of infection (mild, moderate, severe), and so forth. This knowledge base can be queried and reasoned to provide statistical data for the management of and fight against the pandemic by governments and stakeholders. Furthermore, the resulting COVIDonto ontology reuses resources from existing biomedical ontologies on the web; therefore, it can be easily integrated in the linked data infrastructures of these ontologies on the web to provide an open and wider information access on COVID-19 and related biomedical data. Moreover, the proposed COVIDonto can be merged with other existing recent COVID-19 ontologies to build a more comprehensive ontology for the COVID-19 pandemic. Finally, the COVIDonto ontology could serve as a suitable medium for cross-border interoperability of systems of different countries for data sharing in the fight against the COVID-19 pandemic. The future direction of research would focus on expanding the COVIDonto ontology to include concepts related to the treatment of COVID-19, with the introduction of the new vaccines that are currently being administered worldwide.

References

1. Sha, D., et al.: A spatiotemporal data collection of viral cases for COVID-19 rapid response. Big Earth Data **5**, 1–21, (2020)
2. Horgan, D., et al.: Digitalisation and COVID-19: the perfect storm. Biomed. Hub **5**, 1–23 (2020)
3. Makulec, A.: How Is COVID-19 Case Data Collected? The journey from a test swab to a record in a database. https://medium.com/nightingale/how-is-covid-19-case-data-collected-9afd50630c08 Accessed 03 Feb 2021
4. Apps and Covid-19. https://privacyinternational.org/examples/apps-and-covid-19 Accessed 03 Feb 2021
5. Bazzoli, F.: COVID-19 emergency shows limitations of nationwide data sharing infrastructure, https://www.healthcareitnews.com/news/covid-19-emergency-shows-limitations-nationwide-data-sharing-infrastructure Accessed 03 Feb 2021
6. Kachaoui, J., Larioui, J., Belangour, A.: Towards an ontology proposal model in data lake for real-time COVID-19 cases prevention. Int. J. Online Biomed. Eng. (iJOE) **16**, 123–136 (2020)
7. Rawal, R., Goel, K., Gupta, C.: COVID-19: disease pattern study based on semantic-web approach using description logic. In: 2020 IEEE International Conference for Innovation in Technology (INOCON), pp. 1–5 (2020)
8. Tini, G., et al.: Semantic and geographical analysis of COVID-19 trials reveals a fragmented clinical research landscape likely to impair informativeness. Front. Med. **7**, 1–7 (2020)
9. Dutta, B., DeBellis, M.: CODO: an ontology for collection and analysis of COVID-19 data. In: 12th International Conference on Knowledge Engineering and Ontology Development (KEOD), pp. 1–11 (2020)
10. Sargsyan, A., Kodamullil, A.T., Baks, S., Darms, J., Madan, S., et al.: Databases and ontologies, The COVID-19 Ontology. Bioinformatics **2020**, 1–3 (2020)
11. Sherimon, V., Sherimon, P.C., Mathew, R., Kumar, S.M., Nair, R.V., et al.: Covid-19 ontology engineering-knowledge modeling of severe acute respiratory syndrome coronavirus 2 (SARS-CoV-2). Int. J. Adv. Comput. Sci. Appl. **11**, 117–123 (2020)
12. Beverley, J., Babcock, S., Cowell, L., Smith, B.: T he COVID 19 Infectious Disease Ontology. http://www.semantic-web-journal.net/system/files/swj2627.pdf Accessed 05 Feb 2021
13. Suàrez-Figueroa, M.C., Carmen, M., G'omez-Pérez, A., Fernandez-Lopez, M.: The NeOn methodology framework: a scenario-based methodology for ontology development. Appl. Ontology **10**, 107–145 (2015)
14. Krotzsch, M., Simancik, F., Horrocks, I.: A Description Logic Primer. https://arxiv.org/pdf/1201.4089.pdf Accessed 03 Feb 2021
15. Schrim, L.M., et al.: Disease Ontology: a backbone for disease semantic integration. Nucleic Acids Res. **40**, 940–946 (2012)
16. Cowell, L.G., Smith, B.: Infectious disease ontology. In: Vitali Sintchenko, Infectious Disease Informatics, pp. 373–395 (2010)
17. Smith, B., Ashburner, M., Rosse, C., Bard, J., Bug, W., Ceusters, W., et al.: The OBO foundry: coordinated evolution of ontologies to support biomedical data integration. Nature Biotechnol. **25**, 1251–1255 (2007)

18. Arp, R., Smith, B.: Realizable Entities in Basic Formal Ontology, National Center for Biomedical Ontology. University at Buffalo. http://ontology.buffalo.edu/smith/articles/realizables.pdf Accessed 05 Feb 2021

19. He, Y., et al.: CIDO, a community-based ontology for coronavirus disease knowledge and data integration, sharing, and analysis. Sci. Data **7**, 1–5 (2020)

20. Lin, Y., Xiang, Z., He, Y.: Brucellosis ontology (IDOBRU) as an extension of the infectious disease ontology. J. Biomed. Semant. **2**, 1–18 (2011)

21. Coronavirus disease (COVID-19) Q&A. World Health Organization. https://www.who.int/emergencies/diseases/novel-coronavirus-2019/question-and-answers-hub/q-a-detail/coronavirus-disease-covid-19 Accessed 06 Feb 2021

22. Zhu, N., et al.: A novel coronavirus from patients with pneumonia in China, 2019. New England J. Med. **382**, 727–733 (2020)

23. Jin, W., et al.: A rapid advice guideline for the diagnosis and treatment of 2019 novel coronavirus (2019-nCoV) infected pneumonia (standard version). Mil. Med. Res. **7**, 1–23 (2020)

24. Li, Q., et al.: Early transmission dynamics in Wuhan China, of novel coronavirus-infected pneumonia. New England J. Med. **382**, 1199–1207 (2020)

25. Clinical management of COVID-19: interim guidance, World Health Organization. https://apps.who.int/iris/handle/10665/332196 Accessed 06 Feb 2021

26. Horby, P., Wei Shen Lim, W.S., Emberson, J., Mafham, M., Bell, J., et al.: Effect of dexamethasone in hospitalized patients with COVID-19: preliminary report. New England J. Med. (2020)

27. Bezerra, C., Freitas, F., Santana, F.: Evaluating ontologies with competency questions. In: 2013 IEEE/WIC/ACM International Joint Conferences on Web Intelligence (WI) and Intelligent Agent Technologies (IAT), pp. 1–3 (2013)

28. Potoniec, J., Wisniewski, D., Lawrynowicz, A., Keet, C.M.: Dataset of ontology competency questions to SPARQL-OWL queries translations. Data Brief **29**, 1–13 (2020)

K-Nearest Neighbors Classification of Semantic Web Ontologies

Gideon Koech[1] and Jean Vincent Fonou-Dombeu[2(✉)]

[1] Faculty of Applied and Computer Sciences Vaal University of Technology,
Private Bag X021, Andries Potgieter Blvd, Vanderbijlpark 1900, South Africa
[2] School of Mathematics, Statistics and Computer Science University
of KwaZulu-Natal, Private Bag X01, Scottsville, Pietermaritzburg 3209, South Africa
fonoudombeuj@ukzn.ac.za

Abstract. The growing interest in the semantic web technologies in the past years has led to the increase in the number of ontologies on the web. This gives semantic web developers the opportunity to select and reuse these ontologies in new applications. However, none of the existing approaches has leveraged the power of Machine Learning to assist in the choice of suitable ontologies for reuse. In this paper, the k-Nearest Neighbors (KNN) algorithm is implemented to classify ontologies based on their quality metrics. The aim is to group the ontologies that display the same quality properties into classes, thereby, providing some insights into the selection and reuse of these ontologies using a Machine Learning technique. The experiments were carried out with a dataset of 200 biomedical ontologies characterized each by 11 quality metric attributes. The KNN model was trained and tested with 70% and 30% of the dataset, respectively. The evaluation of the KNN model was undertaken with various metrics including accuracy, precision, recall, F-measure and Receiver Operating Characteristic (ROC) curves. For the best value of $k = 5$ the KNN model displayed promising results with an accuracy of 67% and the average precision, recall, and F-measure of 69%, 67%, and 67%, respectively as well as an area under ROC curve of 0.78.

Keywords: Classification · K-Nearest Neighbors · Machine Learning · Ontology · Ontology metrics · Semantic web

1 Introduction

Ontology building has been an active area of interest in Semantic Web development in the past years. This is witnessed by the ever increasing number of ontologies made available to the public on the web today. Three main reasons may explain this proliferation of ontologies on the web, namely, the advent of linked data, ontology libraries and the increase interest in semantic-based applications in various domains. The main benefits of making existing ontologies available on the internet to the public is that they can be reused in other applications and/or in research. Ontology reuse is an active subject of research in

© Springer Nature Switzerland AG 2021
C. Attiogbé and S. Ben Yahia (Eds.): MEDI 2021, LNCS 12732, pp. 241–248, 2021.
https://doi.org/10.1007/978-3-030-78428-7_19

ontology engineering today [1]. Ontology reuse consists of using existing ontologies to build new ones in the process of developing semantic-based applications. This entails integrating and/or merging parts of or the entire existing ontologies to constitute new ontologies.

With the rise of the number of available ontologies on the web today, the users or ontology engineers are presented with many ontologies to choose from. Several criteria may be applied to choose ontologies for reuse. The users or ontology engineers may utilize keywords search or structured queries in an automatic or semi-automatic process to find suitable ontologies that match their desire terms [2]. They may find suitable ontologies from ontologies libraries [3]. Another criterion that may be utilized to choose suitable ontologies for reuse is the quality of the candidate ontologies. Here, a number of metrics that measure the quality of ontologies such as their accuracy, computational efficiency, understandability, conciseness and cohesion [4] may be applied to classify the ontologies describing the same domain into various classes with Machine Learning techniques. However, until to date, the terms "classification of ontology" or "ontology classification" in the semantic web literature has been generally used to refer to the task of computing the subsumption hierarchies for classes and properties in reasoner systems [5–7]. To the best of our knowledge, no previous study has attempted to leverage the power of Machine Learning techniques to classify ontologies into various classes based on their quality metrics.

In this paper, the k-Nearest Neighbors (KNN) algorithm is implemented to classify ontologies based on their quality metrics. The aim is to group the ontologies that display the same quality properties into classes, thereby, providing some insights into the selection and reuse of these ontologies using a Machine Learning technique.

The structure of the rest of this paper is summarized as follows. Section 2 discusses related studies. The materials and methods are presented in Sect. 3. Section 4 reports the experiments and results, and the last section concludes the study.

2   Related Work

As mentioned earlier, the terms "classification of ontology" or "ontology classification" in the semantic web literature has so far been generally used to refer to the task of computing the subsumption hierarchies for classes and properties in reasoner systems, to measure the satisfiability and consistency of ontologies [6,7], classify proteins in molecular biology [5], and so forth. The exception to this trend has been the study in [8] where the authors carried out a review of existing ontologies describing the software engineering activities and performed a manual classification of these ontologies according to the engineering phase their are related to, the type of information they model and the scope of their application.

Another recent study in [9] implemented Machine Learning algorithms including KNN to ontology alignments based on changes that may occur in ontologies overtime. Beside this, KNN is a widely used Machine Learning classifier. It has been successfully applied for loan forecasting [10], in medicine for classifying heart diseases [11], in data mining for classification of medical literature [12], and many more. To the best of our knowledge, no previous study has applied Machine Learning techniques to classify ontologies into various groups/categories based on their quality metrics attributes as it done in this study.

3   Materials and Methods

In this section, the quality metrics of ontologies adopted from the OntoMetrics ontology evaluation framework as well as the mapping of these metrics to four quality dimensions/criteria [4] that form the basis for the KNN classification in this study are presented.

3.1   Quality Metrics of Ontologies

Ontology evaluation is an active research topic in ontology engineering [4,13]. To this end, many approaches have been proposed to address ontology evaluation [13]. The quality metrics of ontologies considered in this study are adopted from OntoMetrics. OntoMetrics is the *state-of-the-art* tool for automated ontology evaluation to date [4].

In this study, 11 quality metrics that fall under the schema, knowledge base and graph categories of OntoMetrics suite of metrics have been selected as the attributes of the ontologies in the dataset, based on their correlation with the abovementioned quality dimensions including accuracy, understandability, cohesion, computational efficiency and conciseness [4]. The schema metrics selected include the attribute richness (AR), inheritance richness (IR) and relationship richness (RR). The knowledge base metrics category encompasses the average population (AP) and class richness (CR). The absolute root cardinality (ARC), absolute leaf cardinality (AC), average depth (AD), maximum depth (MD), average breadth (AB) and maximum breadth (MD) form the graph metrics in OntoMetrics. These ontology quality metrics are linked to quality evaluation criteria in the next subsection.

3.2   Correlating Quality Metrics and Evaluation Criteria

It is further demonstrated [4] that the quality metrics above correlate with four quality criteria/dimensions as follows:

- **Accuracy** - This indicates to which extend the ontology under evaluation represents the real world domain it is modeling. The schema metrics AR, IR and RR and the graph metrics AD, MD, AB and MB are the indicators that can be used to measure this criterion [4].
- **Understandability** - This evaluates the comprehension of the constituents of the ontologies such as the concepts, relationships/properties, and their meanings [4]. The quality metric AC is the indicator for measuring this quality criterion [4].

- **Cohesion** - This indicates the degree of relatedness amongst the constituents of the ontologies. In other words, it judges the degree to which the classes of the ontology are related to one another. Two graph metrics ARC and AC are the indicators that can be utilized to measure this quality dimension/criterion [4].
- **Conciseness** - This measures the degree of usefulness of the knowledge in the ontology [13]. The knowledge base metrics AP and CR are the indicators for measuring this quality criterion [4].

3.3 K-Nearest Neighbors Algorithm

The K-Nearest neighbours algorithm can be used for classification, estimation or prediction and is an example of instance-based learning. The algorithm first loads all the data points that are in the dataset in its memory or makes a plotting of the data in an n-dimensional space [14]. Each and every point on the plotting area represents a label in the dataset. A test sample, which represents a set of training data, shall also be plotted within the same n-dimensional space. It will then make a search for its k nearest neighbours on the basis of a specific distance measure out of the training samples [15].

3.4 Performance Measures

Various techniques are used to evaluate the performance of the KNN algorithm in this study; these include the accuracy, precision, recall, F-measure and ROC curve.

4 Experiments and Discussions

4.1 Dataset

The dataset used in the experiments comprises of 200 ontologies and 11 quality metrics attributes (Eqs. 1 to 11) for each ontology. The ontologies were obtained from the BioPortal [21] repository which provides access to ontologies of the biomedical domain. After randomly downloading the ontologies, their quality metrics attributes were computed using the OntoMetrics platform [4] and loaded into a CSV file.

The four quality criteria/dimensions of accuracy, understandability, cohesion, and conciseness are adopted as the decision classes of ontologies in the KNN model in this study. Therefore, four class labels 0 for accuracy, 1 for understandability, 2 for cohesion, and 3 for conciseness are used in building the KNN model. These class labels appear on the results in Figs. 1, 2 and 4.

	precision	recall	f1-score	support
0	0.59	0.71	0.65	14
1	0.50	0.59	0.54	17
2	0.75	0.43	0.55	14
3	0.93	0.93	0.93	15
accuracy			0.67	60
macro avg	0.69	0.67	0.67	60
weighted avg	0.69	0.67	0.66	60

Fig. 1. Performance of the KNN Classifier for k = 5

4.2 Software Environment

The experiments have been done using a HP Pavilion dv6 computer, Intel(R) Core(TM) i5-2410 M, CPU 2.30 GHz, 6.00 GB RAM, 64-bit Windows 10 operating system, x64-based processor and 500 GB hard drive. Python software, version 3.7.0 and Jupyter Notebook version 5.0.0 have been used in running the experiments.

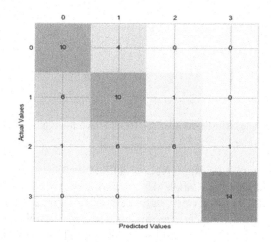

Fig. 2. Confusion matrix for KNN Classier for k = 5

4.3 Results and Discussion

The *KNeighborsClassifier* Python function was used to build the KNN model with the following parameters: algorithm = 'auto', leaf_size = 30, metric = 'minkowski', metric_params = None, n_jobs = None, n_neighbors = 5, p = 2 and weights = 'uniform'. A ratio of 0.3 was used to split the dataset into training and testing sets, that is, the KNN model was trained with 140 ontologies

(70% of dataset) and tested with 60 ontologies (30% of dataset). Different values of k were used to check the most suitable value that produced the highest accuracy score. The results obtained for various k values revealed that the accuracy scores decreased with the increase of k values. For the best $k = 5$, the KNN model displayed an accuracy of 67% and the average precision, recall, and F-Measure scores of 69%, 67%, and 67%, respectively, as shown in Fig. 1. The support column (rightmost column) in Fig. 1 displays the total number of ontologies that were predicted per class. These number of predictions are confirmed in each row of the confusion matrix in Fig. 2 by counting the number of predictions per row/class.

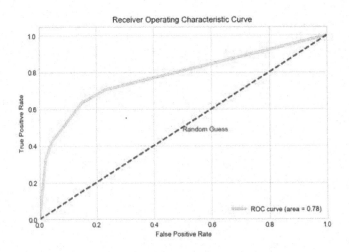

Fig. 3. Macro-average ROC curve for KNN

The confusion matrix for the KNN model for $k = 5$ is shown in Fig. 2, where, the actual values of the class labels 0, 1, 2 and 3 that were predicted correctly are 10, 10, 6 and 14, respectively (right diagonal of Fig. 2). The values on the other cells were misclassified by the model. The total number of values in the confusion matrix in Fig. 2 is 60, which is the size of the testing set, obtained from the 30% split ratio.

The ROC curves were also used to evaluate the performance of the KNN model. Two sets of ROC curves were generated including the macro average curve for the four class labels and the ROC curves for the individual class labels. The curves are as shown in Figs. 3 and 4. The ROC curves evaluate a model by checking the area under the curve. It is shown in Figs. 3 and 4 that the macro average area under the curve for all the classes is 0.78. The area under the curves for the class labels 0, 1, 2, and 3 are 0.78, 0.68, 0.69 and 0.96, respectively in Fig. 4. Since the ROC curves evaluate the scores in the range of 0 to 1, the scores obtained (Figs. 3 and 4) indicate that the KNN model is a good classifier for semantic web ontologies.

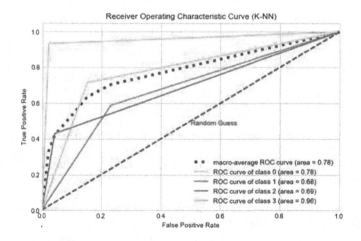

Fig. 4. ROC curves for each class labels for KNN

5   Conclusion

In this paper, the KNN algorithm was implemented to classify ontologies. A dataset of 200 biomedical ontologies along with 11 quality metrics attributes for each ontology was used. The decision classes for the KNN model consisted of four quality criteria/dimensions of ontologies including accuracy, understandability, cohesion and conciseness, labelled 0, 1, 2 and 3, respectively. The performance of KNN model was evaluated using various metrics and the results were promising with an accuracy of 67% for $k = 5$. This showed that KNN is a good classifier for semantic web ontologies. The future direction of research would be to implement more Machine Learning algorithms to classify ontologies and compare their performances.

References

1. Trokanas, N., Cecelja, F.: Ontology evaluation for reuse in the domain of process systems engineering. Comput. Chem. Eng. **85**, 177–187 (2016)
2. Alani, H., Brewster, C., Shadbolt, N.: Ranking ontologies with AKTiveRank. In: 5th International Conference on the Semantic Web, pp. 1–15. Athens, Greece (2006)
3. dAquin, M., Noy, N.F.: Where to publish and find ontologies? A survey of ontology libraries. Web Semant. Sci. Serv. Agents World Wide Web **11**, 96–111 (2012)
4. Lantow, B.: OntoMetrics: putting metrics into use for ontology evaluation. In: Proceedings of the 8th International Joint Conference on Knowledge Discovery, Knowledge Engineering and Knowledge Management (IC3K 2016), pp. 186–191, Joaquim Filipe, Portugal (2016)
5. Wolstencroft, K., Lord, P., Tabernero, L., Brass, A., Stevens, R.: Protein classification using ontology classification. Bioinformatics **22**, e530–e538 (2006)

6. Glimm, B., Horrocks, I., Motik, B., Shearer, R., Stoilos, G.: A novel approach to ontology classification. J. Seman. Web **14**, 84–101 (2012)

7. Wang, C., Feng, Z., Zhang, X., Wang, X., Rao, G., Fu, D.: ComR: a combined OWL reasoner for ontology classification. Front. Comput. Sci. **13**, 139–156 (2019)

8. Zhao, Y., Dong, J., Peng, T.: Ontology classification for semantic-web-based software engineering. IEEE Trans. Serv. Comput. **2**, 303–317 (2009)

9. Jurisch, M., Igler, B.: RDF2Vec-based classification of ontology alignment changes. In: 1st Workshop on Deep Learning for Knowledge Graphs and Semantic Technologies (DL4KGS) co-located with the 15th Extended Semantic Web Conerence (ESWC 2018) (2018)

10. Fan, G.F., Guo, Y.H., Zheng, J.M., Hong, W.C.: Application of the weighted K-nearest neighbor algorithm for short-term load forecasting. Energies **12**, 1–19 (2019)

11. Jabbar, M.A., Deekshatulu, B.L., Chandra, P.: Classification of heart disease using K-nearest neighbor and genetic algorithm. Procedia Technol. **10**, 85–94 (2013)

12. Luschow, A., Wartena, C.: Classifying medical literature using k-nearest-neighbours algorithm. In: NKOS Workshop, pp. 1–13 (2017)

13. Sharman, R., Rao, H.R., Raghu, T.S. (eds.): WEB 2009. LNBIP, vol. 52. Springer, Heidelberg. https://doi.org/10.1007/978-3-642-17449-0

14. Harrison, O: Machine learning basics with the K-nearest neighbors algorithm. Towards Data Sci. (2018)

15. Denoeux, T.: A K-Nearest Neighbor Classification Rule Based on Dempster-Shafer Theory. International Journal on Transactions on Systems, Man and Cybernetics. **25**, 804–813 (1995)

16. Hand, D., Mannila, H., Smyth, P.: Principles of Data mining, pp. 242–267. The MIT Press, Cambridge (2001)

17. Zhang, M., Zhou, Z.: A k-nearest neighbor based algorithm for multi-label classification. In: IEEE International Conference on Granular Computing, pp. 718–721 (2005)

18. Potamias, M., Bonchi, F., Gionis, A., Kollios, G.: K-nearest neighbors in uncertain graphs. J. VLDB Endowment **3**, 997–1008 (2010)

19. Clare, A., King, R.D.: Knowledge discovery in multi-label phenotype data. In: De Raedt, L., Siebes, A. (eds.) PKDD 2001. LNCS (LNAI), vol. 2168, pp. 42–53. Springer, Heidelberg (2001). https://doi.org/10.1007/3-540-44794-6_4

20. Myles, J., Hand, D.: The multi-class metric problem in nearest neighbor discrimination rules. Int. J. Patt. Recogn. **23**, 1291–1297 (1990)

21. Noy, N.F., et al.: BioPortal: ontologies and integrated data resources at the click of a mouse. In: International Conference on Biomedical Ontology, p. 197. New York, USA (2009)

Automated Generation of Datasets from Fishbone Diagrams

Brian Sal, Diego García-Saiz, and Pablo Sánchez[✉]

Software Engineering and Real-Time, University of Cantabria, Santander, Spain
{brian.sal,garciasad,p.sanchez}@unican.es

Abstract. The analysis of the data generated in manufacturing processes and products, also known as *Industry 4.0*, has gained a lot of popularity in last years. However, as in any data analysis process, data to be processed must be manually gathered and transformed into tabular datasets that can be digested by data analysis algorithms. This task is typically carried out by writing complex scripts in low-level data management languages, such as SQL. This task is labor-intensive, requires hiring data scientists, and hampers the participation of industrial engineers or company managers. To alleviate this problem, in a previous work, we developed *Lavoisier*, a language for dataset generation that focuses on what data must be selected and hides the details of how these data are transformed. To describe data available in a domain, Lavoisier relies on object-oriented data models. Nevertheless, in manufacturing settings, industrial engineers are most used to describe influences and relationships between elements of a production process by means of fishbone diagrams. To solve this issue, this work presents a model-driven process that adapts Lavoisier to work directly with fishbone diagrams.

Keywords: Data selection · Industry 4.0 · Fishbone diagrams · Ishikawa diagrams · Cause-effect diagrams

1 Introduction

Production lines of manufacturing industries are nowadays comprised of multiple interconnected elements that gather a wide range of data about how they work [4]. This vast amount of data can then be analyzed, by using *Big Data* and *Data Science* technologies, to improve both the manufacturing process and the manufactured products. For instance, machine learning techniques might be used to predict when a piece of a production line must be replaced before it breaks, helping to reduce maintenance standstills [6]. This idea is a key element of what is currently known as *Industry 4.0* [7], the 4th Industrial Revolution.

So, Industry 4.0 applications are often built on data analysis algorithms. Despite recent data science advances, these algorithms still exhibit an important shortcoming: they only accept as input data arranged in a very specific tabular format, known in the data science community as a *dataset*. Nevertheless, data is rarely found directly in this format. Therefore, data scientists need

© Springer Nature Switzerland AG 2021
C. Attiogbé and S. Ben Yahia (Eds.): MEDI 2021, LNCS 12732, pp. 249–263, 2021.
https://doi.org/10.1007/978-3-030-78428-7_20

to write complex and long scripts in languages like SQL to transform data available in a domain into datasets. These scripts concatenate several low-level operations, such as *joins* and *pivots*, to perform the required transformations. So, data scientists need to work at a low abstraction level, which makes the dataset creation process labor-intensive and prone to errors. Moreover, domain experts, like industrial engineers, can hardly create or modify these scripts, so data scientists need to be hired to execute this task. Since data scientists are expensive and scarce, development time and costs of Industry 4.0 applications increase.

To alleviate these problems, in a previous work, we developed *Lavoisier* [15], a language that provides high-level constructs to build datasets. These constructs are processed by the Lavoisier interpreter, who automatically transforms them into chains of low-level operations. Using Lavoisier, complexity of scripts for dataset creation can be reduced in average by ~60%.

To represent domain data, Lavoisier relies on object-oriented data models, but these models are rarely found in manufacturing settings. However, industrial engineers are used to deal with *fishbone diagrams*, a kind of model to represent cause-effect relationships that might be used to specify relationships between domain data. Therefore, in an Industry 4.0 context, it could be more desirable that Lavoisier worked with these fishbone diagrams instead of object-oriented domain models.

To achieve this goal, this work presents a model-driven process for dataset generation from fishbone diagrams for which we have created new models and languages. To build this process, data-oriented fishbone models were firstly designed to connect causes in a fishbone diagram with domain data. Next, a new language called *Papin* was created to select which concrete causes should be included in a specific dataset. This language interpreter processes this selection and automatically generates the corresponding dataset. This way, industrial engineers can generate datasets directly from domain models. Expressiveness of these languages has been evaluated by applying them to four external case studies [1,2,8,10].

After this introduction, this work is structured as follows: Sect. 2 details the motivation behind this work, using a running example. Section 3 describes our solution. Section 4 comments on related work. Finally, Sect. 5 summarizes our achievements and outlines future work.

2   Background and Motivation

2.1   Running Example: Falling Band

As running example throughout this paper, we will use the production of *drive half-shafts* in a supplier company for the automotive sector. This example is taken from the literature [2] and based on a real case study. A drive half-shaft is a piece of the car transmission system that joins the engine with a wheel. It is composed of a rotating bar of metal, the *shaft*, with two articulated ends, that allow the wheel to move freely as required by the steering and suspension systems at the same time it continues receiving the engine force.

The articulated ends are protected by a flexible piece of rubber known as a *housing*. The housings are fixed to the shaft by means of two metal bands. To produce these drive half-shafts, an operator inserts a shaft in an assembly machine, mounts each housing on the shaft and fastens the bands around the housings using a pneumatic pliers. Once finished, these pieces are sent to car manufacturers, who check they are compliant with their requirements. During these checks, car manufacturers detected that, sometimes, bands on the side of the wheel unfastened, originating different problems. To solve this issue, the company analyzed data gathered during the drive half-shafts production process in order to find the causes of this problem.

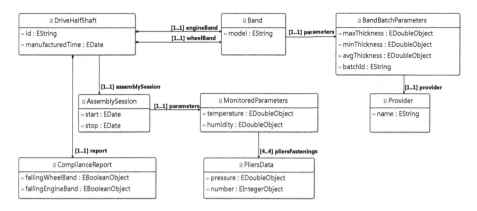

Fig. 1. A fragment of the domain model for the production of drive half-shafts

2.2 Lavoisier and the Dataset Formatting Problem

As commented in the introduction, most of data analysis algorithms require their input data to be arranged in a specific tabular format, where all the information related to an instance of the entities being analyzed must be placed in a single table row. In the following, we will refer to this requirement as the *one entity, one row* constraint. In our running example, this constraint implies that all data related to a specific drive half-shaft must be placed in a fixed-length vector.

Figure 1 shows a fragment of an object-oriented data model for our running example and Listing 1.1 provides an example of data, in JSON format, for a concrete half-shaft, conforming to that model. The data model is in Ecore [11] notation, which, roughly speaking, can be considered as a subset of UML 2.0. As it can be seen, information about a drive half-shaft is comprised of linked and nested data, that we would need to transform so that it can be placed in a single row of a dataset. To avoid writing complex scripts to execute this task, we developed *Lavoisier*.

We illustrate how Lavoisier works using the example of Listing 1.2. In this example, the dataset will contain just data from the compliance reports and the assembly session in which each drive half-shaft was created. When creating a dataset, we must specify first its name, which is *drive-HalfShaft_AssemblyParameters* in our example (Listing 1.2, line 1). Then, we must specify, using the *mainclass* keyword, what are the main entities that will be analyzed. It is worth to remember that, according to the *one entity, one row* constraint, all the information about each instance of these entities must be placed in a single row of the output dataset. In our case, *DriveHalfShafts* are these main entities (Listing 1.2, line 2).

```
1   [{'id':   '5209','manufacturedTime':'2021−02−12  11:27:13',
2     'wheelBand':{'model':'SNR390',
3        'parameters':{'batchId':'2020/2436832',
4           'maxThickness':1.67,'minThickness':1.64,
5           'avgThickness':1.66,'provider':{'name':'ACME'}}},
6        ...,
7     'assemblySession':  {'start':'2021−02−12  11:14:58',
8        'stop':'2021−02−12  11:14:58',
9        'parameters':  {'temperature':20.4,'humidity':53.8,
10          'pliersFastenings':[
11            {'number':1,'pressure':7.9},
12            {'number':2,'pressure':7.9},
13            {'number':3,'pressure':7.8},
14            {'number':4,'pressure':7.9}]}}},
15   {...}]
```

Listing 1.1. Data for a drive half-shaft instance

```
1   dataset driveHalfShaft_AssemblyParameters {
2      mainclass DriveHalfShaft [id] {
3         include report
4         include assemblySession {
5            include parameters {
6               include pliersFastenings by number
7      }}
8   }}
```

Listing 1.2. An example of Lavoisier specification

When a class is selected to be included in a dataset, Lavoisier adds all attributes of that class to the output dataset by default, whereas references to other classes are excluded. This behavior can be modified according to our needs. A concrete subset of attributes of a class can be selected by listing them between square brackets after the class name (e.g., Listing 1.2, line 2).

Regarding references, we can add them using the *include* keyword. This primitive works differently depending on the reference multiplicity. For single-bounded references, i.e., references with upper bound lower or equal to one, we just need to specify the name of the reference to be added. For example, Listing 1.2, line 3 incorporates the compliance report of each half-shaft, retrieved through the *report* reference, to the output dataset.

For each included reference, the default Lavoisier behavior is applied again. As before, this behavior can be modified by specifying again a list of attributes between square brackets and new *include* statements that add references of the reference. For instance, Listing 1.2, line 5 adds the *parameters* reference of the *assemblySession* reference. This recursive inclusion can continue as deeper as we need.

To process reference inclusion, the Lavoisier interpreter relies on a set of transformation patterns [14] that automatically reduce a set of interconnected classes into a single class from which a tabular dataset is directly generated. In the case of single-bounded references, Lavoisier simply adds the attributes of the referenced class to the class holding the reference. For example, in Listing 1.2, line 3, the attributes of the *ComplianceReport* class are added to the *Drive-HalfShaft* class, which would equivalent to performing a *left outer join* between these classes. To handle nested references, the Lavoisier interpreter processes the deepest reference first and continues this way until reaching the main class.

The case of multibounded references, i.e., references with upper bound lower or equal to one, is more challenging, since each instance of the referencing class is related to several instances of the referenced class. So, to satisfy the *one entity, one row* constraint, we need to find a mechanism to spread data of each one of these instances over a single row. To achieve this goal, Lavoisier creates a specific set of columns for each kind of instance that might appear in a multibounded reference. This way, each instance will be able to find a well-defined set of columns to place its data.

This strategy requires that Lavoisier can distinguish between different kinds of referenced instances. For this purpose, we must use one or more attributes of the referenced class as identifiers for the instance kinds. For this purpose, *include* statements involving multibounded reference add a *by* keyword after which these identifiers are specified. As an example, Listing 1.2, line 6 specifies that the *pliersFastenings* reference must be added to the output dataset. This reference contains four measures of the pressure in the pliers pneumatic system (see Listing 1.1, lines 11–14). Each measure is associated to a concrete fastening, identified by a number, performed during an assembly session.

id	fallingWheelBand	...	humidity	temperature	1_pressure	2_pressure	3_pressure	4_pressure
5209	0	...	56.7	20.4	7.9	7.9	7.8	7.9
6556	1	...	58.0	19.8	7.2	7.1	7.1	7.1
...	...	...	...	...	...	...	...	...

Fig. 2. A tabular dataset for drive half-shaft analysis

To process these multibounded references, the Lavoisier interpreter carries out the following actions: First, all distinct values for the identifier attributes are calculated. In our example, these values are $\{1, 2, 3, 4\}$. These values represent all potential kinds of instances in the multibounded reference. For each instance kind, a set of columns containing the non identifier attributes of the referenced class is created. Moreover, each one of these columns is prefixed with its corresponding identifier value. In our case, these sets of columns are $\{\{1\_pressure\},\{2\_pressure\},\{3\_pressure\},\{4\_pressure\}\}$. This structure is illustrated in Fig. 2. Then, each referenced instance is placed in the columns corresponding to its kind. In our case, each measure is placed in the column corresponding to its fastening number. It is worthwhile to mention that the processing of multibounded references contains a lot of picky details that we are omitting here for the sake of simplicity. We refer the interested reader to our previous work [13–15].

By using Lavoisier, the total number of operations required to generate a dataset decreases by ∼40% and script size reduces by ∼60% as compared to other languages used for this purpose, such as SQL. Therefore, we considered Lavoisier might be helpful for data selection in Industry 4.0. However, we soon realized that object-oriented data models are not commonly used in manufacturing settings. Fortunately, we discovered that industrial engineers are quite familiar with a kind of diagram, called *fishbone diagrams*, that might be used to represent relationships between domain data. These diagrams are described in next section.

2.3   Fishbone Diagrams

Fishbone Diagrams [5], also known as *Cause-Effect, Ishikawa* or *Fishikawa diagrams*, aim to identify causes that might lead to a certain effect. They were formally proposed by Kaoru Ishikawa and they are acknowledged nowadays as one of the seven basic tools for quality control and process improvement in manufacturing settings [12].

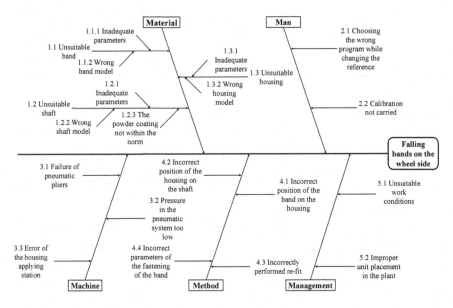

Fig. 3. A fishbone diagram for the drive half-shaft running example (taken from [2])

Figure 3 shows a fishbone diagram, taken from the literature [2], for our running example. As it can be observed, these diagrams are called fishbone diagrams because of its shape, which resembles a fish skeleton. They are elaborated as follows: first, the effect to be analyzed is placed on the right, commonly inside a rectangle or a circle, representing the fish head. In our case, *falling bands on the wheel side* would be the problem, or effect, to be studied. Then, an horizontal line from right to left, representing the fish main bone, is added. This line is used to connect causes with the effect.

Then, we would start to identify causes that might lead to the effect. For each identified cause, we would try to find subcauses that might lead to that cause, repeating iteratively the process with the subcauses until reaching a decomposition level that can be considered as adequate. To help to start the process, a set of predefined main causes, known as *categories*, are used in each domain. These categories are attached as ribs to the main fishbone. In the case of the manufacturing domain, these predefined categories, known as the *5 Ms model* [12], are *Material, Man, Machine, Method* and *Management*, as seen in Fig. 3.

Material specifies causes related to the source materials used during the manufacturing process. For instance, a wheeling band might fall because an unsuitable band has been used (Fig. 3, cause 1.1). A band might be unsuitable either because a wrong model is used or because its parameters are not adequate. For instance, the band might be thicker than usual. *Man* collects causes related to people participating in the process, such as the operator selecting a wrong assembly program (Fig. 3, cause 2.1). *Machine* analyses elements of the machines and tools, such as, for instance, a low pressure in the pliers pneumatic system (Fig. 3, cause 3.2). Finally, *Method* refers to the procedure used to manufacture an item, whereas *Management* considers general aspects of the process. It is worthwhile to comment that more categories might be added if it is considered helpful.

2.4   Problem Statement

As commented, whereas object-oriented data models are rarely found in manufacturing settings, fishbone diagrams are a common and well-known tool for quality control. These diagrams establish cause-effect relationships between domain elements, but they can also be used to identify influence relationships between domain data, helping to decide what concrete data might be useful to place together in a same dataset. This way, it would be desirable that Lavoisier worked on fishbone diagrams instead of object-oriented data models. Nevertheless, fishbone diagrams were not designed to represent data. So, we need to adapt them for this purpose. Moreover, we also need a mechanism to connect fishbone diagrams with domain data, so that datasets can be automatically generated from them.

3   Dataset Generation from Fishbone Diagrams

3.1   Solution Overview

Figure 4 shows the general scheme for our solution. The inputs are a fishbone diagram (Fig. 4, label 1) and a object-oriented data model (Fig. 4, label 2). This fishbone diagram is a classical fishbone diagram, such as the one depicted in Fig. 3, created for quality control. Because of this reason, we will call it *Quality Control Fishbone diagram* (QCF). This fishbone diagram does not need to conform to any metamodel, and it can be created using any tool, including sketching tools or whiteboards.

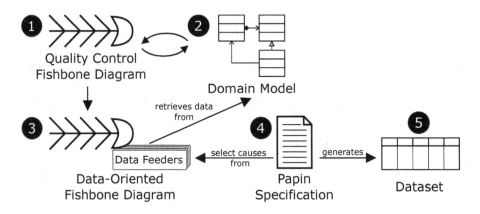

Fig. 4. Overview of our solution

The object-oriented data model is created by data scientists to fulfill two objectives: (1) to provide a well-formed description of data available in a manufacturing setting; and, (2) to supply an access layer to retrieve domain data. This model must conform to the Ecore metamodel [11], so that it can be accessed and navigated using model management tools. In our case, Epsilon is used for this purpose.

We encourage industrial engineers to review domain models and data scientists to check fishbone diagrams, as it is indicated by the go and back arrows between these elements. By inspecting domain models, industrial engineers might discover new causes that should be added to the fishbone diagrams. On the other hand, data scientists could find elements in fishbone diagrams that might be included in the data model.

Using these models, industrial engineers create a new kind of fishbone diagram, which we have called *Data-Oriented Fishbone diagram* (DOF) (Fig. 4, label 3). A DOF reshapes a quality-control fishbone diagram by replacing causes with data that measure these causes. The connections between causes and domain data are established using special blocks of code that we have called *data feeders*. A *data feeder* specifies, using Lavoisier-like primitives, what data of a domain model must be retrieved to characterize a certain cause. Therefore, a DOF represents a set of influence relationships between domain data. For defining these models, we have created a metamodel in Ecore plus a concrete textual syntax in xText.

Finally, industrial engineers specify what causes must be included in a specific dataset using a language called *Papin*, which we designed for this purpose.

Papin specifications (Fig. 4, label 4) are automatically processed by the *Papin* interpreter, which, using the information in the *data feeders*, retrieves and transforms domain data to automatically create a dataset (Fig. 4, label 5).

3.2 Data-Oriented Fishbone Models

Listing 1.3 shows a fragment of a data-oriented fishbone diagram for the quality-oriented fishbone diagram depicted in Fig. 3. This model conforms to the metamodel illustrated in Fig. 5.

```
1  effect FallingBand is DriveHalfShaft
2       include report
3  category Material
4    cause Bands realizes 'Unsuitable band' contains {
5      cause WheelBand contains {
6        cause Model realizes 'Wrong band model' is
7          wheelBand.model
8        cause Parameters realizes 'Inadequate parameters' is
9          wheelBand.parameters { include provider }
10       } -- WheelBand
11       cause EngineBand contains { ... }
12     } -- Bands
13     cause Shaft realizes 'Unsuitable shaft' contains {...}
14   category Machine
15     cause PneumaticPliersPressure realizes
16       'Pressure in the pneumatic system too low' is
17       assemblySession.parameters.plierFastenings by number
18     ...
19   category Management
20     cause WorkingConditions
21       realizes 'Unsuitable Work Conditions' notMapped
```

Listing 1.3. A data-oriented fishbone model for our running example

First of all, to associate an effect with domain data, it must be taken into account that, for analyzing a phenomenon, data analysis algorithms often require to be fed with data of entities that satisfy that phenomenon and entities that do not. For instance, to find patterns that lead to falling bands, these algorithms might need to compare drive half-shafts with and without falling bands. Therefore, in a DOF, the effect is associated to the domain entity whose instances might exhibit it.

The association of an element of a DOF with a domain element is always performed by means of a *data feeder*. The most simple data feeder is a reference to a class. In this case, the data of that class will be used to characterize the element of the DOF. As in Lavoisier, when referencing a class, all its attributes are considered as output data by default, whereas all the references are excluded. As in Lavoisier, this default behavior can be modified using lists of attributes between square brackets and *include* statements.

A DOF must always start with the declaration of an effect. Listing 1.3, lines 1–2 shows an example of an effect declaration. This declaration contains a name for the effect, *FallingBand*, and, after the *is* keyword, a data feeder referencing a class from the data model. In our case, falling bands are analyzed by inspecting instances of *DriveHalfShafts*. Moreover, each half-shaft is characterized by its compliance report, which is added by including the *report* reference.

After the effect declaration, we must specify *category blocks* (Listing 1.3, lines 3–13, 14–17 and 19–21). These blocks simply mimic the category structure of the associated quality-oriented fishbone diagram. Each category block is composed of causes. In a DOF, three kind of causes can be created: (1) compound causes; (2) data-linked causes; and, (3) not mapped causes.

Compound causes represent high-level causes containing one or more sub-causes. Data characterizing a compound cause is obtained by the combination of data coming from its subcauses. *Data-linked causes* represent atomic causes that are linked to domain data by means of data feeders. Finally, *notMapped* causes specify causes that are neither further decomposed nor associated to domain data. They serve to specify causes for which domain data is not available, but that we want to keep to preserve traceability between a DOF and its associated QOF. Each cause in a data-oriented fishbone model has a name. Moreover, each cause can have an optional *realizes* statement, which is used to link data-oriented causes with quality-oriented causes.

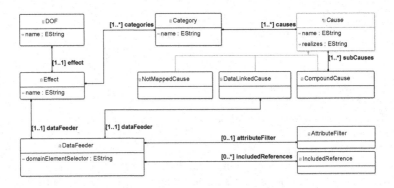

Fig. 5. Metamodel for data-oriented fishbone models

Listing 1.3, lines 4–12 show an example of a *compound cause*, whose name is *Bands* and that realizes the quality-oriented cause *Unsuitable Band* (see Fig. 3, cause 1.1). This cause is comprised of two subcauses: *WheelBand* (Listing 1.3, lines 5–10) and *EngineBand* (Listing 1.3, line 11), which also are compound causes. As it can be observed, these causes are not in the quality-oriented fishbone model, but they are added here to make the DOF fit in with the domain model.

Listing 1.3, lines 6–7 provide an example of *data-linked cause*, which is named as *Model*. In data-linked causes, an expression referencing a domain model element must be specified after the *is* keyword. This expression must provide a path that, starting in the class associated to the effect, reaches a domain element, which will be used to characterize the data-oriented cause. This target element can be: (1) a single attribute; (2) a class; or, (3) a collection.

In Listing 1.3, lines 6–7, the pointed element is an attribute, which means this *Model* cause will be characterized by a single value associated to that attribute. Listing 1.3, lines 8–9 show an example of data-linked cause where the target element is a class. Moreover, in this case, it is specified that the *provider* reference of the pointed class must be also used to characterize this cause. Finally, Listing 1.3, lines 15–17 illustrate a case where a data feeder points to a collection. Collections in a DOF are handled as multibounded references in Lavoisier. This means we need to identify instances in this collection, so that data of each one of them can be spread over a set of well-defined columns. In our concrete case, each measure of the pressure in the pliers pneumatic system is identified by its associated fastening number.

Finally, Listing 1.3, lines 19–21 contain an example of a not mapped cause. Obviously, working conditions, such as, for example, dusty or noisy environments, might be a cause of falling bands. However, we do not have data about these issues, so we cannot link this cause with domain data and it must remain as a not mapped cause. To keep these causes is useful to maintain traceability between fishbone models and to record hints about what new elements might be incorporated to improve domain data.

3.3   Papin: Dataset Specification by Cause Selection

```
1   dataset  FallingBand_MaterialsAndPressures
2     using FallingBand {
3       include Material
4       include Machine {
5           include PneumaticPliersPressure
6   }}
```

Listing 1.4. An example of Papin specification

After building a DOF, *Papin* is used to specify which causes, along with effect data, should be included in a dataset for a specific analysis. Listing 1.4 provides an example of *Papin specification*. A Papin specification starts defining a name for the dataset to be created (Listing 1.4, line 1), which is *FallingBand_MaterialsAndPressures* in our case. Then, we must provide the name of the DOF from which causes will be selected (Listing 1.4, line 2). In our example, this fishbone model is the one depicted in Listing 1.3, which is identified as *FallingBand*. Then, categories to be included in the output dataset must be explicitly listed. In our case, two categories, *Material* and *Machine*, are selected.

By default, all the causes included in a category are added to the output dataset when a category is selected. This behavior also applies to compound causes. For example, in Listing 1.4, line 3, all causes inside the *Material* category are implicitly added to the target dataset. If we wanted to select just a subset of causes of a compound element, we must explicitly list the concrete subcauses to be selected inside an inclusion block after the element name. Listing 1.4, lines 4–6 show an example of an inclusion block. In this case, just the *PneumaticPliersPressure* cause of the *Machine* category is added to the output dataset.

These specifications are processed by the Papin intepreter as follows: first of all, domain data associated to the selected data-linked causes are retrieved using the data feeders associated to these causes. As a result, for each data-linked cause, we get a tabular structure with one column per each attribute that characterizes that cause, plus one column for the identifier of the instances being analyzed[1]. For example, the *Model* cause, contained in the *WheelBand* cause (Listing 1.3, lines 6–7), when selected, generates a tabular structure with *id* and *model* as columns, where *id* is the identifier for a drive half-shaft and *model* is the model number of its band in the wheel side, respectively. Using these data, this cause can be characterized, or measured, for each drive half-shaft.

Since these tabular structures are computed using the same reduction patterns that the *Lavoisier* interpreter uses, each one of them satisfies the *one entity, one row* constraint. So, each one of these tabular structures can be seen as a projection of the desired dataset for viewing just data associated to a specific cause. Therefore, to generate the complete dataset, we would just need to combine these tabular structures. This task can be easily executed by performing *joins*.

4   Related Work

To the best of our knowledge, this is the first work that enriches fishbone diagrams with data so that datasets can be automatically generated from them. Nevertheless, there exists in the literature other work that connect fishbone diagrams with data models.

For example, Xu and Dang [16] develop an approach to generate fishbone diagrams from databases containing problem reports. In these reports, causes for each problem are already identified. These reports are automatically analyzed using natural language processing and machine learning techniques to generate as output sets of causes that might lead to each reported problem. Using this information, fishbone diagrams are automatically constructed. This approach requires that causes for each reported problem are identified, whereas, in our work, we want to build datasets to feed data analysis algorithms that can help to discover these causes.

Shigemitsu and Shinkawa [9] propose using fishbone diagrams as part of the requirements engineering process for developing systems where the problem to be solved is known, but the causes of that problem are unknown and, therefore, the solutions to be implemented for these causes are also unknown. Fishbone diagrams are used to decompose the problem into several subproblem. Then, each atomic subproblem is associated to a software function, and, following some guidelines, a UML class diagram is generated. However, these diagrams contain just functions and no data, so they are not useful to specify domain data.

Yurin et al. [18] present a model-driven process for the automated development of rule-based knowledge bases from fishbone diagrams. In this process,

---

[1] If entities being analyzed required being identified by several values, a column per each one of these values would be added to this tabular structure.

rules of a knowledge base are firstly modeled as cause-effect relationships of a fishbone model. By using fishbone diagrams, domain experts can have a system-wide focus of these rules. As in Xu and Dang [16], cause-effect relationships are perfectly known, since they are expert knowledge. Contrarily, in our approach, we build datasets to find cause-effect relationships that can be used as expert knowledge. On the other hand, similarly to Yurin et al., we use fishbone diagrams to provide a global view of influences between data, so that decisions about what must be in a dataset can be more easily made.

Gwiazda [3] addresses the inclusion of quantitative information in fishbone diagrams. In order to do it, *weighted Ishikawa diagrams* are proposed. These diagrams are fishbone diagrams in which weights are added to the causes. To calculate them, weights of the categories are established using a form of *Saaty matrix*. Then, the weights of the causes and sub-causes are calculated by dis-tributing the category weight between them.

Finally, Yun et al. [17] use fishbone diagrams to control quality of data min-ing processes. In these diagrams, ribs and bones represent steps of a data mining process. Issues that might affect to quality of the results are identified as causes and attached to the corresponding step. These fishbone diagrams help to visu-alize a data analysis process, but they cannot be used to generate datasets.

5   Conclusions and Future Work

This work has presented a model-driven process that aims to automatize the generation of datasets from fishbone models. To build this process, two new lan-guages were designed. This process helps to decrease the time required to create a dataset, which would help to reduce development time, and, therefore, costs, of Industry 4.0 applications. Moreover, some empirical experiments showed this kind of language might be used by people without expertise on data science [13]. Therefore, it is reasonable to expect that industrial engineers can create data-oriented fishbone models and Papin specifications by themselves. However, from a scientific point of view, this hypothesis should be empirically checked. This will be part of our future work.

A benefit of industrial engineers using these languages by themselves would be that the dependence on data scientists, whose fees are often expensive and whose availability might be scarce, would be reduced. This would lead to a reduction of both development costs and time to market of Industry 4.0 appli-cations. Data scientists would be still needed to build and maintain the domain model. Nevertheless, multiple data-oriented fishbone models might be created for a single domain model without the participation of data scientists and several datasets might be generated from each one of these fishbone models. Therefore, our approach does not remove completely the dependence on data scientists, but helps to reduce it.

So, summarizing, our work might help to reduce time required to create a dataset as well as dependence on data scientists. As future work, we are planning a more empirical assessment of our approach and to extend the languages with new features, such as functions to include aggregated values in datasets, like, for example, the average value of a set of measures.

Acknowledgements. Funded by the Spanish Government under grant TIN2017-86520-C3-3-R.

References

1. Dave, N., Kannan, R., Chaudhury, S.K.: Analysis and prevention of rust issue in automobile industry. Int. J. Eng. Res. Technol. **4**(10), 1–10 (2018)
2. Dziuba, S.T., Jarossová, M.A., Gołębiecka, N.: Applying the Ishikawa diagram in the process of improving the production of drive half-shafts. In: Borkowski, S., Ingaldi, M. (eds.) Toyotarity. Evaluation and Processes/Products Improvement, chap. 2, pp. 20–23. Aeternitas (2013)
3. Gwiazda, A.: Quality tools in a process of technical project management. J. Achievements Mater. Manuf. Eng. **18**(1–2), 439–442 (2006)
4. Haverkort, B.R., Zimmermann, A.: Smart industry: how ICT will change the game! IEEE Internet Comput. **21**(1), 8–10 (2017)
5. Ishikawa, K.: Guide to Quality Control. Asian Productivity Organization (1976)
6. Lee, S.M., Lee, D., Kim, Y.S.: The quality management ecosystem for predictive maintenance in the Industry 4.0 era. Int. J. Qual. Innov. **5**(1), 1–11 (2019)
7. Lu, Y.: Industry 4.0: A survey on technologies, applications and open research issues. J. Indus. Inf. Integr. **6**, 1–10 (2017)
8. Piekara, A., Dziuba, S., Kopeć, B.: The use of Ishikawa diagram as means of improving the quality of hydraulic nipple. In: Borkowski, S., Selejdak, J. (eds.) Toyotarity. Quality and Machines Operating Conditions, chap. 15, pp. 162–175 (2012)
9. Shigemitsu, M., Shinkawa, Y.: Extracting class structure based on fishbone diagrams. In: Proceedings of the 10th International Conference on Enterprise Information Systems (ICEIS), vol. 2, pp. 460–465 (2008)
10. Siwiec, D., Pacana, A.: The use of quality management techniques to analyse the cluster of porosities on the turbine outlet nozzle. Prod. Eng. Arch. **24**(24), 33–36 (2020)
11. Steinberg, D., Budinsky, F., Paternostro, M., Merks, E.: EMF: Eclipse Modeling Framework, 2 edn. Addison-Wesley Professional (2008)
12. Tague, N.R.: The Quality Toolbox. Rittenhouse, 2 edn. (2005)
13. de la Vega, A.: Domain-Specific Languages for Data Mining Democratisation. Phd thesis, Universidad de Cantabria (2019). http://hdl.handle.net/10902/16728
14. de la Vega, A., García-Saiz, D., Zorrilla, M., Sánchez, P.: On the automated transformation of domain models into tabular datasets. In: Proceedings of the ER Forum. CEUR Workshop Proceedings, vol. 1979, pp. 100–113 (2017)
15. de la Vega, A., García-Saiz, D., Zorrilla, M., Sánchez, P.: Lavoisier: A DSL for increasing the level of abstraction of data selection and formatting in data mining. J. Comput. Lang. **60**, 100987 (2020)
16. Xu, Z., Dang, Y.: Automated digital cause-and-effect diagrams to assist causal analysis in problem-solving: a data-driven approach. Int. J. Prod. Res. **58**(17), 5359–5379 (2020)

17. Yun, Z., Weihua, L., Yang, C.: The study of multidimensional-data flow of fishbone applied for data mining. In: Proceedings of the 7th International Conference on Software Engineering Research, Management and Applications (SERA), pp. 86–91 (2009)
18. Yurin, A., Berman, A., Dorodnykh, N., Nikolaychuk, O., Pavlov, N.: Fishbone diagrams for the development of knowledge bases, In: Proceedings of the 41st International Convention on Information and Communication Technology, Electronics and Microelectronics (MIPRO) pp. 967–972 (2018)

GPU-Based Algorithms for Processing the k Nearest-Neighbor Query on Disk-Resident Data

Polychronis Velentzas[1]($^{(\boxtimes)}$) iD, Michael Vassilakopoulos[1] iD,
and Antonio Corral[2] iD

[1] Data Structuring and Engineering Laboratory, Department of Electrical
and Computer Engineering, University of Thessaly, Volos, Greece
{cvelentzas,mvasilako}@uth.gr
[2] Department of Informatics, University of Almería, Almería, Spain
acorral@ual.es

Abstract. Algorithms for answering the k Nearest-Neighbor (k-NN) query are widely used for queries in spatial databases and for distance classification of a group of query points against a reference dataset to derive the dominating feature class. GPU devices have much larger numbers of processing cores than CPUs and faster device memory than the main memory accessed by CPUs, thus, providing higher computing power for processing demanding queries like the k-NN one. However, since device and/or main memory may not be able to host an entire, rather big, reference dataset, storing this dataset in a fast secondary device, like a Solid State Disk (SSD) is, in many practical cases, a feasible solution. We propose and implement the first GPU-based algorithms for processing the k-NN query for big reference data stored on SSDs. Based on 3d synthetic big data, we experimentally compare these algorithms and highlight the most efficient algorithmic variation.

Keywords: k Nearest-Neighbor Query · GPU · SSD · Spatial-queries algorithms · Plane-sweep · Max Heap · Parallel computing

1 Introduction

Processing of big spatial data is demanding, and it is often assisted by parallel processing. GPU-based parallel processing has become very popular during last years [1]. In general, GPU devices have much larger numbers of processing cores than CPUs and device memory which is faster than main memory accessed by CPUs, providing high computing capabilities even to commodity computers.

GPU devices can be utilized for efficient parallel computation of demanding spatial queries, like the k Nearest-Neighbor (k-NN) query, which is widely used

---

Work of M. Vassilakopoulos and A. Corral funded by the MINECO research project [TIN2017-83964-R].

C. Attiogbé and S. Ben Yahia (Eds.): MEDI 2021, LNCS 12732, pp. 264–278, 2021.
https://doi.org/10.1007/978-3-030-78428-7_21

for spatial distance classification in many problems areas. We consider a set of query points and a set of reference points. For each query point, we need to compute the k-NNs of this point within the reference dataset. This permits us to derive the dominating class among these k-NNs (in case the class of each reference point is known).

Since GPU device memory is expensive, it is very important to take advantage of this memory as much as possible and scale-up to larger datasets and avoid the need for distributed processing which suffers from excessive network cost, sometimes overcoming the benefits of distributed parallel execution. However, since device and/or main memory may not be able to host an entire, rather big, reference dataset, storing this dataset in a fast secondary device, like a Solid State Disk (SSD) is, in many practical cases, a feasible solution.

In this paper,

- We propose and implement the first (Brute-force and Plane-sweep) GPU-based algorithms for processing the k-NN query on big reference data stored on SSDs.
- We utilize either an array-based, or a max-Heap based buffer for storing the distances of the current k nearest neighbors, which are combined with Brute-force and Plane-sweep techniques, deriving four algorithmic variations.
- Based on 3d synthetic big data, we present an extensive experimental comparison of these algorithmic variations, varying query dataset size, reference dataset size and k and utilizing reference data files which are either presorted in one of the dimensions, or unsorted in all dimensions.
- These experiments highlight that Plane-sweep, combined with either an array or a max-Heap buffer and applied on unsorted reference data, is the performance winner.

The rest of this paper is organized as follows. In Sect. 2, we review related material and present the motivation for our work. Next, in Sect. 3, we present the new algorithms that we developed for the k-NN GPU-based processing on disk-resident[1] data and in Sect. 4, we present the experimental study that we performed for analyzing the performance of our algorithms and for determining the performance winner among four algorithmic variations tested on presorted and unsorted big reference data. Finally, in Sect. 5, we present the conclusions arising from our work and discuss our future plans.

2   Related Work and Motivation

Recent trend in the research for parallelization of nearest neighbor search is to use GPUs. Parallel k-NN algorithms on GPUs can be usually implemented by employing a *Brute-force* (BF) method or by using *indexing data structures*. In the first category, k-NN on GPUs using a Brute-force method applies a two-stage scheme: (1) the computation of distances and (2) the selection of the nearest

---

[1] We used an SSD and in the rest of the text "SSD" instead of "disk" is used.

neighbors. For the first stage, a distance matrix is built grouping the distance array to each query point. In the second stage, several selections are performed in parallel on the different rows of the matrix. There are different approaches for these two stages and the most representative ones can be found in: [7] (each distance matrix row is sorted using radix sort), [2] (insertion sort), [16] (truncated sort), [6] (quick sort), [9] (truncated merge sort), [17] (use heuristics to minimize the reference points near a query point), [18] (use of symmetrical partitioning technique with k-NN list buffer), etc. See [18] for a more complete explanation of the Brute-force approaches. In the second category, we can find effective k-NN GPU implementations based on known indexing methods: k-d tree-based [3], grid-based [8], R-tree-based [11], LSH-based [12], etc.

Flash-based Solid State Drives (SSDs) have been widely used as secondary storage in database servers because of their improved characteristics compared to Hard Disk Drives (HDDs) to manage large-scale datasets [10]. These characteristics include smaller size, lighter weight, lower power consumption, better shock resistance, and faster reads and writes. In these secondary devices, read operations are faster than writes, while difference exist among the speeds of sequential and random I/Os as well. Moreover, the high degree of internal parallelism of latest SSDs substantially contributes to the improvement of I/O performance [14].

To address the necessity of fast nearest neighbor searches on large reference datasets stored in fast secondary devices (SSD), in this paper, we design and implement efficient k-NN GPU-based algorithms.

3 kNN Disk Algorithms

A common practice to handle big data is data partitioning. In order to describe our new algorithms, we should firstly present the mechanism of data partition transfers to device memory. This step is identical in all our methods. Each reference dataset is partitioned in N partitions containing an equal number of reference points. If the total reference points is not divided exactly with N, the Nth partition contains the remainder of the division. Initially the host (the computing machine hosting the GPU device) reads a partition from SSD[2] and loads it into the host memory. The host copies the in-memory partition data into the GPU device memory.

Another common approach in all our four methods is the GPU thread dispatching. Every query point is assigned to a GPU thread. The GPU device starts the k-NN calculation simultaneously for all threads. If the number of query points is bigger than the total available GPU threads, then the execution progresses whenever a block of threads finishes the previously assigned query points calculation. The thread dispatching consists of 4 main steps:

---

[2] Reading from SSD is accomplished by read operations of large sequences of consecutive pages, exploiting the internal parallelism of SSDs, although our experiments showed that reading from SSD does not contribute significantly to the performance cost of our algorithms.

1. The kernel function requests N threads.
2. The requested N threads are assigned to N query points.
3. Every thread carries out the calculation of reference point distances to its query point and updates the k-NN buffer holding the current (and eventually the final) nearest neighbors of this point.
4. The final k-NN list produced by each algorithm is populated with the results of all the query points.

In the next sections we will describe our new methods. These methods are based on two new main algorithms, "Disk Brute-force" and "Disk Plane-sweep". In both of them we have implemented two k-NN buffer variations resulting in a total of four new methods (algorithmic variations).

3.1   Disk Brute-Force Algorithm

The Disk Brute-force algorithm (denoted by DBF) is a Brute-force algorithm enhanced with capability to read SSD-resident data. Brute-force algorithms are highly efficient when executed in parallel. The algorithm accepts as inputs a reference dataset R consisting of m reference points $R = \{r_1, r_2, r_3..r_m\}$ in 3d space and a dataset Q of n query points $Q = \{q_1, q_2, q_3..q_n\}$ also in 3d space. The host reads the query dataset and transfers it in the device memory. The reference dataset is partitioned in equally sized bins and each bin is transferred to the device memory (Algorithm 1; note that the notation $<<< b, t >>>$ denotes execution using b blocks with t threads each). For each partition, we apply the k-NN Brute-force computations for each of the threads.

For every reference point within the loaded partition, we calculate the Euclidean distance (Algorithm 2) to the query point of the current thread. The first k distances are added to the k-NN buffer of this query point. Every other calculated distance is compared with the current largest one and, if it is smaller, it replaces the current largest one in the k-NN buffer.

The organization and implementation of the k-NN buffer is essential for the effective k-NN calculation, because by using it we elude sorting large distance arrays. The sorting step is extremely demanding, regarding GPU computations. Depending on the algorithm, the CUDA profiler revealed that 90% (or more in large datasets) of the GPU computation may be dedicated to sorting [17]. We will use and compare two alternative k-NN buffer implementations, presented in Subsects. 3.3 and 3.4.

Algorithm 1. Brute-force Host algorithm

Input: NN cardinality=K, Reference filename=RF, Query filename=QF, Partition size=S
Output: Host k-NN Buffer=HostKNNBufferVector
1: HostQueryVector ← readFile(QF);
2: queryPoints ← HostQueryVector.size();
3: DeviceQueryVector ← HostQueryVector;
4: createEmptyHostVector(HostKNNBufferVector,queryPoints*K);
5: DeviceKNNBufferVector ← HostKNNBufferVector;
6: **while** not end-of-file RF **do**
7:     HostReferencePartition ← readPartition(RF,S);
8:     DeviceReferencePartition ← HostReferencePartition;
9:     runKNN<<<(queryPoints-1)/256 +1, 256>>>(DeviceReferencePartition,
       DeviceQueryVector,DeviceKNNBufferVector,K);   // 256 cores assumed
10: HostKNNBufferVector ← DeviceKNNBufferVector;

Algorithm 2. Brute-force Device Kernel algorithm (runKNN)

Input: NN cardinality=K, Partition Reference array=R, Query array=Q, Partition size=S
Output: Device k-NN Buffer array=DKB
1: qIdx ← blockIdx.x*blockDim.x+threadIdx.x;
2: knnBufferOffset ← qIdx*K;
3: **for** i ← 0 to S-1 **do**
4:     dist ← $\sqrt[2]{(R[i].x - Q[qIdx].x)^2 + (R[i].y - Q[qIdx].y)^2 + (R[i].z - Q[qIdx].z)^2}$
5:     insertIntoBuffer(DKB,knnBufferOffset,i,qIdx,dist);

3.2   Disk Plane-Sweep Algorithm

An important improvement for join queries is the use of the Plane-sweep technique, which is commonly used for computing intersections [13]. The Plane-sweep technique is applied in [4] to find the closest pair in a set of points which resides in main memory. The basic idea, in the context of spatial databases, is to move a line, the so-called sweep-line, perpendicular to one of the axes, e.g., X-axis, from left to right, and process objects (points, in the context of this paper) as they are reached by this sweep-line. We can apply this technique for restricting all possible combinations of pairs of objects from the two datasets. The Disk Plane-sweep algorithm (denoted as DSP) incorporates this technique which is further enhanced with capability to read SSD-resident data.

Like DBF, DSP accepts as inputs a reference dataset R consisting of m reference points $R = \{r_1, r_2, r_3..r_m\}$ in 3d space and a dataset Q of n query points $Q = \{q_1, q_2, q_3..q_n\}$ also in 3d space. The host reads the query dataset and transfers it in the device memory. The reference dataset is partitioned in equally sized bins, each bin is transferred to the device memory and sorted by the x-values of its reference points. (Algorithm 3). For each partition we apply the k-NN Plane-sweep technique (Fig. 1).

Starting from the leftmost reference point of the loaded partition, the sweep-line moves to the right. The sweep-line hops every time to the next reference point until it approaches the x-value of the query point (Fig. 1). Using the x-value of the query point, a virtual rectangle is created. This rectangle has a length of $2 * l$, where l is the currently largest k-NN distance in the k-NN buffer of the query point of the current thread.

For every reference point within this rectangle, we calculate the Euclidean distance (Algorithm 4) to this query point. The first k distances are added to its

k-NN buffer. Every subsequent calculated distance is compared with the largest one in the k-NN buffer and if it is smaller, it replaces the largest one in the k-NN buffer.

In Fig. 1, we observe that all the reference points located on the right of the right rectangle limit are not even processed. The reference points located on left of the left rectangle limit are only processed for comparing their x-axis value. The costly Euclidean distance calculation is limited within the rectangle.

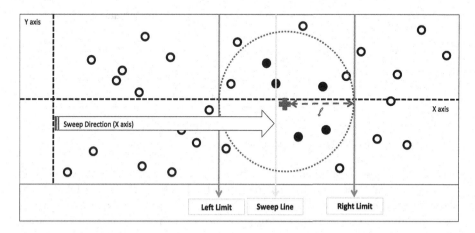

Fig. 1. Plane-sweep k-NN algorithm. Cross is the Query point, selected reference points in solid circles and not selected reference points in plain circles.

Algorithm 3. Plane-sweep Host algorithm

Input: NN cardinality=K, Reference Filename=RF, Query filename=QF, Partition size=S
Output: Host k-NN Buffer=HostKNNBufferVector
1: HostQueryVector←readFile(QF);
2: queryPoints ← HostQueryVector.size();
3: DeviceQueryVector←HostQueryVector;
4: createEmptyHostVector(HostKNNBufferVector,queryPoints*K);
5: DeviceKNNBufferVector←HostKNNBufferVector;
6: **while** not end-of-file RF **do**
7:     HostReferencePartition←readPartition(RF,S);
8:     DeviceReferencePartition←HostReferencePartition;
9:     cudaSort(DeviceReferencePartition);
10:     runKNN<<<(queryPoints-1)/256 +1, 256>>>(DeviceReferencePartition,
        DeviceQueryVector,DeviceKNNBufferVector,K);   // 256 cores assumed
11: HostKNNBufferVector←DeviceKNNBufferVector;

Algorithm 4. Plane-sweep Device Kernel algorithm (runKNN)

Input: NN cardinality=K, Partition Reference array=R, Query array=Q, Partition size=S, Largest distance in k-NN buffer=maxknnDistance
Output: Device k-NN Buffer array=DKB
1: qIdx ← blockIdx.x*blockDim.x+threadIdx.x;
2: knnBufferOffset ← qIdx*K;
3: xSweepline ← R[0].x;
4: i←0;
5: **while** xSweepline<Q[qIdx].x and $i < S$ loop **do**
6:    inc(i);
7:    xSweepline ← R[i].x;
8: leftIdx ← i-1;
9: **while** $(R[leftIdx].x - Q[qIdx].x) < maxknnDistance$ and $leftIdx > 0$ **do**
10:    dec(leftIdx);
11: rightIdx=i;
12: **while** $(R[rightIdx].x - Q[qIdx].x) < maxknnDistance$ and $leftIdx > 0$ **do**
13:    inc(rightIdx);
14: **for** i ← leftIdx to rightIdx **do**
15:    dist ← $\sqrt[2]{(R[i].x - Q[qIdx].x)^2 + (R[i].y - Q[qIdx].y)^2 + (R[i].z - Q[qIdx].z)^2}$
16:    insertIntoBuffer(DKB,knnBufferOffset,i,qIdx,dist);

3.3 k-NN Distance List Buffer

In our methods we implemented two different k-NN list buffers. The first one is the k-NN Distance List Buffer (denoted by KNN-DLB). KNN-DLB is an array where all calculated distances are stored (Fig. 2). KNN-DLB array size is k per thread, resulting in a minimum possible device memory utilization. When the buffer is not full, we append the calculated distances. When the buffer is full, we compare every newly calculated distance with the largest one stored in KNN-DLB. If it is smaller, we simply replace the largest distance with the new one. Therefore, we do utilize sorting. The resulting buffer contains the correct k-NNs, but not in an ascending order. The usage of KNN-DLB buffer is performing better than sorting a large distance array [18].

	New Distance	1	2	3	4	5	6	7	8	9	10
First 10 distances are	5,1	5,1									
appended to the list	2,7	5,1	2,7								
	4,0	5,1	2,7	4,0							
	2,8	5,1	2,7	4,0	2,8						
	11,2	5,1	2,7	4,0	2,8	11,2					
	1,7	5,1	2,7	4,0	2,8	11,2	1,7				
	3,5	5,1	2,7	4,0	2,8	11,2	1,7	3,5			
	0,6	5,1	2,7	4,0	2,8	11,2	1,7	3,5	0,6		
	0,1	5,1	2,7	4,0	2,8	11,2	1,7	3,5	0,6	0,1	
	7,1	5,1	2,7	4,0	2,8	11,2	1,7	3,5	0,6	0,1	7,1
Distances smaller that	8,5	5,1	2,7	4,0	2,8	8,5	1,7	3,5	0,6	0,1	7,1
the maximum distance,	6,9	5,1	2,7	4,0	2,8	6,9	1,7	3,5	0,6	0,1	7,1
replace it	1,6	5,1	2,7	4,0	2,8	6,9	1,7	3,5	0,6	0,1	1,6
	5,8	5,1	2,7	4,0	2,8	5,8	1,7	3,5	0,6	0,1	1,6

KNN Distance List Buffer, k=10

Fig. 2. k-NN Distance List Buffer update.

3.4 k-NN Max-Heap Distance List Buffer

The second list buffer that we implemented is based on a max-Heap (a priority queue represented by a complete binary tree which is implemented using an array). max-Heap array size is $k+1$ per thread, because the first array element is occupied by a sentinel. The sentinel value is the largest value for double numbers (for C++ language, used in this work, it is the constant DBL_MAX). KNN-DLB is adequate for smaller k values, but when k value increases performance deteriorates, primarily due to KNN-DLB O(n) insertion complexity. On the other hand, max-Heap insertion complexity is O(log(n)) and for large enough k max-Heap implementations are expected to outperform KNN-DLB ones.

4   Experimental Study

We run a large set of experiments to compare the application of our proposed algorithms. All experiments query at least 500M reference points. We did not include less than 500M reference points because we target reference datasets that do not fit in the device memory. The largest dataset that could fit in device memory in our previous work was 300M [18]. Furthermore, we increased the points accuracy representation from single precision numbers to double precision (Algorithm 5) to be able to discriminate among small distance differences.

Algorithm 5. Point Structure

```
       // Point record structure, used in reference datasets. Record size 32 bytes
    1: record point_struct begin
    2:     id,        // Point ID, type unsigned long long, 8 bytes
    3:     x, y, z    // 3 Dimensions, type double, 8 bytes per dimension
    4: end;
```

All the datasets were created using the SpiderWeb [5] generator. This generator allows users to choose from a wide range of spatial data distributions and configure the size of the dataset and its distribution parameters. This generator has been successfully used in research work to evaluate index construction, query processing, spatial partitioning, and cost model verification [19].

Table 1 lists all the generated datasets. For the reference dataset, we created four datasets using the "Bit" distribution (Fig. 3 right), with file sizes ranging from 16 GB to 64 GB. The reference points dataset size ranges from 500M points to 2G points. For the query points dataset we created one "Uniform" dataset (Fig. 3, center) of 10 points and five "Gaussian" datasets (Fig. 3, left) ranging from 10K to 50K points.

Three different sets of experiments were conducted. In the first one, we scaled the reference dataset size, in the second one we scaled the query dataset size and in the last one we scaled the number of the nearest neighbors, k. For every set, we used presorted and unsorted reference datasets to evaluate their effect on the

Table 1. SpiderWeb Dataset generator parameters.

Distribution	Size	Seed	File size	Dataset usage
Bit	500M	1	16 GB	Reference
Bit	1G	2	32 GB	Reference
Bit	1.5G	3	48 GB	Reference
Bit	2G	4	64 GB	Reference
Uniform	10	5	32 B	Query
Gaussian	10K	6	320 KB	Query
Gaussian	20K	7	640 KB	Query
Gaussian	30K	8	960 KB	Query
Gaussian	40K	9	1,3 MB	Query
Gaussian	50K	10	1,6 MB	Query

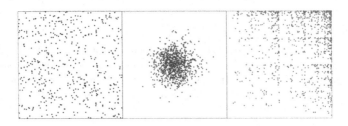

Fig. 3. Experiment distributions, Left=Uniform, Middle=Gaussian, Right=Bit.

methods' performance. We also evaluated the performance of the two alternative list buffers to clarify the pros and cons of using KNN-DLB and max-Heap buffer.

All experiments were performed on a Dell G5 15 laptop, running Ubuntu 20.04, equipped with a six core (12-thread) Intel I7 CPU, 16 GB of main memory, a 1 TB SSD disk used and a NVIDIA Geforce 2070 (Mobile Max-Q) GPU with 8 GB of device memory. CUDA version 11.2 was used.

We run experiments to compare the performance of k-NN queries regarding execution time, as well as memory utilization. We tested a total of four algorithms, listed in the following.

1. DBF, Disk Brute-force using KNN-DLB buffer
2. DBF Heap, Disk Brute-force using max Heap buffer
3. DPS, Plane-sweep using KNN-DLB buffer
4. DPS, Plane-sweep using max Heap buffer

To the best of our knowledge, these are the first methods to address the k-NN query on SSD-resident data.

4.1 Reference Dataset Scaling

In our first series of tests, we used the "Bit" distribution synthetic datasets for the reference points. The size of the reference point dataset ranged from 500M points to 2G points. Furthermore, we used a small query dataset of 10 points, with "Uniform" distribution and a relatively small k value, 10, in order to focus only on the reference dataset scaling.

In Fig. 4, we can see the presorted dataset results in blue and the unsorted dataset results in stripped yellow. In the presorted experiment, we notice that the execution time of all methods is quite similar, for each reference dataset size. For example, for the 500M dataset the execution time is 172 sec. for DBF, 171 s for DBF Heap, 181 for DPS and 178 sec. for DPS Heap. The execution times increase proportionally to the reference dataset size. As expected, we get the slowest execution times for the 2G dataset, 691 sec. for both DBF, 683 sec. for DBF Heap, 730 sec. for DPS and 729 for DPS Heap.

In the unsorted dataset experiments, we observe that all execution times are smaller, especially for the Plane-sweep methods. The Brute-force methods are slightly faster in the unsorted dataset experiments than in the presorted ones. For the 500M unsorted dataset, the execution time for DBF is 154 sec., for DBF Heap is 156 sec., for DPS 67 sec. and for DPS Heap just 68 sec.. Once again, the execution times increase proportionally to the reference dataset size. For the 2G unsorted dataset, we get 638 sec. for DBF, 640 sec. for DBF Heap, 262 sec. for DPS and 262 for DPS Heap.

The reference dataset scaling experiments reveal that all the methods performed better for the unsorted dataset. For the unsorted dataset, the Brute-force methods performed slightly better, but the Plane-sweep methods performed exceptionally better than for the presorted dataset. Furthermore, the Plane-sweep methods were more than 1.7 times faster than Brute-force ones, in the unsorted dataset experiments.

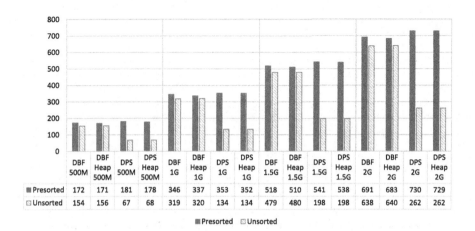

Fig. 4. Reference scaling experiment (Y-axis in sec.). (Color figure online)

4.2 Query Dataset Scaling

In our second set of experiments, we used between 10K and 50K query points with "Gaussian" distribution. For the reference points we used a 500M "Bit" distribution synthetic dataset. These experiments also used a relatively small k value of 10, in order to focus only on query dataset scaling.

In Fig. 5, we can see the presorted dataset results in blue and the unsorted dataset results in stripped yellow. In the presorted experiments, we notice that the execution time of the Brute-force algorithms is always larger. Depending on the query dataset size, we observe that the execution time gradually increases. For the 10K dataset, the execution time is 585 sec. for DBF, 618 s for DBF Heap, 512 for DPS and 520 sec. for DPS Heap. The slowest execution times were recorded for the 50K dataset, 1882 sec. for DBF, 3475 sec. for DBF Heap, 1504 sec. for DPS and 1584 for DPS Heap.

In the unsorted experiments, we observe once more that all execution times are smaller, especially for the Plane-sweep methods. The Brute-force methods are slightly faster for the unsorted dataset than for the presorted one. For the 10K unsorted query dataset, the execution time for DBF is 579 sec., for DBF Heap is 616 sec., for DPS 81 sec. and for DPS Heap also 81 sec.. Once again, the execution times increase proportionally to the query dataset size. For the 50K unsorted query dataset, we get 1846 sec. for DBF, 3377 sec. for DBF Heap, 218 sec. for DPS and 225 for DPS Heap.

The results of the query dataset scaling experiments conform with the ones of the reference scaling experiments. All the methods performed better with the unsorted dataset. For the unsorted dataset, the Brute-force methods performed slightly better, but the Plane-sweep methods performed once again exceptionally better than for the presorted dataset. Furthermore, the Plane-sweep methods were more than 7 to 15 times faster than Brute-force ones, in the unsorted dataset experiments.

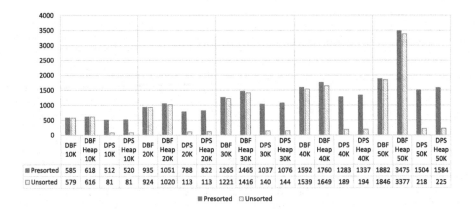

	DBF 10K	DBF Heap 10K	DPS 10K	DPS Heap 10K	DBF 20K	DBF Heap 20K	DPS 20K	DPS Heap 20K	DBF 30K	DBF Heap 30K	DPS 30K	DPS Heap 30K	DBF 40K	DBF Heap 40K	DPS 40K	DPS Heap 40K	DBF 50K	DBF Heap 50K	DPS 50K	DPS Heap 50K
Presorted	585	618	512	520	935	1051	788	822	1265	1465	1037	1076	1592	1760	1283	1337	1882	3475	1504	1584
Unsorted	579	616	81	81	924	1020	113	113	1221	1416	140	144	1539	1649	189	194	1846	3377	218	225

■ Presorted ◻ Unsorted

Fig. 5. Query scaling experiment (Y-axis in sec.). (Color figure online)

4.3   *k* Scaling

The *k* scaling is our last set of experiments. In these tests we used *k* values of 10,20,50 and 100. For the reference points we used the 500M "Bit" distribution synthetic dataset and a small query group of 10 points, with "Uniform" distribution, in order to focus only on the *k* scaling.

In Fig. 6, we can see the presorted dataset results in blue and the unsorted dataset results in stripped yellow. In the presorted experiments, we notice that the execution time of the Brute-force algorithms is slightly smaller than the Plane-sweep ones. Depending on the *k* value, we observe that the execution time increases slightly for larger *k* values. For *k* equal to 10 the execution time is 172 sec. for DBF, 171 s for DBF Heap, 181 for DPS and 178 sec. for DPS Heap. The slowest execution times were recorded for *k* equal to 100, 180 sec. for DBF, 181 sec. for DBF Heap, 210 sec. for DPS and 208 for DPS Heap.

In the unsorted experiments, we observe once more that the execution is faster, especially for the Plane-sweep methods. The Brute-force methods are slightly faster for the unsorted dataset than for the presorted one. For *k* equal to 10 and the unsorted query dataset, the execution time for DBF is 158 sec., for DBF Heap is 156 sec., for DPS 68 sec. and for DPS Heap also 63 sec.. Once again, the execution times increase proportionally to the *k* value. For the $k = 100$, we get 168 sec. for DBF, 166 sec. for DBF Heap, 98 sec. for DPS and 93 for DPS Heap.

The results of the *k* scaling experiments also conform with the results of the previous experiments. All the methods performed better with the unsorted dataset. For the unsorted dataset, the Brute-force methods performed slightly better, but the Plane-sweep methods performed once again exceptionally better than for the presorted dataset. Furthermore, the Plane-sweep methods were about 2 times faster than Brute-force ones, in the unsorted dataset experiments. Although, the two *k*-NN list buffer methods were shown equal, for even larger *k* values than the ones studied in this paper, the *k* max-Heap list buffer is expected to outperform the KNN-DLB one.

4.4   Interpretation of Results

Exploring why the application of the Plane-sweep algorithms on unsorted reference data is significantly more efficient, we observed that, when the reference dataset is presorted, each partition contains points that fall within a limited *x*-range and in case the query point under examination is on the right side of this partition regarding *x*-dimension, most of the reference points of this partition will likely replace points already included in the current set of *k*-NNs for this query point (Fig. 7 left), since partitions are loaded from left to right and previous partitions examined were less *x*-close to this query point. However, when the reference dataset is unsorted, each partition contains points that cover a wide *x*-range and it is likely that many of the reference points of this partition will be rejected by comparing their *x*-distance to the distance of the *k*-th NN found so far (Fig. 7 right).

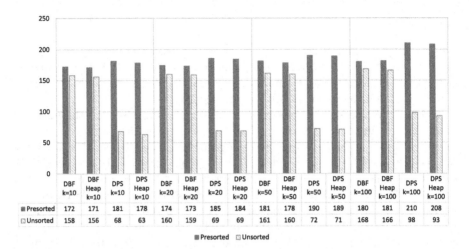

The chart below shows Presorted versus Unsorted values:

	DBF k=10	DBF Heap k=10	DPS k=10	DPS Heap k=10	DBF k=20	DBF Heap k=20	DPS k=20	DPS Heap k=20	DBF k=50	DBF Heap k=50	DPS k=50	DPS Heap k=50	DBF k=100	DBF Heap k=100	DPS k=100	DPS Heap k=100
■ Presorted	172	171	181	178	174	173	185	184	181	178	190	189	180	181	210	208
▨ Unsorted	158	156	68	63	160	159	69	69	161	160	72	71	168	166	98	93

■ Presorted ▨ Unsorted

Fig. 6. k scaling experiment (Y-axis in sec.). (Color figure online)

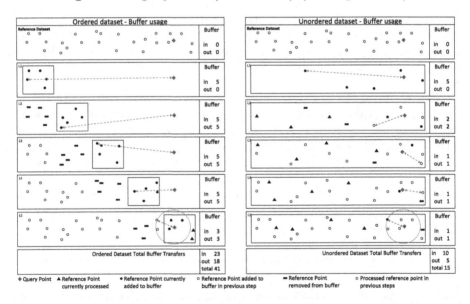

Fig. 7. Presorted versus unsorted reference dataset buffer update.

5   Conclusions and Future Plans

In this paper, we presented the first GPU-based algorithms for parallel processing the k-NN query on reference data stored on SSDs, utilizing the Brute-force and Plane-sweep techniques. These algorithms exploit the numerous GPU cores, utilize the device memory as much as possible and take advantage of the speed and storage capacity of SSDs, thus processing efficiently big reference datasets. Through an experimental evaluation on synthetic datasets, we highlighted that

Plane-sweep on unsorted reference data (with either an array or a max-Heap buffer for organizing the current k-NNs) is a clear performance winner.

Our future work plans include:

- development of k-NN GPU-based algorithms for big SSD resident data which exploit the use of indexes [15] to further speed-up processing,
- implementation of join queries (like K-closest pairs), based on techniques utilized in this paper.

References

1. Barlas, G.: Multicore and GPU Programming: An Integrated Approach. 1 edn, Morgan Kaufmann, Amsterdam (2014)
2. Garcia, V., Debreuve, E., Nielsen, F., Barlaud, M.: K-nearest neighbor search: Fast gpu-based implementations and application to high-dimensional feature matching. In: ICIP Conference, pp. 3757–3760 (2010)
3. Gieseke, F., Heinermann, J., Oancea, C.E., Igel, C.: Buffer k-d trees: Processing massive nearest neighbor queries on GPUs. In: ICML Conference, pp. 172–180 (2014)
4. Hinrichs, K.H., Nievergelt, J., Schorn, P.: Plane-sweep solves the closest pair problem elegantly. Inf. Process. Lett. **26**(5), 255–261 (1988)
5. Katiyar, P., Vu, T., Eldawy, A., Migliorini, S., Belussi, A.: Spiderweb: a spatial data generator on the web. In: SIGSPATIAL Conference, pp. 465–468 (2020)
6. Komarov, I., Dashti, A., D'Souza, R.M.: Fast k-NNG construction with GPU-based quick multi-select. PloS ONE **9**(5), 1–9 (2014)
7. Kuang, Q., Zhao, L.: A practical GPU based KNN algorithm. In: SCSCT Conference, pp. 151–155 (2009)
8. Leite, P.J.S., Teixeira, J.M.X.N., de Farias, T.S.M.C., Reis, B., Teichrieb, V., Kelner, J.: Nearest neighbor searches on the GPU - a massively parallel approach for dynamic point clouds. Int. J. Parallel Program. **40**(3), 313–330 (2012)
9. Li, S., Amenta, N.: Brute-force k-nearest neighbors search on the GPU. In: SISAP Conference, pp. 259–270 (2015)
10. Mittal, S., Vetter, J.S.: A survey of software techniques for using non-volatile memories for storage and main memory systems. IEEE Trans. Parallel Distributed Syst. **27**(5), 1537–1550 (2016)
11. Nam, M., Kim, J., Nam, B.: Parallel tree traversal for nearest neighbor query on the GPU. In: ICPP Conference, pp. 113–122 (2016)
12. Pan, J., Lauterbach, C., Manocha, D.: Efficient nearest-neighbor computation for GPU-based motion planning. In: IROS Conference, pp. 2243–2248 (2010)
13. Preparata, F.P., Shamos, M.I.: Computational Geometry - An Introduction. Texts and Monographs in Computer Science, Springer, New York (1985) https://doi.org/10.1007/978-1-4612-1098-6
14. Roh, H., Park, S., Kim, S., Shin, M., Lee, S.: B+-tree index optimization by exploiting internal parallelism of flash-based solid state drives. Proc. VLDB Endow. **5**(4), 286–297 (2011)
15. Roumelis, G., Velentzas, P., Vassilakopoulos, M., Corral, A., Fevgas, A., Manolopoulos, Y.: Parallel processing of spatial batch-queries using xbr$^+$-trees in solid-state drives. Clust. Comput. **23**(3), 1555–1575 (2020)

16. Sismanis, N., Pitsianis, N., Sun, X.: Parallel search of k-nearest neighbors with synchronous operations. In: HPEC Conference, pp. 1–6 (2012)
17. Velentzas, P., Vassilakopoulos, M., Corral, A.: In-memory k nearest neighbor GPU-based query processing. In: GISTAM Conference, pp. 310–317 (2020)
18. Velentzas, P., Vassilakopoulos, M., Corral, A.: A partitioning gpu-based algorithm for processing the k nearest-neighbor query. In: MEDES Conference. pp. 2–9 (2020)
19. Vu, T., Migliorini, S., Eldawy, A., Belussi, A.: Spatial data generators. In: Spatial-Gems - SIGSPATIAL International Workshop on Spatial Gems, pp. 1–7 (2019)

Revisiting Data Compression
in Column-Stores

Alexander Slesarev⬤, Evgeniy Klyuchikov⬤, Kirill Smirnov⬤,
and George Chernishev$^{(\boxtimes)}$⬤

Saint-Petersburg University, Saint-Petersburg, Russia

Abstract. Data compression is widely used in contemporary column-oriented DBMSes to lower space usage and to speed up query processing. Pioneering systems have introduced compression to tackle the disk bandwidth bottleneck by trading CPU processing power for it. The main issue of this is a trade-off between the compression ratio and the decompression CPU cost. Existing results state that light-weight compression with small decompression costs outperforms heavy-weight compression schemes in column-stores. However, since the time these results were obtained, CPU, RAM, and disk performance have advanced considerably. Moreover, novel compression algorithms have emerged.

In this paper, we revisit the problem of compression in disk-based column-stores. More precisely, we study the I/O-RAM compression scheme which implies that there are two types of pages of different size: disk pages (compressed) and in-memory pages (uncompressed). In this scheme, the buffer manager is responsible for decompressing pages as soon as they arrive from disk. This scheme is rather popular as it is easy to implement: several modern column and row-stores use it.

We pose and address the following research questions: 1) Are heavy-weight compression schemes still inappropriate for disk-based column-stores?, 2) Are new light-weight compression algorithms better than the old ones?, 3) Is there a need for SIMD-employing decompression algorithms in case of a disk-based system? We study these questions experimentally using a columnar query engine and Star Schema Benchmark.

Keywords: Query execution · Compression · PosDB

1 Introduction

The fact that DBMSes can benefit from data compression has been recognized since the early 90's [12, 16, 25]. Using it allows to reduce the amount of disk space occupied by data. It also allows to improve query performance by 1) reducing the amount of data read from disk, which may decrease the run time of a particular query if it is disk-bound, 2) operating on compressed data directly [2, 13, 18], thus allowing to speed up execution in compression ratio times minus overhead. Compression is applied to other database aspects as well, such as: results transferred

© Springer Nature Switzerland AG 2021
C. Attiogbé and S. Ben Yahia (Eds.): MEDI 2021, LNCS 12732, pp. 279–292, 2021.
https://doi.org/10.1007/978-3-030-78428-7_22

between the DBMS and the client [24], indexes [11,19], intermediate results [9], etc. Nowadays, data compression is used in almost all contemporary DBMSes.

Column-stores stirred up the interest in data compression in DBMSes. These systems store and handle data on a per-column basis, which leads to better data homogeneity. It allows to achieve better compression rates while simultaneously making simpler compression algorithms worthy of adoption.

Early experiments with column-stores [2,30] have demonstrated that a special class of compression algorithms (light-weight) should be employed for data compression in this kind of systems. However, almost fifteen years have passed since the publication of these works, and many changes have arisen:

- CPU, RAM, and disk performance have considerably advanced;
- novel compression algorithms have appeared;
- SIMD-enabled versions of existing algorithms have appeared as well.

These factors call for a reevaluation of the findings described by the founders. There are several recent papers [3,7,8,20,29] that examine the performance of classic and novel compression algorithms in a modern environment. However, these studies are insufficient, since they can not be used to answer questions related to performance of compression algorithms during query processing. In order to do so, these methods should be integrated into a real DBMS.

In this paper we study the impact of compression algorithms on query processing performance in disk-based column-stores. Despite the focus shift to in-memory processing, disk-based systems are still relevant. Not all workloads can be handled by pure in-memory systems, regardless of the availability and decreasing costs of RAM. This is especially true for analytical processing.

The exact research questions studied in this paper are:

- RQ1: Are heavy-weight compression schemes still inappropriate for disk-based column-stores?
- RQ2: Are new light-weight compression algorithms better than the old ones?
- RQ3: Is there a need for SIMD-employing decompression algorithms in case of a disk-based system?

These questions are studied experimentally using a columnar query engine and Star Schema Benchmark.

2   Background and Related Work

There are two approaches to implementing compression inside DBMSes [30]: I/O-RAM and RAM-CPU. The idea of the former is the following: data is stored on disk as a collection of compressed pages, which are decompressed as soon as they are loaded into the buffer manager. Therefore, there are two types of pages in the system: disk and in-memory. The second approach uses a single page type throughout the whole system. Therefore, a buffer manager stores compressed pages and when a data request comes, data is decompressed on demand (Fig. 1).

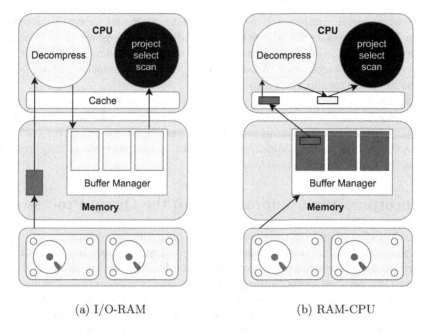

(a) I/O-RAM   (b) RAM-CPU

Fig. 1. Compression implementing approaches, adapted from [30]

The RAM-CPU approach is considered a superior option, especially for in-memory systems due to the higher performance it allows to achieve. At the same time, the I/O-RAM approach is still very popular since it is easy to implement in existing systems. Many classic systems rely on the I/O-RAM scheme, such as MySQL [4], SybaseIQ [30], and Apache Kudu [22]. Apache Druid has also used I/O-RAM compression for a long time before developing a more complex hybrid approach [21].

Next, two types of compression algorithms can be used in databases: light-weight and heavy-weight. Light-weight algorithms are usually characterized as simple algorithms that require little computational resources to decompress data. At the same time, the compression ratio they offer is relatively low: it is rarely higher than 2–3. On the contrary, heavy-weight compression schemes require significant effort to perform decompression while offering significantly higher compression ratios.

The following algorithms are considered light-weight in literature [1,14]: RLE, bit-vector, dictionary [5], frame-of-reference, and differential encoding. Examples of heavy-weight compression schemes [14] are BZIP and ZLIB.

Pioneering column-stores argued in favor of light-weight compression algorithms since they allowed to operate on compressed data directly and led to negligible decompression overhead. Another motivating point was the fact that light-weight algorithms worked well in their contemporary environment (i.e., engine implementation, hardware, OS, etc.), unlike heavy-weight ones.

However, novel compression algorithms have appeared recently, alongside with a trend of SIMD-ing algorithms (including compression). Furthermore, Google has released the Brotli library, which can be considered a novel heavy-weight compression technique. All of this calls for the reevaluation of approaches used to integrate compression into DBMSes.

Note that in this paper we consider "old-school" compression techniques, i.e. techniques that: 1) operate not on a set of columns (as, for example in a study [28]), but on each column individually, and 2) do not search for patterns in data to perform its decomposition, like many of the most recent compression studies [10,17,23] for column-stores do.

3   Incorporating Compression into the Query Processor

We have decided to address the posed research questions by performing an experimental evaluation. For this, we have implemented compression inside PosDB — a distributed column-store that is oriented towards disk-based processing. Before starting this work, it had a buffer pool which stored uncompressed pages that were the same as the pages residing on disk. In this study, we have implemented a generalized I/O-RAM compression scheme in which the compression algorithm is a parameter which we can change. Below, we present a general overview of the system and describe the architecture of our solution.

3.1   PosDB Fundamentals

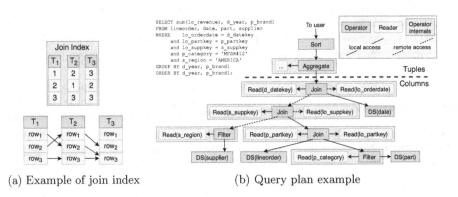

(a) Example of join index   (b) Query plan example

Fig. 2. PosDB internals

Query plans in PosDB consist of two parts: position- and tuple-based (separated by a dotted line in Fig. 2b). Operators in the bottom part pass over blocks that consist of positions — references to qualifying tuples (e.g. the ones that conform to query predicates). These positions are represented as a generalized join index [27] which is shown in Fig. 2a. Here, the second row of T_1 was joined with the first row of T_2 and then the resulting tuple was joined with the second row of T_3.

The upper part of any plan consists of operators that work with tuples, similarly to classic row-stores. In any plan, there is a materialization point (or points) where position-based operator(s) transform join indexes into tuples of actual values. Such materializing operators can be operators performing some useful job (e.g. aggregation or window function computation) or dedicated materialization operators.

When positional operators (e.g. Join, Filter) require data values they invoke auxiliary entities called readers. PosDB offers several types of readers that form a hierarchy, for example:

- ColumnReader retrieves values of a single attribute,
- SyncReader encapsulates several simpler readers in order to provide values of several attributes synchronously.

In their turn, readers invoke access methods — low-level entities that interact with data that is stored in pages residing on disk. There are three such entities: AccessRange, AccessSorted, and AccessJive. The first one assumes that positions are sequential and dense, the second one relies only on the monotonicity of positions (gaps are possible), and the last one is intended for arbitrary position lists. This type of organization is necessary to ensure efficiency of disk operations by relying on sequential reads as much as possible.

Readers interact with data pages that reside in main memory instead of disk. Therefore, pages are read from disk, stored in main memory, and pushed back (if, for example, a page is not required anymore) during query execution. This process of handling pages is governed by the buffer manager [15]. The PosDB buffer manager is built according to the classic guidelines.

A detailed description of PosDB's architecture can be found in paper [6].

3.2   Architecture of the Proposed Solution

Each column in PosDB can be stored in one or more files with the following structure: the file starts with a PageIndex that contains metadata such as the total number of pages. Next, it contains the data itself as pages (see Fig. 3a).

In the uncompressed form, all pages on disk are of equal size. However, implementing a compression subsystem in accordance to the I/O-RAM scheme required us to support pages of different size since each page may have different compression ratio, depending on its data. Therefore, we had to store extra information on compressed page offsets separately (see Fig. 3b). The physical parameters of all column files are stored in the catalog file.

After being loaded from disk to the buffer manager, a page is represented as a structure called ValBlock, which consists of a header and data buffer. This data buffer has to be decompressed each time a block is loaded from disk, just before it takes its place in the buffer manager slot.

In PosDB compression can be applied on a per-column basis with a specified (fixed) algorithm. During this process, corresponding column files will be changed and the catalog file will be updated. No other changes from the user's point of view will occur.

Query execution process can be represented as interaction of three types of processes: client, worker and I/O. The client passes a query to the queue and waits for the result before passing a new query from its set. The worker takes the query from the queue for execution. The I/O puts pages into buffer when the worker needs them. A page can be loaded immediately before it is needed or, alternatively, it could be preloaded. Therefore, data acquisition and query plan evaluation occur in parallel. All these processes can have several instances, except the query queue, which is uniquely instantiated.

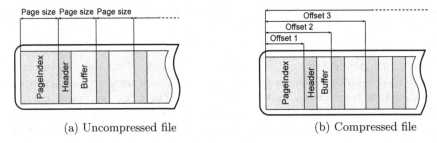

(a) Uncompressed file   (b) Compressed file

Fig. 3. File formats in PosDB

4   Experiments

4.1   Experimental Setup

To answer the research questions posed in Sect. 1, we have implemented the I/O-RAM scheme inside PosDB. The next step was to try a number of different compression algorithms. We have compiled the set of state-of-the-art compression algorithms from several different studies. Light-weight compression schemes were selected from the study by Lemire and Boytsov [20]. While selecting these, we aimed to obtain top-performing algorithms in terms of decoding speed. For heavy-weight compression we have selected Brotli [3] — an open source general-purpose data compressor that is now adopted in most known browsers and Web servers. Brotli is considered a heavy-weight competitor to BZIP.

Overall, we have selected the following algorithms:

1. Light-weight:
    - Regular: PFOR, VByte;
    - SIMD-enabled: SIMD-FastPFOR128, SIMD-BinaryPacking128.
2. Heavy-weight: Brotli (default configuration).

The source code of these implementations was taken from their respective Github repositories[1,2].

Experiments were run on the following hardware: Inspiron 15 7000 Gaming (0798), 8 GiB RAM, Intel(R) Core(TM) i5-7300HQ CPU @ 2.50 GHz, TOSHIBA 1 TB MQ02ABD1. The following software specification was used: Ubuntu 20.04.1 LTS, 5.4.0-72-generic, g++ 9.3.0, PosDB version 0043bba9.

In our experiments, we studied the impact of data compression on query evaluation time. For these purposes, we have employed the Star Schema Benchmark [26] with a scale factor of 50. All 13 SSB queries were evaluated. We have compressed only the integer columns of the LINEORDER table. Overall, ten columns turned out to be suitable for compression. For all results, we also provide measurements without compression.

The following PosDB settings were chosen: 65536 byte pages, 16K pages buffer manager capacity, which approximately equals 1 GB. The mean value of 10 iterations with a 95% confidence interval was presented as the result. Each iteration was performed with two sequential executions of the randomly shuffled query set. The first execution of the query set was needed to fill the buffer manager with pages, so its results were not taken into account. Operating system caches were dropped by writing "3" to /proc/sys/vm/drop_caches and swap was restarted between iterations.

4.2   Results and Discussion

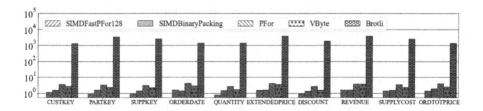

Fig. 4. Compression time (Seconds)

To address the posed research questions, we have run a number of experiments. Their results are presented in Figs. 4, 5, 6, 7, 8 and 9 and Table 1.

- Figure 4 shows the time it took to compress each involved column. Note that we had to use the logarithmic axis due to Brotli's significant compression time.
- Figure 5 shows the size of each compressed column, using each studied method. The overall impact of compression on size is presented in Table 1. Here, "over columns" lists the total sum of sizes of all compressed columns

---

[1] https://github.com/lemire/FastPFor.
[2] https://github.com/google/brotli.

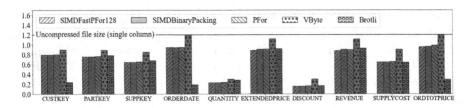

Fig. 5. Compressed column sizes (Gigabytes)

of the LINEORDER table. The "over whole table" presents the same data but over the whole table (including columns which we did not compress as we compress only integer data).

- Figures 6 and 7 present query run times for all SSB queries. These graphs show two different scenarios: "sequential" and "parallel". In "sequential", we feed queries into the system in a sequential fashion, i.e. the next query is run as soon as the previous has returned its answer. This simulates an ideal scenario in which there is no competition over resources. The "parallel" scenario is different: all queries are submitted at once, and the time it takes an individual query to complete is recorded. Additionally, we simulate disk contention by allowing only a single I/O thread in the buffer manager. Furthermore, we tried to break down the whole run time into two parts: query plan and data access. The first part denotes the portion of time the system spent actually executing the query and the second denotes the time it spent accessing data (reading from disk and decompressing).
- Figure 8 shows the breakdown of actions for the I/O thread for the "sequential" scenario.
- Figure 9 visualizes the total volume of data read from disk by each query in the "sequential" scenario.

Table 1. Data sizes in detail (Gigabytes)

Counting	Compressor					
	Raw	SIMDFastPFor	SIMDBinaryPacking	PFor	VByte	Brotli
Over columns	12.0	6.8	6.9	7.0	8.7	5.1
Over the whole table	16.6	11.5	11.5	11.6	13.4	9.7

RQ1: Are Heavy-Weight Compression Schemes Still Inappropriate for Disk-Based Column-Stores? First of all, note the time it takes to compress the data. It is at least two orders of magnitude higher than that of light-weight approaches. Compressing 1.2 GB of raw data (single column) takes about 15 min. However, such low compression speed is compensated by the achieved compression rate, which is almost 30% higher.

Turning to query run time breakdown, we can see that in the "sequential" scenario (Fig. 6), the heavy-weight compression scheme loses to all other methods. This happens due to the overall decompression overhead which is comparable to accessing data from disk. A close study of I/O thread actions (Fig. 8) reveals that decompression can take 10 times more time than reading the data from disk. Nevertheless, it is still safe to say that using this compression method can improve DBMS performance by 10%–20% (Fig. 6), compared to the uncompressed case.

Considering the "parallel" scenario (Fig. 7), where contention is simulated, one may see that the heavy-weight compression scheme is comparable to the best (SIMD-enabled) light-weight compression approaches.

Therefore, the answer to this RQ is largely yes. The only environment where the application of these approaches may be worthwhile is the read-only datasets with disk-intensive workloads that put a heavy strain on disk. In this case, the application of such compression scheme can at least save disk space. Note that this may be not a desirable mode of DBMS operation since the system is clearly overworked: individual queries significantly slow down each other, thus increasing their response time. A better choice may be to postpone some of the queries thus reducing the degree of inter-query parallelism.

RQ2: Are New Light-Weight Compression Algorithms Better Than the Old Ones? From the compression ratio standpoint (see Table 1) there is little to no difference for old PFor. Concerning compression speed, there is a significant difference: older PFor and VByte almost always lose to newer SIMD-enabled versions of classic light-weight approaches. However, VByte also loses to another old compression algorithm — vanilla PFor. Another observation can be derived from Fig. 5: Vbyte failed to compress two columns out of ten and demonstrated the worst performance overall (Table 1).

During query execution in the "sequential" scenario, VByte demonstrated the second worst result on average. Its data access cost can rival that of Brotli (see Fig. 6), and looking into the I/O thread breakdown (Fig. 8) one can see that: 1) depending on the query, VByte's disk reading costs are 5 to 25 times higher than Brotli's (8.5 on average), 2) VByte's decompression costs are approximately 6 times lower, and 3) VByte's decompression takes 23% of its total run time on average, compared to the 93% of Brotli.

The "parallel" query execution scenario shows that VByte is the worst method out of all evaluated. Sometimes (Query 1.1, 1.2, and 1.3) its performance can be even worse than that of running without compression. We believe that this happens due to poor compression rate and high decompression cost: the sum of costs of reading poorly compressed pages and decompressing them is larger than the cost of operating on uncompressed pages.

The light-weight SIMDBinaryPacking algorithm performs comparably to PFor in terms of compression rates, but it is faster. In the "sequential" scenario, this method is mostly superior to all others. In the "parallel" scenario,

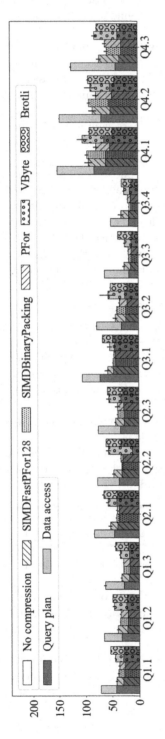

Fig. 6. System run time break down for "sequential" scenario (Seconds)

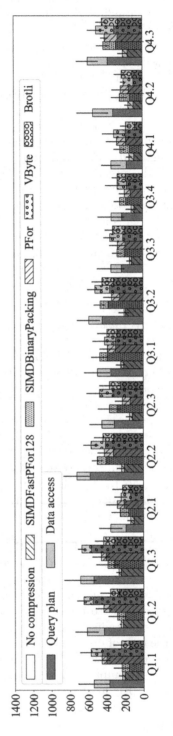

Fig. 7. System run time break down for "parallel" scenario (Seconds)

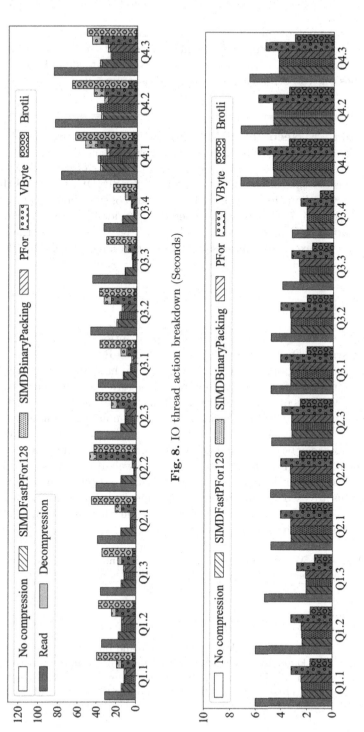

Fig. 8. IO thread action breakdown (Seconds)

Fig. 9. Total volume of data read by query (Gigabytes)

this method loses to another SIMD-enabled algorithm that we tested — SIMD-FastPFor128.

Overall, we cannot definitely conclude that there is progress (beneficial to DBMSes) in light-weight compression schemes, aside from the appearance of SIMD-enabled versions. Furthermore, we believe that VByte should not be used inside DBMSes due to both its compression and decompression speed, as well as poor compression ratios.

RQ3: Is There a Need for SIMD-employing Decompression Algorithms in Case of a Disk-Based System? Experiments demonstrated that SIMD-enabled versions are needed, even in case of a modern disk-based system. First of all, consider the compression time (Fig. 4) for SIMDFastPFor128 that is several times lower than that of vanilla PFor. Next, consider both query execution scenarios (Figs. 6, 7): SIMD-enabled versions provide the best performance.

Looking into the I/O thread action breakdown (Fig. 8), one can see that these algorithms provide excellent compression rates while having negligible decompression costs.

5   Conclusion

In this paper, we have re-evaluated compression options inside a disk-based column-store system. Our work specifically targets I/O-RAM compression architecture, which is still a popular alternative today. We have experimentally tried new light-weight and heavy-weight compression algorithms, including existing SIMD-enabled versions of them. The results indicate that modern heavy-weight compression schemes can be beneficial in a limited number of cases and can provide up to 20% of run time improvement over uncompressed data. However, compression costs may be extremely high and thus, this approach is not appropriate for frequently changing data. Next, novel light-weight compression schemes do not provide significant benefits for in-DBMS usage, except when their SIMD-enabled versions are used. To our surprise, experiments demonstrated that SIMD usage in compression algorithms is absolutely necessary for disk-based DBMSes, even in the case when the workload is disk-bound.

Acknowledgments. We would like to thank Anna Smirnova for her help with the preparation of the paper.

References

1. Abadi, D., Boncz, P., Harizopoulos, S.: The Design and Implementation of Modern Column-Oriented Database Systems. Now Publishers Inc., Hanover (2013)
2. Abadi, D., Madden, S., Ferreira, M.: Integrating compression and execution in column-oriented database systems, SIGMOD 2006, pp. 671–682. ACM, New York (2006)

3. Alakuijala, J., et al.: Brotli: a general-purpose data compressor. ACM Trans. Inf. Syst. **37**(1), (2018). https://doi.org/10.1145/3231935
4. Bains, S.: InnoDB transparent page compression (August 2015). https://mysqlserverteam.com/innodb-transparent-page-compression/
5. Binnig, C., Hildenbrand, S., Färber, F.: Dictionary-based order-preserving string compression for main memory column stores. In: Proceedings of the 2009 ACM SIGMOD International Conference on Management of Data, SIGMOD 2009, pp. 283–296. Association for Computing Machinery, New York (2009). https://doi.org/10.1145/1559845.1559877
6. Chernishev, G.A., Galaktionov, V.A., Grigorev, V.D., Klyuchikov, E.S., Smirnov, K.K.: PosDB: an architecture overview. Program. Comput. Softw. **44**(1), 62–74 (2018)
7. Damme, P., Habich, D., Hildebrandt, J., Lehner, W.: Insights into the comparative evaluation of lightweight data compression algorithms. In: Markl, V., Orlando, S., Mitschang, B., Andritsos, P., Sattler, K., Breß, S. (eds.) Proceedings of the 20th International Conference on Extending Database Technology, EDBT 2017, Venice, Italy, 21–24 March 2017, pp. 562–565. OpenProceedings.org (2017). https://doi.org/10.5441/002/edbt.2017.70
8. Damme, P., Habich, D., Hildebrandt, J., Lehner, W.: Lightweight data compression algorithms: An experimental survey (experiments and analyses). In: Markl, V., Orlando, S., Mitschang, B., Andritsos, P., Sattler, K., Breß, S. (eds.) Proceedings of the 20th International Conference on Extending Database Technology, EDBT 2017, Venice, Italy, 21–24 March 2017. pp. 72–83. OpenProceedings.org (2017). https://doi.org/10.5441/002/edbt.2017.08
9. Galaktionov, V., Klyuchikov, E., Chernishev, G.A.: Position caching in a column-store with late materialization: An initial study. In: Song, I., Hose, K., Romero, O. (eds.) Proceedings of the 22nd International Workshop on Design, Optimization, Languages and Analytical Processing of Big Data co-located with EDBT/ICDT 2020 Joint Conference, DOLAP@EDBT/ICDT 2020, Copenhagen, Denmark, March 30, 2020. CEUR Workshop Proceedings, vol. 2572, pp. 89–93. CEUR-WS.org (2020). http://ceur-ws.org/Vol-2572/short14.pdf
10. Ghita, B., Tomé, D.G., Boncz, P.A.: White-box compression: Learning and exploiting compact table representations. In: 10th Conference on Innovative Data Systems Research, CIDR 2020, Amsterdam, The Netherlands, 12–15 January 2020, Online Proceedings. www.cidrdb.org (2020). http://cidrdb.org/cidr2020/papers/p4-ghita-cidr20.pdf
11. Goldstein, J., Ramakrishnan, R., Shaft, U.: Compressing relations and indexes. In: Proceedings 14th International Conference on Data Engineering, pp. 370–379 (1998). https://doi.org/10.1109/ICDE.1998.655800
12. Graefe, G., Shapiro, L.: Data compression and database performance. In: 1991 Symposium on Applied Computing, pp. 22–27. IEEE Computer Society, Los Alamitos, CA, USA (April 1991). https://doi.org/10.1109/SOAC.1991.143840
13. Habich, D., et al.: MorphStore – in-memory query processing based on morphing compressed intermediates live. In: Proceedings of the 2019 International Conference on Management of Data, SIGMOD 2019, pp. 1917–1920. Association for Computing Machinery, New York (2019). https://doi.org/10.1145/3299869.3320234
14. Harizopoulos, S., Abadi, D., Boncz, P.: Column-Oriented Database Systems, VLDB 2009 Tutorial. (2009). nms.csail.mit.edu/ stavros/pubs/tutorial2009-column_stores.pdf

15. Hellerstein, J.M., Stonebraker, M., Hamilton, J.: Architecture of a database system. Found. Trends Databases **1**(2), 141–259 (2007). https://doi.org/10.1561/1900000002

16. Iyer, B.R., Wilhite, D.: Data compression support in databases. In: Proceedings of the 20th International Conference on Very Large Data Bases, VLDB 1994, pp. 695–704. Morgan Kaufmann Publishers Inc., San Francisco, CA, USA (1994)

17. Jiang, H., Liu, C., Jin, Q., Paparrizos, J., Elmore, A.J.: PIDS: attribute decomposition for improved compression and query performance in columnar storage. Proc. VLDB Endow **13**(6), 925–938 (2020). https://doi.org/10.14778/3380750.3380761

18. Jianzhong L., Srivastava, J.: Efficient aggregation algorithms for compressed data warehouses. IEEE Trans. Knowl. Data Eng. **14**(3), 515–529 (2002). https://doi.org/10.1109/TKDE.2002.1000340

19. Johnson, T.: Performance measurements of compressed bitmap indices. In: Proceedings of the 25th International Conference on Very Large Data Bases, VLDB 1999, pp. 278–289. Morgan Kaufmann Publishers Inc., San Francisco, CA, USA (1999)

20. Lemire, D., Boytsov, L.: Decoding billions of integers per second through vectorization. Softw. Pract. Exper. **45**(1), 1–29 (2015). https://doi.org/10.1002/spe.2203

21. Li, D.: Compressing longs in druid (December 2016). https://imply.io/post/compressing-longs

22. Lipcon, T., Alves, D., Burkert, D., Cryans, J.D., Dembo, A.: Kudu: Storage for fast analytics on fast data (2016). https://kudu.apache.org/kudu.pdf

23. Liu, H., Ji, Y., Xiao, J., Tan, H., Luo, Q., Ni, L.M.: TICC: Transparent intercolumn compression for column-oriented database systems. In: Proceedings of the 2017 ACM on Conference on Information and Knowledge Management, CIKM 2017, pp. 2171–2174. Association for Computing Machinery, New York (2017). https://doi.org/10.1145/3132847.3133077

24. Mullins, C.M., Lim, L., Lang, C.A.: Query-aware compression of join results. In: Guerrini, G., Paton, N.W. (eds.) Joint 2013 EDBT/ICDT Conferences, EDBT 2013 Proceedings, Genoa, Italy, 18–22 March 2013, pp. 29–40. ACM (2013). https://doi.org/10.1145/2452376.2452381

25. O'Connell, S.J., Winterbottom, N.: Performing joins without decompression in a compressed database system. SIGMOD Rec. **32**(1), 6–11 (2003). https://doi.org/10.1145/640990.640991

26. O'Neil, P., Chen, X.: Star Schema Benchmark (June 2009). http://www.cs.umb.edu/~poneil/StarSchemaB.PDF

27. Valduriez, P.: Join Indices. ACM Trans. Database Syst. **12**(2), 218–246 (1987)

28. Wandelt, S., Sun, X., Leser, U.: Column-wise compression of open relational data. Inf. Sci. **457-458**, 48–61 (2018). https://doi.org/10.1016/j.ins.2018.04.074

29. Wang, J., Lin, C., Papakonstantinou, Y., Swanson, S.: An experimental study of bitmap compression vs. inverted list compression. In: Proceedings of the 2017 ACM International Conference on Management of Data, SIGMOD 2017, pp. 993–1008. Association for Computing Machinery, New York (2017). https://doi.org/10.1145/3035918.3064007

30. Zukowski, M., Heman, S., Nes, N., Boncz, P.: Super-scalar RAM-CPU cache compression. In: 22nd International Conference on Data Engineering (ICDE 2006), p. 59 (2006). https://doi.org/10.1109/ICDE.2006.150

Using Multidimensional Skylines
for Regret Minimization

Karim Alami and Sofian Maabout[✉]

Univ. Bordeaux, CNRS, LaBRI, UMR 5800, 33400 Talence, France
{karim.alami,sofian.maabout}@u-bordeaux.fr

Abstract. Skyline and Top-K operators are both multi-criteria preference queries. The advantage of one is a limitation of the other: Top-k requires a scoring function while skyline does not, and Top-k output size is exactly K objects while skyline's output can be the whole dataset. To cope with this state of affairs, regret minimization sets (RMS) whose output is bounded by K and where there is no need to provide a scoring function has been proposed in the literature. However, the computation of RMS on top of the whole dataset is time-consuming. Hence previous work proposed the Skyline set as a candidate set. While it guarantees the same output, it becomes of no benefit when it reaches the size of the whole dataset, e.g., with anticorrelated datasets and high dimensionality. In this paper we investigate the speedup provided by other skyline related candidate sets computed through the structure Negative SkyCube (NSC) such as Top k frequent skylines. We show that this query provides good candidate set for RMS algorithms. Moreover it can be used as an alternative to RMS algorithms as it provides interesting regret ratio.

1 Introduction

Skyline [3] and Top K [4] are two well known preference queries. The Skyline queries are based on the dominance relation. A tuple t is said to be dominated by a tuple t' iff (i) t' is *better* or equal on all dimensions and (ii) t' is strictly *better* on at least one dimension. The Skyline result is then the set of non dominated tuples. Top-K queries are based on scoring functions given users. Often scoring functions are linear, e.g., $f(t) = \sum_{i=1}^{d} w[i] * t[i]$ where w is called the weight vector. In a normalized setting, $0 \leq w[i] \leq 1 \; \forall i \in [1, d]$ and $\sum_{i=1}^{d} w[i] = 1$. The result of Top-K query by considering the scoring function f is K tuples with the best scores.

Example 1. Consider Table 1 that describes Hotels by their price and their distance from the beach. Suppose that cheaper and closer to the beach is better

The Skyline set with respect to this dataset is illustrated in Table 2. Only h_2 does not belong to the Skyline set because it is dominated by t_1. Indeed, t_1 is cheaper and closer to the beach

© Springer Nature Switzerland AG 2021
C. Attiogbé and S. Ben Yahia (Eds.): MEDI 2021, LNCS 12732, pp. 293–304, 2021.
https://doi.org/10.1007/978-3-030-78428-7_23

Table 1. Hotels.

Hotels	Price	Distance
h_1	200	120
h_2	390	140
h_3	465	20
h_4	395	90
h_5	100	300

Table 2. Skyline hotels.

Hotels	Price	Distance
h_1	200	120
h_3	465	20
h_4	395	90
h_5	100	300

Table 3. Top K Hotels.

Hotels – Weight vector	$(0.2, 0.8)$	$(0.5, 0.5)$	$(0.8, 0.2)$
h_1	136	<u>160</u>	184
h_2	190	215	340
h_3	<u>109</u>	242.5	376
h_4	151	242.5	334
h_5	260	200	<u>140</u>

Table 3 represents the hotels' score wrt three linear scoring functions. Note that lower the value the better the hotel. Top-1 hotels score is underlined. h_1 is Top-1 wrt $(0.5, 0.5)$, h_3 is Top-1 wrt $(0.2, 0.8)$ and h_5 is Top-1 wrt $(0.8, 0.2)$

However Skyline queries and Top K queries have some limitations. On one hand, Skyline queries do not bound the results. Indeed the output may be the whole dataset, e.g., in presence of high dimensions and anti-correlated data. On the other hand, Top-K queries require the user to provide weight vector which is not an easy task. To solve these limitations, [8] presented the regret minimization queries. These queries bound the results and they do not require the user to provide the weight vector. Given a family of functions $\mathcal{F}$, they compute a subset $S \subset T$ that minimizes the maximum regret ratio. In a nutshell, the maximum regret ratio of a set S represents how far a user's best tuple in the whole dataset is from the best tuple in S. To simplify, consider the family of 3 functions $\mathcal{F} = \{f_{(0.2,0.8)}, f_{(0.5,0.5)}, f_{(0.8,0.2)}\}$ and consider a set $S = \{h_3, h_1\}$. The maximum regret ratio of S wrt $\mathcal{F}$, i.e., $mrr(S, \mathcal{F})$, is 31.4% which represents the ratio between the best score within T and the best score within S wrt the function

$f_{(0.8,0.2)}$. Concretely, this means that for a user whose scoring function is in $\mathcal{F}$, the best score he can get from S is at most 31.4% less than the best score he can get from T. Further details about the computation of the maximum regret ratio are in Sect. 2.1.

[8] presented the RMS problem: Given a dataset T, the family of all linear functions $\mathcal{L}$, and an integer K, compute a set $S \subseteq T$ of size K such that $mrr(S, L)$ is minimum. This problem has been proven NP-Complete in [5]. Hence, [8] and later work proposed heuristics. Nonetheless the computation is still time-consuming. [8] showed that it is sufficient to consider the Skyline set rather than the whole dataset as input to compute the regret minimization set. However the Skyline becomes with marginal benefit when its size grows, e.g., in anti-correlated setting. [6,7] proposed respectively the Top-K frequent skylines (Top-KF) and Top-K priority (Top-KP) skylines as candidates sets for computing the regret minimization set. They claimed that both operators speed up the RMS computation by up to two orders of magnitude. However, the empirical evaluation in that work are not conclusive. In this paper, we investigate the speed up provided by these two candidates sets. Concretely, we verify wrt several parameters (i) if these sets speed up RMS computation and (ii) impact the output regret. We consider the RMS state of the art algorithm *sphere* [12]. Moreover, we use NSC [1] an indexing structure to compute (i) Skyline, (ii) Top-KF and (iii) Top-KP sets.

2   Background

2.1   Regret Minimization Sets (RMS)

[8] presented the regret minimization queries RMS to avoid the limitations of skyline and Top-k queries. Unlike Top-K queries, RMS do not require scoring functions, and unlike Skyline queries, they bound the result size. The main idea is to select a subset S of a dataset T such that S minimizes the user *regret*. The *regret* represents how far the user's *best* tuple in S is from the user's *best* tuple in T. Specifically, reference [8] addressed the following problem:

> *Problem* **RMS** Given a dataset T, the family of all linear scoring function $\mathcal{L}$, an integer K, compute a set $S \subset T$ of size K that minimizes the maximum regret ratio $mrr(S, \mathcal{L})$.

In the following, we explain the maximum regret ratio of a set S wrt $\mathcal{L}$. Let $f \in \mathcal{L}$ be a scoring function. Given a dataset X, let $f_1(X)$ be the highest score by considering tuples in X. The regret of $S \subseteq T$ wrt a function f is $f_1(T) - f_1(S)$ and the regret ratio is $\frac{f_1(T)-f_1(S)}{f_1(T)}$. The maximum regret ratio is then $mrr(S, \mathcal{L}) = max_{f \in \mathcal{L}} \frac{f_1(T)-f_1(S)}{f_1(T)}$. [5] proved the NP hardness of RMS problem. The regret minimization set has been shown (i) scale-invariant, i.e., the maximum regret ratio remains the same even if the values in the dataset are multiplied by the same factor, and (ii) stable, i.e., the RMS does not change when *weak* tuples (tuples not having the highest score wrt any function) are inserted or deleted from the dataset. Sphere [12] is currently the state of the art heuristic algorithm. It has interesting time complexity and provides theoretical guarantees on the output. Its time complexity is $O(n \cdot e^{O(\sqrt{d \cdot ln(n)})} + n \cdot k^3 \cdot d)$ where n represents the dataset size, d the number of dimensions and k the output size. [8] showed that it suffices to consider the skyline set to compute the RMS rather than the whole dataset. In other words, the optimal solution S^* is only composed of skyline tuples. [9] presented an even smaller and accurate candidate set, namely *Happy* tuples. However, it is time-consuming. Its time complexity is $O(n^2 \cdot d^2)$.

2.2 Multidimensional Skyline

The multidimensional skyline or subspace skyline consists in considering subsets of the set of dimensions for skyline analysis. Given a set of dimensions D and a dataset T. Let $X \subset D$, $Sky(T, X)$ is the set of skyline points by considering only attributes in X. The Skycube [10] has been proposed to optimize the evaluation of the skyline wrt any subspace. It consists simply in materializing the results wrt any subspace. Since it requires an exponential space wrt the number of dimensions, [1,2,11] proposed summarization techniques. We note NSC [1] which stores in an intelligent way, for every tuple, the subspaces where the tuple is dominated. It has been shown as time and memory efficient.

The multidimensional skyline analysis of a dataset gives a useful insight on the best tuples within a dataset. For example, the frequency of a tuple is the number of subspaces in respect to which it belongs to its respective skyline. Let $t \in T$ $Frequency(t) = |\{X \subseteq D \text{ s.t. } t \in Sky(T, X)\}|$. The tuple with the highest frequency may have the best values on the dimensions. Another interesting operator is called the skyline priority which simply is the size of the smallest subspace wrt to which the tuple belongs to its respective skyline. Let $t \in T$ $Priority(t) = min_{X \subseteq D | t \in Sky(X)}(|X|)$. A tuple with low priority may belong to several skylines.

In this paper, we want to investigate the impact of (i) Top-KF a ranking query based on the skyline frequency and (ii) Top-KP a ranking query based on the skyline priority on *sphere* performance, i.e., processing time and output regret. Given K, computing Top-KF and Top-KP requires exponential time wrt the numbers of dimensions d. Hence we use NSC for that purpose.

2.3 The NSC Structure

NSC (Negative SkyCube) stores for each tuple t a list of pairs, each summarizing the subspace where t is dominated. Let $t \in T$ and a let $p = \langle X|Y \rangle$ computed wrt some tuple $t' \in T$. X represents dimensions where t' is strictly better than t, and Y represents dimensions where t and t' are equal. p summarizes the set of subspaces where t' dominates t. This set, denoted by $cover(p)$ is equal to

$\{Z \subset D | Z \subseteq X \cup Y \text{ and } Z \not\subseteq Y\}$. We give an example to illustrate NSC and we refer the interested reader to [1] for more details for this structure.

Example 2. Consider Table 4. We assume that small values are preferred for every dimension. The skyline of T wrt dimension A is $\{t_1, t_2\}$ because these two tuples have the least value of A. The skyline wrt dimensions AD is $\{t_1, t_2, t_3, t_4, t_5\}$. t_6 does not belong to the skyline because it is dominated by t_4: the later has a better value of A and a better value on D.

Table 4. Dataset T

Id	A	B	C	D
t_1	1	1	3	3
t_2	1	1	2	3
t_3	2	2	2	2
t_4	4	2	1	1
t_5	3	4	5	2
t_6	5	3	4	2

By comparing some tuple t to all the others, we obtain a set of *pairs* that summarizes the the subspaces where t is dominated. For example, comparing t_1 to t_2 returns the pair $\langle C | ABC \rangle$: t_2 is better than t_1 in C and these two tuples are equal on ABC. From this pair, we can deduce that, e.g., t_1 doesn't belong to the skyline wrt AC. Table 5 depicts the list of pairs associated to each tuple after comparing it to all the others. Note that pairs $\langle X | Y \rangle$ where $X = \emptyset$ are not stored because they do not bring any dominance information.

Table 5. List of pairs associated to every $t \in T$

Tuples	Pairs
t_1	$\langle C | ABD \rangle, \langle CD | \emptyset \rangle, \langle D | \emptyset \rangle$
t_2	$\langle D | C \rangle, \langle CD | \emptyset \rangle, \langle D | \emptyset \rangle$
t_3	$\langle AB | \emptyset \rangle, \langle AB | C \rangle, \langle CD | B \rangle$
t_4	$\langle AB | \emptyset \rangle, \langle A | B \rangle, \langle A | \emptyset \rangle$
t_5	$\langle ABC | \emptyset \rangle, \langle ABC | D \rangle, \langle BCD | \emptyset \rangle, \langle BC | D \rangle$
t_6	$\langle ABC | \emptyset \rangle, \langle ABC | D \rangle, \langle ABCD | \emptyset \rangle, \langle A | D \rangle$

In Table 5, some pairs can be seen as *redundant*. For example, pair $\langle D | \emptyset \rangle$ associated to t_1 tells that t_1 is dominated wrt D. The same information can be derived from $\langle CD | \emptyset \rangle$ that is associated to t_1 too. Hence, $\langle D | \emptyset \rangle$ without losing any information regarding dominance. The summarized sets of pairs are represented in Table 6. Note that the number of pairs decreases from 20 to 9. This is the NSC associated to the dataset T.

Table 6. NSC of Table T

Tuples	Pairs
t_1	$\langle C|ABD\rangle, \langle CD|\emptyset\rangle$
t_2	$\langle CD|\emptyset\rangle$
t_3	$\langle AB|C\rangle, \langle CD|B\rangle$
t_4	$\langle AB|\emptyset\rangle$
t_5	$\langle ABC|D\rangle, \langle BCD|\emptyset\rangle$
t_6	$\langle ABCD|\emptyset\rangle$

Again, [1] gives all the details about how this summary is obtained, maintained in case of dynamic data and used to speed up skyline queries evaluation.

Algorithm 1 describes the procedure to compute Top-KF through NSC. We compute the subspaces where a tuple t is dominated by computing the *cover* of all pairs related to t (line 4–7). We then compute the score of each tuple and put them in list *Score* (line 8). We sort *Score* and select Top-K tuples (line 9–11). Algorithm for Top-KP is similar to Algortihm 1 with a difference in computing the score (line 8).

Algorithm 1: Top K frequent tuples

Input: NSC, T, K, D
Output: $Top - KF$

1 **begin**
2     $Top - KF \leftarrow \emptyset$
3     $Score \leftarrow []$
4   **foreach** $t \in T$ *in parallel* **do**
5        $E \leftarrow \emptyset$
6   **foreach** $p \in NSC[t]$ **do**
7           $E \leftarrow E \cup cover(p)$
8        $Score.append(t, 2^{|D|} - |E|)$
9     $sort(Score)$
10   **foreach** $i \in [0, K)$ **do**
11       $Top - KF \leftarrow Top - KF \cup Score[i].first$

12 **return** $Top - KF$

---

3   Experiments

In this section, we report on some of the experimental results we obtained so far. We focus our comments on three aspects:

1. We evaluate the speed up of RMS computation provided by considering the Skyline set as a candidate set.
2. We investigate the speed up and output regret of RMS algorithm *sphere* by considering Top-K Frequent and Top-K priority sets as candidates sets for a given K.
3. Given K, we evaluate the regret of Top-K frequent and priority sets.

Hardware & Software. we consider the state of the art algorithm *sphere* [12] for computing regret minimizing sets and the structure NSC [1] for computing (i) skyline, (ii) Top-K frequent and (iii) Top-K priority sets. All the experiments are conducted on a Linux machine equipped with two 2.6 ghz hexacore CPUs and 32GB RAM. Software is in C++ and available on GitHub[1].

Datasets. We consider synthetic datasets generated through the framework in [3]. The parameters considered for these experiments and their (default) values are illustrated in Table 7.

Table 7. Parameters

Parameters	Values
Distribution	Independent (INDE), Anti-correlated (ANTI)
n (dataset size)	**100K**, $1M$
d(number of dimensions)	$4, \mathbf{8}, 12$
k(output size)	$20, \mathbf{30}, 40, 60, 80, 100$

3.1   Speed up with Skyline Set

Here, we evaluate the speed up of *sphere* by considering the Skyline set (S(n)) as input instead of the whole data set (D(n)). Note that the output set and regret are the same whether we consider the skyline set or the whole dataset (Refer [8]). Figures 1 and 2 depict the results. It is also important to note that the reported execution time when the skyline is used as input data set includes the execution time used to obtain this skyline.

The first observation we can make is that using Skyline as input data set enables faster computation of the minimum regret set on all cases. That's, thanks to NSC structure, the skyline computation time is negligible compared to *Sphere*

---

[1] https://github.com/karimalami7/NSC.

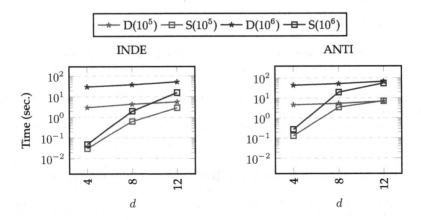

Fig. 1. Speedup of *Sphere* with Skyline as input set by varying dimensionality d

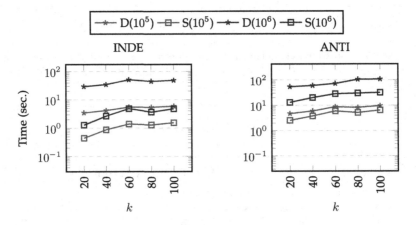

Fig. 2. Speedup of *Sphere* with Skyline as input set by varying the output size k

execution time. The second observation is that when d increases, the benefit of using the skyline as input decreases. This is because the skyline size gets closer to that of the entire data set. Hence, the benefit of *Sphere* gets smaller. For example, for a dataset with 1 million tuples with uniform distribution, the skyline set contains 418 tuples with 4 dimensions and 237726 tuples with 12 dimensions. This behavior is also observed when we compare INDE and ANTI distributions. With anti-correlated data, gain of using the skyline is already negligible when $d = 8$ with anti-correlated data which is not the case with independent uniform data distribution.

It is also interesting to note that the execution time of *Sphere* is almost constant wrt to K (the size of its output) (see Fig. 2).

We conclude that considering the Skyline set as candidate set has a limitation (its size is large when d is large), even if its computation time is negligible. In the next section, we investigate the impact of multidimensional skyline variations related ranking functions, i.e., Top-KF and Top-KP, on *sphere*.

3.2 Speedup and Regret of *Sphere* with Multidimensional Skyline

In this section, we evaluate the speedup of *Sphere* by providing Top-KF and Top-KP sets as input sets. Figure 3 depict the obtained results.

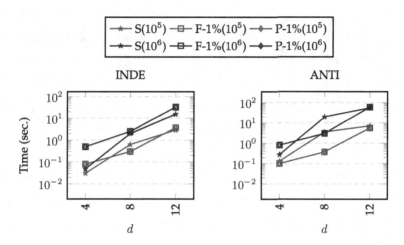

Fig. 3. Computation time of *sphere* with inputs sets (i) Skyline (ii) Top K Frequent and (iii) Top K Priority by varying d

Regarding computation time, we do not see big improvement by providing Top 1% tuples by frequency or priority. Indeed, *Sphere* computation time is improved because the input sets are smaller and have constant sizes (Top 1% instead of the whole skyline) however the computation time of input sets is now large and grows rapidly with increasing dimensionality. Indeed, computing Top-KF tuples requires the computation of an exponential number of skylines. Hence this computation time is not amortized by the fact that the obtained result is small.

3.3 Regret Ratio with Different Input Sets

In this section we analyze the quality of *Sphere* output when using different input data sets by contrast to the previous section where we analyzed just the execution time. Figures 4 and 5 show the results obtained from the same data sets as those used in the previous section. We see that all input sets provide similar regret ratio. We also see that for small k (under 60) when considering Top 1% frequent tuples as input sets, the regret ratio computed by *Sphere* is better than that of the output when the skyline is used as input set. This is explained by the fact that *sphere* is actually a heuristic approach to compute RMS. Indeed, Top 1% frequent tuples discard some *noisy* points that are consequently not selected by *Sphere*.

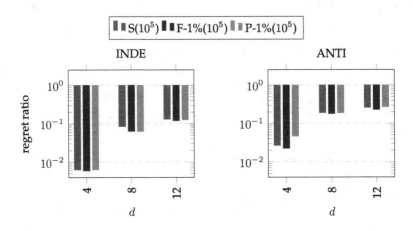

Fig. 4. Regret of *sphere* by input sets (i)Sky (ii) TopKF (iii) TopKP and varying d

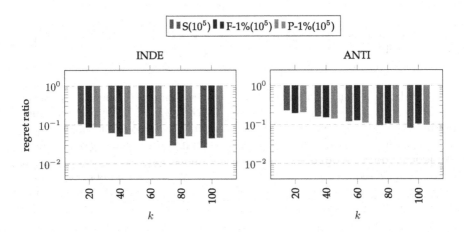

Fig. 5. Regret of *sphere* by input sets (i)Sky (ii) TopKF (iii) TopKP and varying k

3.4 Multidimensional Skyline Metrics as Alternatives to RMS Algorithms

So far, we showed that Top-KF and Top-KP provide good input sets for *sphere*. Now we want to answer the question: Can Top-KF or Top-KP (without *sphere*) compute sets that achieve regret ratio close to that achieved by *sphere*? Concretely, we evaluate the regret ratios of sets of size K computed with (i)*sphere* (ii) TopKF and (ii) TopKP. Figures 6 and 7 depict the results. Globally, we can see that TopKF achieves a good regret ratio when dimensionality gets large and k is small. One possible explanation of this behavior would be the fact that higher dimensionality means a higher number of skylines. This makes tuples better differentiated. Indeed, with lower dimensionality, many tuples share the same score(number of skyline they belong to) which makes them hardly distinguishable.

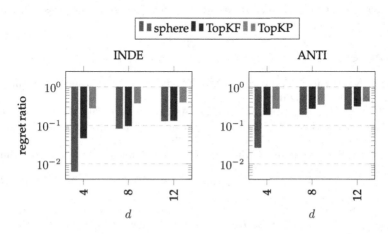

Fig. 6. Regret of (i) *sphere* (ii) TopKF (iii) TopKP by varying k

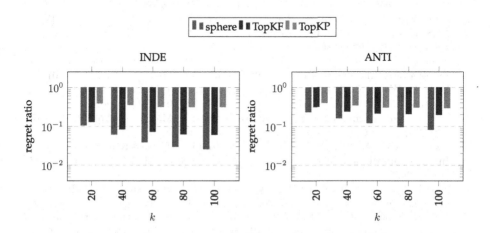

Fig. 7. Regret of (i) *sphere* (ii) TopKF (iii) TopKP by varying k

4 Discussion

The experimental results we obtained show that NSC is useful for RMS algorithms as it computes very efficiently different variants of input data sets that are provided to RMS algorithms such as *Sphere*. Some of these input data sets can be even used as *an approximations* of RMS results. Indeed, our empirical results showed that some of them already provide small regrets. From the practical point of view, this is very important since the *Sphere* application on even small input data sets can be prohibitive when the number of dimensions is large. The encouraging empirical results obtained so far should be pursued to state some theoretical error guarantee bounds wrt chosen input data sets. This is our plan for future research.

References

1. Alami, K., Hanusse, N., Wanko, P.K., Maabout, S.: The negative skycube. Inf. Syst. **88**, 101443 (2020)
2. Bøgh, K.S., Chester, S., Sidlauskas, D., Assent, I.: Template skycube algorithms for heterogeneous parallelism on multicore and GPU architectures. In: Proceedings of SIGMOD Conference, pp. 447–462 (2017)
3. Börzsönyi, S., Kossmann, D., Stocker, K.: The skyline operator. In: Proceedings of ICDE Conference, pp. 421–430 (2001)
4. Chaudhuri, S., Gravano, L.: Evaluating top-k selection queries. VLDB **99**, 397–410 (1999)
5. Chester, S., Thomo, A., Venkatesh, S., Whitesides, S.: Computing k-regret minimizing sets. Proc. VLDB Endowment **7**(5), 389–400 (2014)
6. Han, S., Zheng, J., Dong, Q.: Efficient processing of *k*-regret queries via skyline frequency. In: Meng, X., Li, R., Wang, K., Niu, B., Wang, X., Zhao, G. (eds.) WISA 2018. LNCS, vol. 11242, pp. 434–441. Springer, Cham (2018). https://doi.org/10.1007/978-3-030-02934-0_40
7. Han, S., Zheng, J., Dong, Q.: Efficient processing of *k*-regret queries via skyline priority. In: Meng, X., Li, R., Wang, K., Niu, B., Wang, X., Zhao, G. (eds.) WISA 2018. LNCS, vol. 11242, pp. 413–420. Springer, Cham (2018). https://doi.org/10.1007/978-3-030-02934-0_38
8. Nanongkai, D., Sarma, A.D., Lall, A., Lipton, R.J., Xu, J.: Regret-minimizing representative databases. Proc. VLDB Endowment **3**(1–2), 1114–1124 (2010)
9. Peng, P., Wong, R.C.: Geometry approach for k-regret query. In: Cruz, I.F., Ferrari, E., Tao, Y., Bertino, E., Trajcevski, G. (eds.) IEEE 30th International Conference on Data Engineering, Chicago, ICDE 2014, IL, USA, March 31 - April 4, 2014, pp. 772–783. IEEE Computer Society (2014)
10. Tao, Y., Xiao, X., Pei, J.: Subsky: efficient computation of skylines in subspaces. In: 22nd International Conference on Data Engineering (ICDE 2006), pp. 65–65. IEEE (2006)
11. Xia, T., Zhang, D., Fang, Z., Chen, C.X., Wang, J.: Online subspace skyline query processing using the compressed Skycube. ACM TODS **37**(2), 15:1–15:36 (2012)
12. Xie, M., Wong, R.C., Li, J., Long, C., Lall, A.: Efficient k-regret query algorithm with restriction-free bound for any dimensionality. In: Das, G., Jermaine, C.M., Bernstein, P.A. (eds.) Proceedings of the 2018 International Conference on Management of Data, SIGMOD Conference 2018, Houston, TX, USA, 10–15 June 2018, pp. 959–974. ACM (2018)

Enhancing Sedona (formerly GeoSpark) with Efficient k Nearest Neighbor Join Processing

Francisco García-García[1] , Antonio Corral[1(✉)] , Luis Iribarne[1] , and Michael Vassilakopoulos[2]

[1] Department of Informatics, University of Almeria, Almeria, Spain
{paco.garcia,acorral,liribarn}@ual.es
[2] Department of Electrical and Computer Engineering, University of Thessaly, Volos, Greece
mvasilako@uth.gr

Abstract. *Sedona* (formerly GeoSpark) is an in-memory cluster computing system for processing large-scale spatial data, which extends the core of Apache Spark to support spatial datatypes, partitioning techniques, indexes, and operations (e.g., spatial range, k Nearest Neighbor (kNN) and spatial join queries). k Nearest Neighbor Join Query (kNNJQ) finds for each object in one dataset $\mathbb{P}$, k nearest neighbors of this object in another dataset $\mathbb{Q}$. It is a common operation used in numerous spatial applications (e.g., GISs, location-based systems, continuous monitoring, etc.). kNNJQ is a time-consuming spatial operation, since it can be considered a hybrid of spatial join and nearest neighbor search. Given that *Sedona* outperforms other Spark-based spatial analytics systems in most cases and, it does not support kNN joins, including kNNJQ is a worthwhile challenge. Therefore, in this paper, we investigate how to design and implement an efficient kNNJQ algorithm in *Sedona*, using the most appropriate spatial partitioning technique and other improvements. Finally, the results of an extensive set of experiments with real-world datasets are presented, demonstrating that the proposed kNNJQ algorithm is efficient, scalable and robust in *Sedona*.

Keywords: k nearest neighbor join · Distributed spatial data processing · Sedona · Spatial Query Evaluation

1 Introduction

k Nearest Neighbor Join Query (kNNJQ) is a Distance-based Join Query (DJQ) that obtains from each object in one dataset $\mathbb{P}$, the k objects in another dataset $\mathbb{Q}$ that are closest to it. kNNJQ usually serves as a primitive operation and is

Research of all authors is supported by the MINECO research project [TIN2017-83964-R].

© Springer Nature Switzerland AG 2021
C. Attiogbé and S. Ben Yahia (Eds.): MEDI 2021, LNCS 12732, pp. 305–319, 2021.
https://doi.org/10.1007/978-3-030-78428-7_24

commonly used in many data mining and analytic applications, such as clustering and outlier detection methods [6]. Given that kNNJQ combines nearest neighbor and spatial join queries, it becomes a time-consuming operation, especially if the datasets involved in the join are very large or they follow different data distributions. This type of query is common in several application cases in the context of big spatial data. For example, in mobile location services, with two spatial datasets, locations of shopping centers and positions of possible customers using a smart phone with mobile data and GPS enabled. kNNJQ could *find the 10 nearest possible customers to each shopping center* for sending an advertising SMS about a fashion brand available there.

In the era of *Big Spatial Data*, the research community is focusing on developing new data management systems that can efficiently analyze and process large-scale spatial data. Nowadays, with the development of modern mobile applications, the increase of the volume of available spatial data, from smartphones, sensors, cars, satellites, etc., is huge world-wide. Recent developments of big spatial data systems have motivated the emergence of novel technologies for processing large-scale spatial data on clusters of computers in a distributed environment. Parallel and distributed computing using shared-nothing clusters on large-scale datasets is becoming a dominant trend in the context of data processing and analysis. *Apache Spark*[1] is a fast, reliable and distributed in-memory large-scale data processing framework. It takes advantage of Resilient Distributed Datasets (RDDs), that allow data to be stored transparently in memory and persisted to disk only if it is needed. Hence, Spark can avoid a huge number of disk writes and reads, and it outperforms the Hadoop platform. Since Spark maintains the status of assigned resources until a job is completed, it reduces time consumption in resource preparation and collection.

There are several modern spatial analytics systems for managing and analyzing spatial data [8]. The most representative ones based on Spark are: *SpatialSpark* [11], *Simba* [10], *LocationSpark* [9] and *GeoSpark* [13] (currently called Sedona). An exhaustive experimental evaluation of these Spark-based spatial analytics systems using real-world datasets was performed in [8], and four important conclusions of *GeoSpark* were extracted: (1) it is close to be a complete spatial analytic system because of spatial data types and queries are supported; (2) it also exhibits the best performance in most cases; (3) it is actively under development; and (4) it does not support kNN joins. *Apache Sedona*[2] is an in-memory cluster computing system for processing large-scale spatial data. More accurately, *Sedona*, as a full-fledged cluster computing system, extends the core engine of Apache Spark and SparkSQL to support spatial data types, indexes, and geometrical operations; with the aim of being able to load, process, and analyze large-scale spatial data across machines.

Considering the above analysis, in this paper, we propose a novel kNNJQ distributed algorithm in the *Sedona* architecture. The most important contributions of this paper are the following:

---

[1] Available at https://spark.apache.org/.

[2] Available at http://sedona.apache.org/.

- The design and implementation of a new algorithm in *Sedona* for efficient parallel and distributed kNNJQ processing on large spatial datasets.
- The execution of a set of experiments using real-world datasets for examining the efficiency and the scalability of the kNNJQ distributed algorithm in *Sedona*, considering several performance parameters and measures.

This paper is organized as follows. In Sect. 2, we review related work on kNNJQ algorithms on Spark-based systems and provide the motivation of this paper. In Sect. 3, we present preliminary concepts related to kNNJQ and *Sedona* as a Spark-based spatial analytics system. In Sect. 4, the kNNJQ distributed algorithm in *Sedona* is proposed. In Sect. 5, we present the most representative results of the experiments that we have performed, using real-world datasets. Finally, in Sect. 6, we provide the conclusions arising from our work and discuss related future work directions.

2   Related Work and Motivation

In this section, we review the most representative contributions of kNNJQ algorithms in distributed data management systems, especially Spark-based spatial analytics systems.

kNNJQ processing has been actively investigated in parallel and distributed environments, and here we review the most relevant contributions. The first one was [6], where the problem of answering the kNNJ using MapReduce is studied. This is accomplished by exploiting the Voronoi-Diagram based partitioning method, that divides the input datasets into groups, such that kNNJ can answer by only checking object pairs within each group. Moreover, effective mapping mechanisms and pruning rules for distance filtering are applied to minimize the number of replicas and the shuffling cost. In [14], novel (exact and approximate) algorithms in MapReduce to perform efficient parallel kNNJQ on large datasets are proposed, and they use the R-tree and Z-value-based partition joins to implement them. In [1], the *Spitfire* approach was presented; it improves over the duplication rate of [6] by utilizing grid-based partitioning to divide the data and bound the support set. A new *Data Partitioning* approach, called kNN-DP, is proposed in [15] to alleviate load imbalance incurred by data skewness in the kNNJQ. kNN-DP dynamically partitions data to optimize kNNJQ performance by suppressing data skewness on Hadoop clusters. Finally, in [3], a new kNNJQ MapReduce algorithm, using Voronoi-Diagram based partitioning technique, was developed in SpatialHadoop.

There are very few works in the literature devoted to the design and implementation of efficient kNNJQ algorithms in distributed spatial data management systems based on Spark. As a matter of fact, according to [8] only two Spark-based systems support kNNJQ for spatial data: *Simba* and *Location-Spark*. In *Simba* [10] several variations of the kNNJQ algorithm in Spark were proposed: *BKJSpark-N* (block nested loop kNNJ method), *BKJSpark-R* (block R-tree kNNJ method), *VKJSpark* (Voronoi kNNJ method), *ZKJSpark* (Z-value kNNJ method) and *RKJSpark* (R-tree kNNJ method). As a general conclusion from the experimental results, *RKJSpark* showed the best performance and the best scalability, with respect to both data size and k. In *LocationSpark* [9], five approaches were proposed for kNNJQ distributed algorithm: two indexed nested-loops joins (*nestR-tree* and *nestQtree*) and three block-based joins (*sfcurve*, *pgbjk* and *spitfire* [1]). After an exhaustive performance study of the five approaches, the *nestQtree* kNNJ algorithm for the local workers was the fastest. *nestQtree* is an indexed nested loop join where a Quadtree index is used in the inner dataset, and a kNNQ for each query point of the outer dataset is executed. Moreover, an improved version of *nestQtree* was also implemented with an optimized query plans and the *sFilter*, showing the best performance in all cases.

As we have seen above, there is a lack in the design and implementation of kNNJQ distributed algorithms in Spark-based spatial analytics systems (i.e., kNNJQ is only included in *Simba* and *LocationSpark*). Moreover, *GeoSpark* (currently *Sedona*) exhibits the best performance in most cases for the implemented spatial queries (e.g., spatial and distance joins) [8] and it is actively under development with interesting and recent extensions [2,12], but one of the main weakness is that it does not implement kNNJQ [8]. Motivated by these observations, the design and implementation of a new efficient kNNJQ distributed algorithm in *Sedona* is the main aim of this research work.

3 Preliminaries and Background

In this section, we first present the basic definitions of the kNNQ and kNNJQ, followed by a brief introduction of preliminary concepts of *Sedona*.

3.1 The k Nearest Neighbor Join Query - kNNJQ

We know the kNNJQ retrieves for each point of one dataset, k nearest neighbors of this point in the other dataset. Therefore, we can deduce that the kNNJQ is based on the kNNQ, and we are going to define it.

Given a set of points $\mathbb{P}$, kNNQ discovers the k points of $\mathbb{P}$ that are the closest to a given query point q (i.e., it reports only top-k points of $\mathbb{P}$ from q). It is one of the most important and studied spatial operations. The formal definition of the kNNQ for points is the following:

Definition 1. k Nearest Neighbor query, kNN query
Let $\mathbb{P} = \{p_1, p_2, \cdots, p_n\}$ be a set of points in E^d (d-dimensional Euclidean space). Then, the result of the k Nearest Neighbor query, with respect to a query point q in E^d and a number $k \in \mathbb{N}^+$, is an ordered set, $kNN(\mathbb{P}, q, k) \subseteq \mathbb{P}$, which contains k $(1 \leq k \leq |\mathbb{P}|)$ different points of $\mathbb{P}$, with the k smallest distances from q:
$kNN(\mathbb{P}, q, k) = \{p_1, p_2, \cdots, p_k\} \subseteq \mathbb{P}$, such that $\forall p \in \mathbb{P} \setminus kNN(\mathbb{P}, q, k)$ we have $dist(p_i, q) \leq dist(p, q), 1 \leq i \leq k$

When two datasets ($\mathbb{P}$ and $\mathbb{Q}$) are combined, one of the most studied DJQs is the k Nearest Neighbor Join (kNNJ) query, where, given two points datasets

($\mathbb{P}$ and $\mathbb{Q}$) and a positive number k, finds for each point of $\mathbb{P}$, k nearest neighbors of this point in $\mathbb{Q}$. The formal definition of this DJQ for points is given below.

Definition 2. k Nearest Neighbor Join query, kNNJ query
Let $\mathbb{P} = \{p_1, p_2, \cdots, p_n\}$ and $\mathbb{Q} = \{q_1, q_2, \cdots, q_m\}$ be two set of points in E^d, and a number $k \in \mathbb{N}^+$. Then, the result of the k Nearest Neighbor Join query is a set $kNNJ(\mathbb{P}, \mathbb{Q}, k) \subseteq \mathbb{P} \times \mathbb{Q}$, which contains for each point of $\mathbb{P}$ ($p_i \in \mathbb{P}$) k nearest neighbors of this point in $\mathbb{Q}$:
$$kNNJ(\mathbb{P}, \mathbb{Q}, k) = \{(p_i, q_j) : \forall p_i \in \mathbb{P}, q_j \in kNN(\mathbb{Q}, p_i, k)\}$$

Note that the kNNJQ is asymmetric [3], i.e., $kNNJ(\mathbb{P}, \mathbb{Q}, k) \neq kNNJ(\mathbb{Q}, \mathbb{P}, k)$. Given $k \leq |\mathbb{Q}|$, the cardinality of $kNNJ(\mathbb{P}, \mathbb{Q}, k)$ is $|\mathbb{P}| \times k$, and in this paper, we will assume that $k \leq |\mathbb{Q}|$ ($k \ll |\mathbb{Q}|$).

3.2 Sedona

Sedona exploits the core engine of Apache Spark and SparkSQL, by adding support for spatial data types, indexes, and geometrical operations. *Sedona* extends the Resilient Distributed Datasets (RDDs) concept to support spatial data. It adds two more layers, the Spatial RDD (SRDD) Layer and Spatial Query Processing Layer, thus providing Spark with in-house spatial capabilities. The SRDD layer consists of three newly defined RDDs: PointRDD, RectangleRDD and PolygonRDD. SRDDs support basic geometrical operations, like Overlap, Intersect, etc. SRDDs are automatically partitioned by using different spatial partitioning techniques (e.g., uniform grid, R-tree, Quadtree, kDB-tree, etc.), where the global spatial grid file is split into a number grid cells. The framework implements spatial partitioning in three main steps: building a global spatial grid file, assigning a grid cell ID to each object and re-partitioning SRDD across the cluster. *Sedona* spatial indexes rely on the R-tree or Quadtree data structure. There are three kind of index options: build local indexes, query local indexes, and persist local indexes. The Spatial Query Processing Layer includes spatial range query, distance query, spatial kNN query and range join query. *Sedona* relies heavily on the JTS (Java Topology Suite) and therefore conforms to the specifications published by the Open Geospatial Consortium. It is a robust, well implemented and actively under development cluster computing system for processing large-scale spatial data, since several companies are using it[3]. Moreover, a lot of heterogeneous data sources are supported, like CSV, GeoJSON, WKT, NetCDF/HDF and ESRI Shapefile. *Sedona* does not directly support temporal data and operations.

Sedona visualization extension is supported with the core visualization framework *GeoSparkViz* [12]. *GeoSparkViz* is a large-scale geospatial map visualization framework, and it extends Apache Spark with native support for general cartographic design. *GeoSparkSim* [2] is a scalable traffic simulator which extends Apache Spark to generate large-scale road network traffic datasets with microscopic traffic simulation. *GeoSparkSim* integrates with *GeoSpark*, to deliver a

---

[3] see http://sedona.apache.org/download/features/.

holistic approach that allows data scientists to simulate, analyze and visualize large-scale urban traffic data.

3.3 kNNJQ in Spark-Based Spatial Analytics Systems

As we mentioned above, the only Spark-based spatial analytics systems that implement kNNJQ are *Simba* and *LocationSpark*.

In *Simba* [10], the kNN join algorithms proposed on MapReduce [6,14] had been redesigned and implemented with the RDD abstraction of Spark, resulting the following algorithms: BKJSpark-N, BKJSpark-R, VKJSpark, ZKJSpark and RKJSpark. *BKJSpark-N* and *BKJSpark-R* are block nested loop kNNJ methods. Particularly, *BKJSpark-R* builds a local R-tree index for every bucket of $\mathbb{Q}$, and uses the R-tree for local kNN joins. *VKJSpark* is a Voronoi kNNJ method based on [6], and it executes n local kNN joins by partitioning both $\mathbb{P}$ and $\mathbb{Q}$ into n partitions respectively, where the partition strategy is based on the Voronoi diagram for a set of pivot points selected from $\mathbb{P}$. *ZKJSpark* is a Z-value kNNJ method based on [14], and it exploits Z-values to map multidimensional points into one dimension and uses random shift to preserve spatial locality. This algorithm only reports approximate results and there is an extra cost for returning exact results in a post-processing step. Finally, *RKJSpark* is an R-tree kNNJ method, and it partitions $\mathbb{P}$ into n partitions $\mathcal{P}_i^{\mathbb{P}}$, $1 \leq i \leq n$, using the STR partitioning strategy. It then takes a set of random samples $\mathbb{Q}'$ from $\mathbb{Q}$ and builds an R-tree $\mathcal{T}_{\mathbb{Q}'}$ over $\mathbb{Q}'$. From these $(\mathcal{P}_i^{\mathbb{P}}, \mathcal{T}_{\mathbb{Q}'})$ a distance bound can be derived from each $\mathcal{P}_i^{\mathbb{P}}$ to find a subset $\mathbb{Q}_i \subset \mathbb{Q}$ such that for any point $p \in \mathcal{P}_i^{\mathbb{P}}$, $kNN(p, \mathbb{Q}_i) = kNN(p, \mathbb{Q})$. After a detailed performance study of all kNNJ algorithms [10], *RKJSpark* showed the best performance and scalability, with respect to both data size and k, closely followed by *VKJSpark*.

In *LocationSpark* [9], five approaches were proposed for the kNNJQ distributed algorithm: two indexed nested-loops joins (*nestR-tree* and *nestQtree*) and three block-based joins (*sfcurve*, *pgbjk* and *spitfire*). For the indexed nested-loops kNNJQ algorithms, an index (R-tree or Quadtree) is built on the inner dataset and for each query point in the outer dataset, a kNNQ is executed over the index, obtaining the *nestR-tree* and *nestQtree* versions, respectively. The block-based kNNJQ algorithms partition the outer dataset (query points) and the inner datasets (data points) into different blocks, and find the kNN candidates for queries in the same block. Then, a post-processing refine step computes kNN for each query point in the same block. *sfcurve* is based on Gorder algorithm and the Hilbert-curve, *pgbjk* is based on PGBJ algorithm, and *spitfire* [1] is a parallel kNN self-join algorithm for in-memory data that replicates the possible kNN candidates into its neighboring data blocks. After an exhaustive performance study of the five approaches, *nestQtree* was the fastest. Moreover, an improved version of *nestQtree* was also implemented with an optimized query plans and the *sFilter*, showing the best performance in all cases.

In [8], a performance comparison between the two Spark-based systems was accomplished. For the kNNJQ, the authors fixed points datasets, $k = 5$, *RKJSpark* (Simba) and *nestQtree* (LocationSpark). *LocationSpark* outperformed

Simba for total runtime, join time and shuffling costs; while for peak execution memory, *LocationSpark* showed larger values than *Simba*.

4 kNNJQ Algorithm in *Sedona*

In this section, we present how we have designed and implemented kNNJQ distributed algorithm in *Sedona*. Being a distance-based join, it is an expensive spatial operation in which two datasets are combined, and usually shows an increase in the amount of shuffled data in each of its steps. Using the functionalities and features that *Sedona* architecture [13] provides, the datasets to be joined are spatially distributed using some spatial partitioning technique (e.g., Quadtree) and cached to be reused throughout the algorithm. Furthermore, indices can be used for local data processing if the dataset is indexed using R-tree. Moreover, the use of these partitioning and indexing techniques makes it possible to skip data that are not part of the join, furtherly reducing the size of the shuffled data.

Our kNNJQ algorithm in *Sedona* adapts the general approach proposed in [7], with different strategies and improvements made in [4], and using spatial partitioning techniques for processing the kNNJQ algorithm in a parallel and distributed manner. The proposed kNNJQ algorithm for two datasets $\mathbb{P}$ and $\mathbb{Q}$ consists of the following steps:

1. The **Information Distribution** step partitions the $\mathbb{Q}$ dataset with any of the disjoint partitioning methods provided by *Sedona* (namely uniform Grid, R-tree, Quadtree and kDB-tree) [13] and, optionally, builds an R-tree local index over each partition. Next, $\mathbb{P}$ is re-partitioned over the partitions of $\mathbb{Q}$ and both SRDDs are cached to accelerate their use in following steps.
2. The **Bin kNNJ** step consists of a *Bin Spatial-Join* of the input RDDs, in which the join operand is kNNQ. Like spatial range join in *Sedona*, this step *zips* (combines) the partitions [13] from $\mathbb{P}$ and $\mathbb{Q}$ using their partition IDs. For instance, each point in $\mathbb{P}$ is combined with the partition in $\mathbb{Q}$ where it is located, so that, the local kNNQ of that point is executed with the points of the same partition in $\mathbb{Q}$. Depending on whether we have available indices in the partitions of $\mathbb{Q}$, the local kNNQ is performed in the two following ways:
   - **No Index.** This step runs a partition-level local kNNQ join on each pair of zipped partitions. If no R-tree index exists on the $\mathbb{Q}$ partition, a plane-sweep kNNQ algorithm for each point of the $\mathbb{P}$ partition is executed and a kNN list for each of them is returned.
   - **R-tree Index.** If an R-tree index exists on the $\mathbb{Q}$ partition, an index-nested loop kNNQ algorithm is employed. For instance, a kNNQ is executed for each point of the $\mathbb{P}$ partition over the Sedona's built-in index to get a candidate kNN list for each one. Therefore, the number of operations and calculations is reduced, resulting in a spatial operation with higher performance.

   To end this step, a completeness check is performed over the previous kNN lists to verify whether they contain less than k points as a result for each point p in $\mathbb{P}$ and also whether neighboring partitions are overlapping with the circular range centered on p and radius the distance to the current k-th nearest neighbor. Therefore, the kNN lists that pass these validations are considered final results for the kNNQ.

3. In the **kNNJ on Overlapping Partitions** step, the points from $\mathbb{P}$ with kNN lists that are not considered part of the final result are further processed through a spatial range join using each distance to their k-th nearest neighbor in which only k elements are considered. Notice that since each point can overlap with more than one partition, the result of this step can include multiple kNN lists per point.

4. Finally, the **Merge Results** step, creates a combined RDD from the results of the *Bin kNNJ* and *kNNJ on Overlapping Partitions*. To do this, first, the non-final kNN lists from the *Bin kNNJ* step are combined with the results of the *kNNJ on Overlapping Partitions* step employing an *aggregation* operation that generates a new RDD. Finally, the definitive results of the kNNJQ are obtained by joining this last RDD with the kNN lists considered final in the *Bin kNNJ* step by using the *union* operation of Spark.

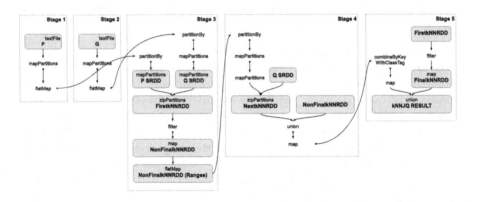

Fig. 1. Spark DAG for kNNJQ algorithm in *Sedona*.

Figure 1 shows the *Spark DAG* (Directed Acyclic Graph) or *Execution Plan* of kNNJQ following the stages of the multi-step algorithm discussed above. In Stage 1 and Stage 2, the two datasets $\mathbb{P}$ and $\mathbb{Q}$ are partitioned according to a given spatial partitioning method. In Stage 3, the initial kNN lists (*FirstkN-NRDD*) are calculated, using a local kNNQ algorithm using the *zip partitions* transformation. Moreover, the resulting RDD is cached to be reused in the following stages. Stage 4 collects the points where a final result has not been obtained (*NonFinalkNNRDD*) as ranges, using each distance to their k-th nearest neighbor and, then performs a spatial range join. Next, in Stage 5, the results of the previous stage (*NextkNNRDD*) are combined with the non-final results of *NonFinalkNNRDD* by an *aggregation* transformation. Lastly, the final kNN lists from *FirstkNNRDD* (*FinalkNNRDD*) are combined with the previous results (*FinalNextkNNRDD*) using the *union* operator of Spark. Notice that the use of wide dependencies has been minimized, prioritizing the appearance of narrow dependencies. The wide dependencies that appear in the DAG and that give rise to the different Stages are mainly due to the need to partition the initial datasets and the SRDD of ranges from the *NonFinalkNNRDD* for the spatial range join. The main use of operations like *zip partitions*, *aggregation* and *union* only produce narrow dependencies. Also, caching the *FirstkNNRDD* increases the performance of Stage 5.

5   Experimental Results

In this section, we present the most representative results of our experimental evaluation. We have used real-world 2d point datasets to test our $kNNJQ$[4] algorithm in *Sedona*. We have used datasets from OpenStreetMap[5]: *LAKES* (*L*) which contains 8.4M records (8.6 GB) of boundaries of water areas (polygons), *PARKS* (*P*) which contains 10M records (9.3 GB) of boundaries of parks or green areas (polygons), *ROADS* (*R*) which contains 72M records (24 GB) of roads and streets around the world (line-strings), and *BUILDINGS* (*B*) which contains 115M records (26 GB) of boundaries of all buildings (polygons). To create sets of points from these four spatial datasets, we have transformed the MBRs of line-strings into points by taking the center of each MBR. In particular, we have considered the centroid of each polygon to generate individual points for each kind of spatial object. The main performance measures that we have used in our experiments have been: (1) the *Total Execution Time*, that stands for the total run-time the system takes to execute a given task or stage, (2) the *Total Shuffled Data*, that represents the amount of information that is redistributed across partitions and that may or may not cause moving data across processes, executors or nodes, and (3) the *Peak Execution Memory*, that aggregates the highest execution memory of all the tasks of a specific job.

All experiments were conducted on a cluster of 7 nodes on an OpenStack environment. Each node has 8 vCPU with 64GB of main memory running Linux operating systems, Spark 2.4.7 with Hadoop 2.7.1.2.3. The cluster setup features a Master Node and 6 Slaves Nodes, with a total of 12 executors and has been tuned following parameter guidelines from [5]. Finally, we used the latest code available in the repositories of *Sedona*[6].

The first experiment aims to test the proposed $kNNJQ$ distributed algorithm, considering different spatial partitioning techniques (i.e., *Grid* (G), *R-tree* (R), *Quadtree* (Q) and *kDB-tree* (KD)), by using different combinations of datasets and varying k. In Fig. 2, the left chart shows the execution time of $kNNJQ$ for several dataset combinations (i.e., $LAKES \times PARKS$ ($L \times P$), $LAKES \times ROADS$ ($L \times R$), $LAKES \times BUILDINGS$ ($L \times B$)), fixing $k = 25$. The first conclusion

---

[4] Available at https://github.com/acgtic211/incubator-sedona/tree/KNNJ.

[5] Available at http://spatialhadoop.cs.umn.edu/datasets.html.

[6] Available at https://github.com/apache/incubator-sedona.

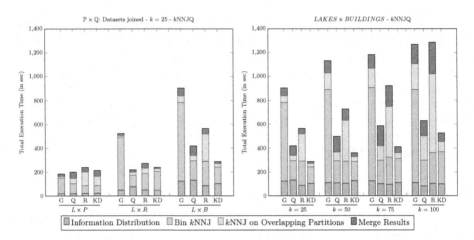

Fig. 2. Total execution time of kNNJQ considering different spatial partitioning techniques, varying the dataset combinations (left) and the k values (right).

is that the execution times for all the partitioning techniques grow as dataset size increases. For *Grid*, the increase is much higher because it is a uniform partitioning, which behaves well when joining uniformly distributed data. However, for real spatial datasets, it presents important skew problems as we can observe in the significant growth of the *Bin kNNJ* step time. Next, *R-tree* partitioning shows a growth in its execution time when the dataset size is increased, especially for $L \times B$. Furthermore, the *Bin kNNJ* step times are similar to those of Quadtree and kDB-tree, but R-tree has much higher time values in the *kNNJ on Overlapping Partitions* step. This is due to the creation of non-regular partitions and the use of an overflow data partition that cause points that do not fall in any cell of R-tree partitioning, to overlap with a higher number of partitions. Finally, *Quadtree* and k*DB-tree* behave similarly and provide the best performance in terms of run-time. This is because of the way they generate the partitions, that is, Quadtree is a space-based partitioning technique and kDB-tree is a data-based one. For the combination of the biggest datasets $(L \times B)$, kDB-tree has the smallest execution time because it has more balanced partitions than Quadtree (e.g., min/median/max values of the partitions sizes of *BUILDINGS* for kDB-tree are 5.9MB/6.3MB/12.1MB and Quadtree 0MB/3.5MB/12.4MB). The main reason is how they generate the partitions, while Quadtree divides it based on its spatial properties (i.e., a rectangle is split into four equal rectangles), kDB-tree uses the number of points present in the current rectangle to generate two different sized regions.

In Fig. 2, the right chart shows the execution times of kNNJQ for the $L \times B$ combination and several k values (i.e., 25, 50, 75 and 100). We can observe that the larger the k value, the larger the times of *Bin kNNJ*, *kNNJ on Overlapping Partitions* and *Merge Results* are, since the number and size of the results to be processed increases too. First, *Grid* shows stable growth of the execution time, highlighting a small increment when using $k = 50$. Being a uniform partitioning, the number of overlapping partitions tends to increase in steps when a certain threshold in the generated ranges is reached. Next, *R-tree* shows a higher increase in the time of the *kNNJ on Overlapping Partitions* step than in the *Bin $k$textitNNJ* step. This indicates that the number of overlaps grows significantly with the increase of k, producing a larger number of non-final kNN lists distributed over a greater number of partitions. Moreover, the increase of execution time in the *Merge Results* step confirms that more partial kNN lists need to be combined. Then, *Quadtree* and *kDB-tree* again show similar behavior by showing the smallest execution times, especially in the case of *kDB-tree* (it adapts better to the data distribution). As a result, the number of overlapping partitions is reduced as we can see in the small increase of the execution time of the *kNNJ on Overlapping Partitions* and *Merge Results* steps. However, for *Quadtree*, the time of the *Merge Results* step increases faster than *kNNJ on Overlapping Partitions*, meaning that the number of these partitions increases for each point with a non-final kNN list. Finally, we can conclude that *kDB-tree* is the spatial partitioning technique which shows the best performance, for both different dataset combinations and distinct k values.

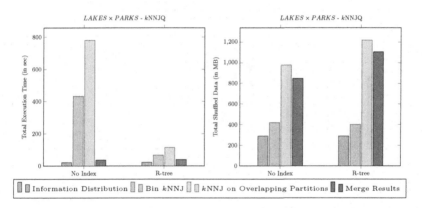

Fig. 3. kNNJQ cost per step, considering the use of indices on the $LAKES \times PARKS$ combination. Execution time in seconds (left) and shuffled data in MBytes (right).

Our second experiment studies the use of indices with the proposed kNNJQ algorithm, using the best spatial partitioning method from previous experiment (kDB-tree) for the $LAKES \times PARKS$ combination and fixing $k = 25$. In Fig. 3, the left chart shows the execution time of kNNJQ, for each of the algorithm steps for *non-indexed* and *R-tree indexed* datasets. First, for both algorithms, the execution time of the *Information Distribution* and *Merge Results* steps is similar, since they use the same input datasets and generate a similar number of final and non-final kNN lists. Small variations can be due to the random sampling of the data at the time of partitioning. Furthermore, the time spent in the *Bin kNNJ* and *kNNJ on Overlapping Partitions* steps is proportional between them. More-over, note that the use of indices causes an important performance improvement of the algorithm with a speedup of 6×. Although a plane-sweep algorithm has a higher performance over a nested loop approach, it is lower to an in-memory R-tree index algorithm where only the necessary data are accessed directly. As shown in Fig. 3, right chart, the Total Shuffled Data for each step of the *k*NNJQ algorithm for both local *k*NN options is very similar. On the one hand, the values in the *Information Distribution* and *Bin kNNJ* steps are almost the same, because the partitioning of the datasets and the number of generated *k*NN lists are very similar. On the other hand, for the *kNNJ on Overlapping Partitions* and *Merge Results* steps, the use of a customized plane-sweep algorithm allows to reduce the number of *k*NN lists and therefore, the shuffled data decreases. For the *R-tree index* variant, due to the fact that it uses the built-in spatial range join, it is not possible to add the same optimizations. Finally, the main conclusion is that the use of the *R-tree index* obtains a significant performance improvement compared to not using indexes, and at the same time there is no considerable increase in the size of the shuffled data.

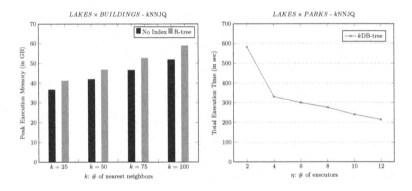

Fig. 4. *k*NNJQ cost with respect to the *peak execution memory* usage (left) and the number of executors η (right).

The third experiment, represented in the left chart of Fig. 4, shows the Peak Execution Memory of the *k*NNJQ algorithm, for $L \times B$ combination, using the *k*DB-tree partitioning method varying k. The first conclusion is that the memory requirement increases linearly with the increase of k, since the size of the *k*NN lists depends directly on that value. Moreover, this increase of Peak Execution Memory is not high (i.e., approximately 180 MBytes per unit of k for *R-tree*) and allows to use higher k values without consuming many cluster resources. In general, *No Index* consumes less memory than *R-tree* because the later needs the tree nodes information in-memory for the index to work (i.e., MBRs of the internal/leaf nodes and pointers to the children). Therefore, the left chart shows an increment of approximately 13% when using the index variant. Finally, given the input dataset sizes of 8.6 GBytes (L) and 26 GBytes (B), the overhead needed for partitioning, indexes, and the kNN lists is restrained.

The fourth and last experiment aims to study the speedup of the kNNJQ varying the number of executors η. The right chart of Fig. 4 shows the impact of different number of executors on the performance of the kNNJQ algorithm, for $L \times B$ with $k = 25$. From this chart, we can deduce that a better performance would be obtained if more executors are added. For instance, notice how doubling the number of executors from 2 to 4, reduces the execution time in the same amount. However, for higher values of η, the performance gain does not increase in the same way, mainly due to data skew issues. The time required to execute the most expensive task is 155 s, while the median is 11.4 s. Therefore, to solve this type of problems, the current spatial partitioning techniques should be improved (i.e., using bulk-loading methods for building the partition trees) or the algorithm must apply some repartitioning techniques [4].

By analyzing the previous experimental results, we can extract several interesting conclusions that are shown below:

- We have experimentally demonstrated the *efficiency* (in terms of total execution time) and the *scalability* (in terms of k values, sizes of datasets and number of executors (η)) of the proposed parallel and distributed kNNJQ algorithm in *Sedona*.
- kDB-tree partitioning technique shows the best performance thanks to its regular data-based subdivision and its more balanced partitions.
- The use of *R-tree*, as an in-memory local index, significantly increases the performance of the kNNJQ algorithm when performing the local kNNQ compared to other non-indexed methods such as a plane-sweep algorithm.
- Memory requirements (in terms of *Peak Execution Memory* and *Shuffled Data*) are measured and, they increase linearly with the value of k, allowing the use of higher k values without consuming many cluster resources.
- The performance of the kNNJQ algorithm improves as the number of executors (η)) increases, although there are skew problems that prevent further improvements.

6   Conclusions and Future Work

Sedona is a recent in-memory cluster computing system for processing large-scale spatial data. Moreover, *Sedona* is actively under development and several companies are currently using it. kNNJQ is a common distance-based join operation used in numerous spatial applications, and it is time-consuming query (i.e., combination of kNNQ and spatial join). In this paper, we have proposed a new kNNJQ distributed algorithm in *Sedona*. The performance of the proposed algorithm has been experimentally evaluated on large spatial real-world datasets, demonstrating its efficiency (in terms of total execution time and memory requirements) and scalability (in terms of k values, sizes of datasets and number of executors). As part of our future work, we are planning to extend our current research in several directions: (1) implement the kNNJQ using Quadtree as local index; (2) extend the algorithm to perform kNNJQ between other spatial data types supported by *Sedona*, like point-rectangle, rectangle-polygon, etc.; (3) compare the kNNJQ distributed algorithm implemented in *Sedona* with the same join operation included in *LocationSpark* [9].

References

1. Chatzimilioudis, G., Costa, C., Zeinalipour-Yazti, D., Lee, W., Pitoura, E.: Distributed in-memory processing of all k nearest neighbor queries. IEEE Trans. Knowl. Data Eng. **28**(4), 925–938 (2016). https://doi.org/10.1109/TKDE.2015. 2503768

2. Fu, Z., Yu, J., Sarwat, M.: Demonstrating geosparksim: A scalable microscopic road network traffic simulator based on apache spark. In: SSTD Conference, pp. 186–189 (2019). https://doi.org/10.1145/3340964.3340984

3. García-García, F., Corral, A., Iribarne, L., Vassilakopoulos, M.: Improving distance-join query processing with voronoi-diagram based partitioning in spatialhadoop. Future Gener. Comput. Syst. **111**, 723–740 (2020). https://doi.org/ 10.1016/j.future.2019.10.037

4. García-García, F., Corral, A., Iribarne, L., Vassilakopoulos, M., Manolopoulos, Y.: Efficient distance join query processing in distributed spatial data management systems. Inf. Sci. **512**, 985–1008 (2020). https://doi.org/10.1016/j.ins.2019.10.030

5. Gounaris, A., Torres, J.: A methodology for spark parameter tuning. Big Data Res. **11**, 22–32 (2018). https://doi.org/10.1016/j.bdr.2017.05.001

6. Lu, W., Shen, Y., Chen, S., Ooi, B.C.: Efficient processing of k nearest neighbor joins using mapreduce. PVLDB **5**(10), 1016–1027 (2012). https://doi.org/10. 14778/2336664.2336674

7. Nodarakis, N., Pitoura, E., Sioutas, S., Tsakalidis, A.K., Tsoumakos, D., Tzimas, G.: kdann+: a rapid aknn classifier for big data. Trans. Large-Scale Data Knowl. Centered Syst. **24**, 139–168 (2016). https://doi.org/10.1007/978-3-662-49214-7_5

8. Pandey, V., Kipf, A., Neumann, T., Kemper, A.: How good are modern spatial analytics systems? PVLDB **11**(11), 1661–1673 (2018). https://doi.org/10.14778/ 3236187.3236213

9. Tang, M., Yu, Y., Mahmood, A.R., Malluhi, Q.M., Ouzzani, M., Aref, W.G.: Locationspark: In-memory distributed spatial query processing and optimization. Front. Big Data **3**, 30 (2020). https://doi.org/10.3389/fdata.2020.00030

10. Xie, D., Li, F., Yao, B., Li, G., Zhou, L., Guo, M.: Simba: efficient in-memory spatial analytics. In: SIGMOD Conference, pp. 1071–1085 (2016). https://doi.org/ 10.1145/2882903.2915237

11. You, S., Zhang, J., Gruenwald, L.: Large-scale spatial join query processing in cloud. In: ICDE Workshops, pp. 34–41 (2015). https://doi.org/10.1109/ICDEW. 2015.7129541

12. Yu, J., Zhang, Z., Sarwat, M.: Geosparkviz: a scalable geospatial data visualization framework in the apache spark ecosystem. In: SSDBM Conference, pp. 15:1–15:12 (2018). https://doi.org/10.1145/3221269.3223040

13. Yu, J., Zhang, Z., Sarwat, M.: Spatial data management in apache spark: the GeoSpark perspective and beyond. Geo Informatica **23**(1), 37–78 (2018). https:// doi.org/10.1007/s10707-018-0330-9

14. Zhang, C., Li, F., Jestes, J.: Efficient parallel kNN joins for large data in MapReduce. In: EDBT Conference, pp. 38–49 (2012). https://doi.org/10.1145/2247596.2247602
15. Zhao, X., Zhang, J., Qin, X.: knn-dp: handling data skewness in kNN joins using mapreduce. IEEE Trans. Parallel Distrib. Syst. **29**(3), 600–613 (2018). https://doi.org/10.1109/TPDS.2017.2767596

Author Index

Printed in the United States
by Baker & Taylor Publisher Services